The Mechanics of Solder Alloy Wetting and Spreading

The Mechanics of Solder Alloy Wetting and Spreading

Edited by

F. G. Yost
F. M. Hosking
D. R. Frear

VNR VAN NOSTRAND REINHOLD
New York

Chapter 4 was prepared by government employees and is public domain.

Softcover reprint of the hardcover 1st edition 1993

Library of Congress Catalog Card Number 93-23151

ISBN-13: 978-1-4684-1442-4 e-ISBN-13: 978-1-4684-1440-0
DOI: 10.1007/978-1-4684-1440-0

Van Nostrand Reinhold
115 Fifth Avenue
New York, NY 10003

International Thomson Publishing GmbH
Konigswinterer Str. 518
5300 Bonn 3
Germany

International Thomson Publishing
Berkshire House, 168-173
High Holborn, LondonWC1V 7AA
England

International Thomson Publishing Asia
38 Kim Tian Rd., #0105
Kim Than Plaza
Singapore 0316

Thomas Nelson Australia
102 Dodds Street
South Melbourne 3205
Victoria, Australia

International Thomson Publishing Japan
Kyowa Building, 3F
2-2-1 Hirakawacho
Chiyada-Ku, Tokyo 102
Japan

Nelson Canada
1120 Birchmount Road
Scarborough, Ontario
M1K 5G3, Canada

16 15 14 13 12 11 10 9 8 7 6 5 4 3 2 1

Library of Congress Cataloging-in-Publication Data

Yost, Frederick G.
The mechanics of solder alloy wetting and spreading / F. G. Yost, F. M. Hosking, D. R. Frear
p. cm.
Includes bibliographical references and index.

1. Solder and soldering. 2. Wetting. I. Hosking, F. M. II. Frear, D. R. III. Title.
TS610.Y67 1993
671.5'6--dc20

Contents

Foreword

In 1992 Congress passed the Defense Manufacturing Engineering Education Act with the intent of encouraging academic institutions to increase their emphasis on manufacturing curricula. The need for this incentive to integrate the academic and industrial communities was clear: gaps in manufacturing science were inhibiting the evolution of new manufacturing technologies that are required for the U.S. to maintain a competitive posture in the world marketplace.

The Army Research Laboratory and Sandia National Laboratories sought to contribute to the congressional intent by initiating a new series of graduate level college textbooks. The goal was to focus next-generation scientists onto issues that were common to the needs of the commercial market, the affordability of DoD weapons systems, and the mobilization readiness of the U.S. Armed Forces.

The textbook *The Mechanics of Solder Wetting and Spreading* was written in this spirit by nationally renowned scientists for academe and industry. Researchers using the book are encouraged to formulate programs that will establish scientific correlations between manufacturing process controls and product reliability. Such correlations are essential to the building of a new electronics industry which is based upon the futuristic concepts of Virtual Factories, Prototyping, and Testing.

George K. Lucey
U.S. Army MANTECH Manager
for Soldering Technology

Acknowledgments

The editors want to especially acknowledge the professional effort put forth by the team of authors that made this book. We asked a great deal from each of them to write a seamless critical appraisal of the subject of wetting and spreading of solders. Each of the authors went above and beyond what is normally expected and responded in a timely and enthusiastic manner. We would also like to thank the many reviewers whose excellent comments improved the quality of the book.

Much of the technical insight resulted from both the formal and informal discussions at the Solder Alloy Wetting and Spreading Workshop that was held in Santa Fe, New Mexico in June of 1992. We acknowledge the attendees whose participation helped stimulate the ideas and concepts that are summarized in this book.

The enthusiastic support of George Lucey of the U. S. Army Research Laboratories brought this project to fruition. George strongly believes in the concept of bringing fundamental knowledge into the manufacturing arena to improve quality and it is our hope that with this book we have supported his efforts.

This book was reviewed, and the quality was greatly improved, by the professional editors of Technically Write. We would like to acknowledge the great efforts put in by Jennifer Scofield, who proofread and tirelessly edited the manuscripts, and Linda and Jim Innes, who spent many hours collating and compositing the book. We would also like to thank the people at Van Nostrand Reinhold for believing in this project and helping with the many details in putting a volume of this size together. Specifically we would like to mention the efforts of Marjorie Spencer who coordinated the effort at VNR, with the assistance of Dawn Wechsler and Chris Grisonich.

Finally, the editors want to acknowledge the direct, and indirect support of the U. S. Army Research Laboratories and Sandia National Laboratories without which this book could never have been written.

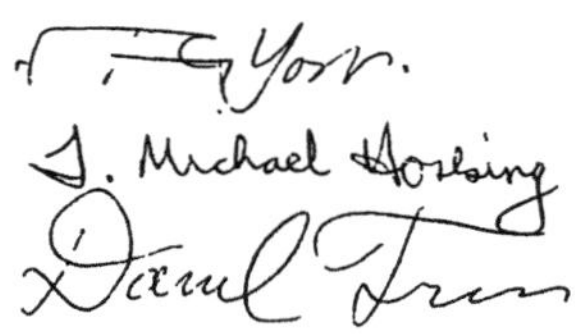

Albuquerque, New Mexico

Chapter Authors & Affiliations

W. J. Boettinger
NIST
A153/223
Gaithersburg, MD 20899

D. R. Frear
Sandia National Laboratories
Center for Solder Science & Technology
Department 1832
P.O. Box 5800
Albuquerque, NM

J. Golden
U. S. Laser Corporation
825 Windham Court North
Wyckoff, NJ 07481

C. A. Handwerker
NIST
A153/223
Gaithersburg, MD 20899

W. B. Hampshire
Tin Research Institute
1353 Perry Street
Columbus, Ohio 43201

F. M. Hosking
Sandia National Laboratories
Center for Solder Science & Technology
Department 1831
P.O. Box 5800
Albuquerque, NM 87185

J. D. Hunt
Department of Metallurgy
Parks Road
Oxford University
Oxford, OX1 3PH, England

J. L. Jellison
Sandia National Laboratories
Center for Solder Science & Technology
Department 1833
P.O. Box 5800
Albuquerque, NM 87185

K. L. Jerina
Washington University
Dept. of Metallurgy & Materials Science
Campus Box 1185
One Brookings Drive
St. Louis, MO 63130

U. R. Kattner
NIST
A153/223
Gaithersburg, MD 20899

D. M. Keicher
Sandia National Laboratories
Center for Solder Science & Technology
Department 1831
P.O. Box 5800
Albuquerque, NM 87185

G. Kuo
Washington University
Dept. of Metallurgy & Materials Science
Campus Box 1185
One Brookings Drive
St. Louis, MO 63130

C. A. MacKay
MCC
12100 Technology Blvd.
Austin, TX 78727

D. L. Millard
Renssalaer Polytechnic Institute
C11 9015
Troy, NY 12180-3590

K. Rajan
Materials Engineering Department
Rensselaer Polytechnic Institute
Troy, NY 12180-3590

S. M. L. Sastry
Washington University
Dept. of Metallurgy & Materials Science
Campus Box 1185
One Brookings Drive
St. Louis, MO 63130

D. Morgan Tench
Rockwell International Science Center
1049 Camino Dos Rios
Thousand Oaks, CA 91358

R. Trivedi
Ames Laboratory
USDOE
Dept. of Materials Science and Engg.
Iowa State University
Ames, Iowa 50010

M. Wolverton
Texas Instruments
P.O. Box 655474, MS409
Dallas, TX 75265

F. G. Yost
Sandia National Laboratories
Center for Solder Science & Technology
Department 1831
P.O. Box 5800
Albuquerque, NM 87185

The Mechanics of Solder Alloy Wetting and Spreading

1

Introduction: The Mechanics of Solder Alloy Wetting and Spreading

F. G. Yost, F. M. Hosking, and D. R. Frear

1.1 SOLDERING IN ELECTRONICS

The challenge of keeping pace with emerging microelectronic device technologies has become increasingly difficult, and the fundamental limits of existing printed wiring board (PWB) technologies are becoming increasingly apparent. As a result, the domestic PWB industry now faces serious foreign competition from large, well-financed firms. The 1990 worldwide market for PWBs was \$25.2 billion and has grown by just under 10% per year since 1988. North America's demand is 28.5%, Japan's is 27.6%, and Europe's is 20.7%. The market is expected to grow to over \$35 billion in 1993. The PWB industry in the U.S. consists of approximately 930 large companies. Of those, 600 have revenues of less than \$5M/year and 200 have revenues of between \$5M and \$25M/year. PWB technology is now approaching limitations that will have a profound effect on the ability to provide a means of utilizing the ever-increasing functionality that integrated circuits possess and promise. In addition, these limitations and increasing foreign competition will force many U.S. PWB companies out of business. One of the limiting processes referred to above is the age-old method of soldering.

The soldering process has followed every microelectronic revolution from hard wired boards to multichip modules (MCM) with no foreseeable threat to seriously replace it. The high density MCM market is expected to rise to $800M by 1995 and exceed $1 trillion by 1997. The biggest users of MCM technology will be those having high throughput, such as computer and telecommunication manufacturers, and the defense industry. Defense applications of MCMs for smart systems are driven by the 10–1 volume reduction they offer but will require agile manufacturing processes such as solder bump technology. Despite a lack of semantic agreement as to what a multichip module actually is, there is no disputing the difficulties encountered in their assembly. To begin assembly, it is important to know that each and every integrated circuit (IC) to be mounted in the module is a good IC, since the cost of rejection or repair of the module is too high. This notion of "known good die" can be implemented by solder bump technology since any scarring of bond pads, due to probing during electrical test, can be eliminated by solder reflow. Solder bumps also allow rework after assembly. Although any chip mounting and bonding-out process can be used to build these modules, solder bumps accomplish both in one process.

Solder bump technology has also been chosen for the attach method by several optoelectronic system manufacturers. Photonic devices, emitters and detectors, must be aligned and attached to substrates that become part of a larger assembly. Accurate placement of the devices is a technology requirement. Other photonic packaging applications require a GaAs chip to be attached to a substrate that must be aligned to one or more optical fibers. The optical signal coupling efficiency from the device to the fiber is severely affected by misalignment. Solder bumps, and self alignment by associated capillary forces, offer a unique method to accomplish this technology.

To make a sound joint, solder must come into contact with the base metal. The clean surface of most metals is very active and will therefore readily react with many ambient gases. Since most metal solders and substrates are oxidized or otherwise coated, a fluxing action is required to remove these stabilizing surface layers. At the present time, liquid chemical fluxes are commonly used to strip native oxides from metals that are to be soldered. These fluxes typically contain corrosive compounds such as zinc chloride and oleic acid in a rosin medium that makes them difficult to remove after soldering. Traditionally, fluxes have been removed with one or more chlorofluorocarbon or other halogenated solvent rinses. These chemicals have come under increasing scrutiny (The Montreal Protocol and the U.S. Clean Air Act of 1990) because of their effect on the ozone layer; they will soon be banned from manufacturing.

Increased pressure has also been applied to all products containing lead, including most solder alloys, which is a known toxin and heavy metal poison. Environmental groups have exerted their influence on state and federal politicians and regulatory agencies such as the Environmental Protection Agency and the Centers for Disease Control. This has resulted in proposed legislation introduced by the 102nd Congress: S. 391 – Lead Exposure Act of 1991, and four related bills (H.R. 2922, H.R. 1750, H.R. 870, and S. 398). If passed, these laws will regulate, ban, or tax all U.S.-manufactured lead products and imported foreign lead products. This action will not affect world-wide marketing by other industrialized nations and would put U.S. industrial competitiveness in the world market at a serious disadvantage. The many problems presented by such environmentally driven legislation require a fundamental understanding of the fluxing and soldering process to prevent serious economic damage to the multi-billion dollar PWB industry.

The future of soldering as a technology seems to be secure, so it remains the responsibility of the technologist to provide the most reliable and cost-effective processes available. The process of soldering an electronic assembly involves a number of complex metallurgical reactions. In this book we describe the operative mechanisms and processes that occur when a solder joint is formed. After the joint is formed, ways to evaluate the solder joint and improve reliability are discussed. This volume represents a summary assessment of the state-of-the-art understanding of the mechanism of solder alloy wetting and spreading. The book is derived from the *Second International Workshop: A Solder Wetting and Spreading Assessment* (the second in a series of workshops on the mechanics of solder[1]). Present at the workshop were scientists and engineers recognized for their expertise in wetting behavior. The goal of the meeting was to determine what was known, what was not well understood, and what needed further study in the area of solder wetting and spreading. This book represents the general consensus of the workshop attendees.

This book focuses on wetting. A brief introduction to the wetting problem is illustrative of the complex issues that are involved in making a sound solder joint.

1.2 THE WETTING PROBLEM

Wetting is a problem that has received a great deal of theoretical and fundamental study for liquids on solids. The bulk of this work has been performed on simple systems like water and oil on a non-reacting surface (e.g. glass). Even in these

[1] The first workshop, *The First International Workshop on Materials and Mechanics Issues of Solder Alloy Applications*, was held in June of 1990 and covered all areas of soldering science and technology. That workshop was summarized in the volume *Solder Mechanics: A State of the Art Assessment* [TMS publications (1991)]. This book is derived from the *Second International Workshop: A Solder Wetting and Spreading Assessment* held in June of 1992. The third workshop was held in September of 1992 and dealt with the mechanics and modeling issues of solder joints and is summarized in the book *The Mechanics of Solder Alloy Interconnections* [Van Nostrand Reinhold (1993)].

"simple" systems, a description of the wetting behavior is complex. Consider a droplet, small enough to be essentially spherical, having a height, h, and radius, r, spreading on a smooth, flat, chemically homogeneous substrate. It will be assumed that the substrate does not dissolve or otherwise react with the droplet. The capillary driving force for the uniform advance of the droplet triple point line (TPL) can be written [Lau et al. 1972]:

$$F_d = \gamma_l (\cos\theta_y - \cos\theta) \tag{1-1}$$

while the viscous force resisting the driving force is [Levinson et al. 1988; de-Gennes 1985]:

$$F_v = \left(\frac{\eta r}{h}\right)^{-1} \frac{dr}{dt} \tag{1-2}$$

where γ_l is liquid surface energy, θ_y is the equilibrium contact angle, θ is the temporal contact angle, η is liquid viscosity, h is droplet height, and t is time. Equating these forces yields the equation of motion for the TPL:

$$\frac{dr}{dt} = \frac{h}{\eta r}\gamma_l (\cos\theta_e - \cos\theta) \tag{1-3}$$

Assuming that θ is zero and θ_e is small, this equation can be simplified and solved, which gives the following:

$$r(t) = (V_0)^{\frac{1}{3}} \left(\frac{\gamma_l}{\eta} t\right)^{\frac{1}{10}} \tag{1-4}$$

where V_0 is the constant liquid volume. Alternatively, Eq. (1-3) can be solved numerically with the spherical assumption and constant liquid volume condition, which give

$$\cos\theta = \frac{1 - \left(\frac{h}{r}\right)^2}{1 + \left(\frac{h}{r}\right)^2} \tag{1-4a}$$

and respectively

$$\frac{dh}{dr} = -\frac{2rh}{r^2 + h^2} \tag{1-4b}$$

The kinetics given by Eq. (1-4) are followed by many different kinds of non-interacting liquid-solid systems. If there is no interaction or coupling of any kind between solder and substrate, spreading should follow these kinetics. Since the values for viscosity and surface tension are known for solder, Eq. [1-3] can be solved and compared with experimental solder spreading kinetics. Figure 1-1 shows a solution obtained by assuming that the equilibrium contact angle is 15 degrees, the surface tension is 460 ergs/cm^2, and the viscosity is 1.5 centipoise. Typical experimental results for lead-tin solder spreading on a copper substrate are also shown in Figure 1-1. Note that there is a large discrepancy between experiment and theory. This discrepancy is due to the added metallurgical complexity in solder wetting and spreading.

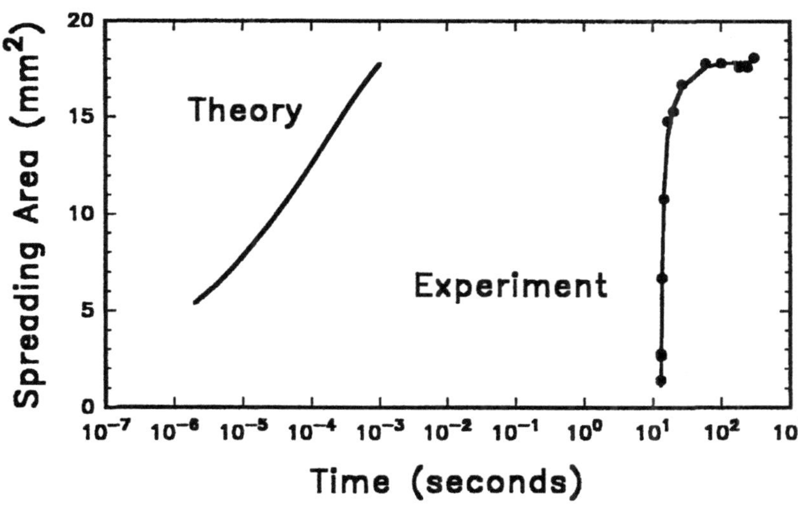

Figure 1-1 A comparison of contemporary spreading theory with experimental solder spreading results.

The above model would become more complicated for solder because it has long been recognized in the solder industry that most wetting systems exhibit either partial substrate dissolution or substrate reaction. In practical solder-substrate systems, an intermetallic compound is formed at the reaction interface. The initial compound layer is thin and does not thicken appreciably during the soldering process. When a droplet of liquid is placed on a wettable solid surface, it first experiences extremely rapid and dynamic spreading. The process then slows, and for many liquids, the radius of the droplet expands as a power fractional of the spreading time. In summary, the real soldering process involves a number of complex reactions because the molten metal does react with the solid metal substrate. The reactivity between the solder and the metal surface results in the formation of intermetallic compounds. The reactive wetting process and intermetallic formation are described in Chapter 4.

To be able to use the relations described above (or derivations thereof), a number of variables must be determined. The surface energy of the molten solder, γ, and the contact angles, θ, are needed. Unfortunately, there is not an extensive data base of surface energies, particularly for solder alloys. Chapter 7 of this book describes what these surface energies mean and how they can be measured. Chapter 2 describes how contact angles may be determined and some of the limitations on values derived from experimental measurements. Before a solder can spread across a surface and react with the metal, the solid surface must be clean. Solderable systems of interest are metal substrates that are reactive with air and oxidize. Chapter 6 describes the surface reactions that form these oxide coatings and some novel techniques that can be used to characterize the oxides. Typical solders will not react with oxide coatings, so a liquid flux is required to remove contaminants. Chapter 3 describes what a flux is, and the reactions that are involved in removing a contaminant layer are described fundamentally.

Under ideal conditions, the molten solder reacts with a metal substrate and a good metallurgical bond results. Occasionally, however, the joint wets and then dewets, resulting in poor solderability and rejected parts. The process of dewetting is described in Chapter 5.

The solder wetting process is one-half physical metallurgy and one-half process metallurgy. The physical metallurgy is the reaction of the molten solder to a metal substrate. The process metallurgy is how to make the solder molten. Heating an assembly to form a solder joint is becoming an important aspect of the soldering process with the advent of smaller, more complex electronic assemblies. Chapter 8 is a treatment of advanced soldering processes that covers new processes, such as laser soldering and fluxless soldering, that will be needed to be able to make more complex assemblies in the future.

After a solder joint is made it is essential that its quality be assessed. In the past this assessment was simply a measure of how shiny the solder joint was. This is not a satisfactory measure of reliability because it cannot determine anything about the bulk of the joint. Chapter 9 treats a number of novel and advanced inspection technologies that are available to assess the quality of solder joints in electronic assemblies.

The reliability of a solder joint does not end after assembly. Often the electronic assemblies must survive extreme environments where strain and temperature are imposed upon the solder joints. The solder alloys used in electronic assemblies are susceptible to low-cycle fatigue failures under these severe conditions. Chapter 10 describes new composite solder alloys that show promise in alleviating problems due to cyclical temperature and strain. The purpose of this book is to give designers and engineers who are responsible for electronic assembly manufacture background on the wetting behavior of solder alloys. The issue is complex, and knowledge of all aspects of the soldering process is needed. This book is a step in that direction; it is an assessment and it is a picture in time. The subject of solder wetting is constantly changing. However, the current state of knowledge that is summarized here goes a long way toward setting the stage for soldering capabilities in the next century.

REFERENCES

deGennes, P. G. 1985. Rev. Mod. Phys. 57:827.

Frear, D. R., W. B. Jones, and K. R. Kinsman. 1991. Solder Mechanics: A State of the Art Assessment. Warrendale, PA.: TMS Publications.

Frear, D. R., S. N. Burchett, H. S. Morgan, and J. H. Lau. 1993. The Mechanics of Solder Alloy Interconnections. New York: Van Nostrand Reinhold.

Lau, W. Y., and C. M. Burns. 1972. Surface Science. 30:497.

Levinson, P., A. M. Cazabat, M. A. Cohen Stuart, F. Heslot, and S. Nicolet. 1988. Revue Phys. Appl. 23:1009.

2

Solderability Testing

Bill Hampshire and Mike Wolverton

2.1 INTRODUCTION

Solderability testing is used to predict the success of the soldering process before the large investment of assembly has taken place. Once several parts costing several hundreds of dollars or more are assembled and soldering is attempted, it is very expensive (and makes no sense) to discover that the unit will not function because one individual part did not solder. Hence, solderability testing is performed prior to assembly on the individual components to be joined to show the probability of soldering success.

It is important to keep this goal in mind because we sometimes forget why we test. If a factor will have no bearing on soldering yields, then it is unlikely to be meaningful for solderability testing. As we look at the various solderability test schemes and practices, we will see that some of these attempt to simulate actual joining, while others seem to be far removed from typical, practical situations. The only thing that counts for a solderability test, in the end, is its ability to predict soldering success.

2.1.1 Definition of Solderability

One of the problems surrounding solderability testing is the definition of terms. Solderability has been defined simply as "the ability of a surface to be wet by molten solder." This simple definition was perhaps adequate when the electronics industry was typically using wave soldering with rosin mildly activated (RMA) fluxes. As the range of process variation has expanded, so has the need for a better definition of solderability.

One aspect of this problem is that the definition of solderability usually ends up being one of "wettability." Wettability is what is usually measured by solderability testing, so we define what we can measure. Solderability depends on achieving success in the specific environment of the actual solder joint, but we certainly cannot measure the actual joint in advance of making it, so we must compromise by measuring wettability. We must also remember that having good wettability (which we measure) does not automatically translate into good solderability.

Solder technologists have attempted to circumvent this definition problem by defining separate solderabilities [DeVore 1978, 1984a,b,c; 1990], such as functional solderability, which depends not only on material properties, but on the specific environment in which soldering is attempted; and inherent solderability, which is actually wettability. DeVore and others have also suggested a different approach by defining "soldering ability," which takes into account the local physical and thermal conditions at the actual joint area. There seems to be a tendency in industry to confuse the concepts of solderability and soldering ability.

A set of definitions based on probability considerations has been offered by one authority (Appendix 2-1). This is a useful contribution because it reminds us that we must apply results to a less than perfect world.

2.1.2 Why Test for Solderability

As indicated above, we cannot afford to create an expensive assembly of parts and then find out that we cannot get the desired functionality because one part failed to accept solder properly. In one study [Fazekas 1987], solderability defects accounted for over two-thirds of the total defects found (Figure 2-1). Because of the definition difficulties, this probably overstates the actual situation. Any drive to better quality (whether it be the older "zero defects" philosophy or the more recent "six sigma" version) must take a very careful look at solderability and solderability testing.

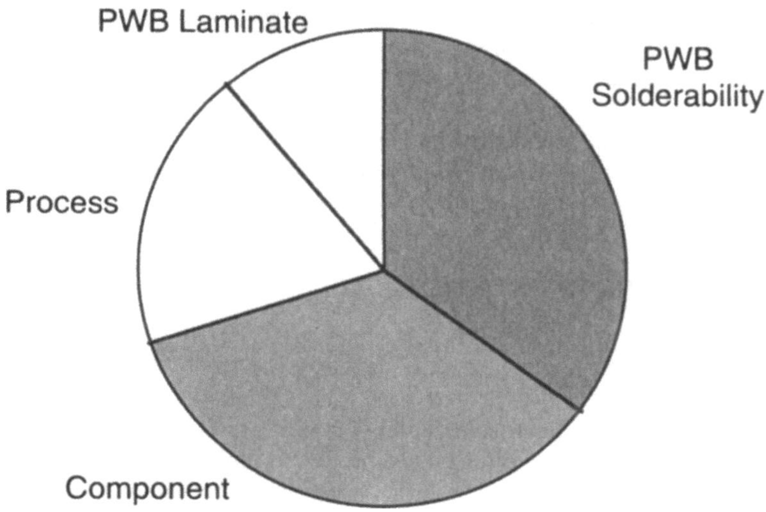

Figure 2-1 Underlying causes of loss of solderability

Rework is done after soldering to correct visible solder joint defects, but rework is a potential reliability risk. Most soldering experts agree that the first joint made is likely to be the best. The heat cycle of reworking tends to degrade the materials, resulting in a generally poorer joint, even if the joint is visually repaired. This aspect is also part of the push to better quality.

Quality is more than just initial defects, of course. A perfectly functioning piece of equipment must continue to function, so reliability is very much an issue. Solderability defects may have a long term, hidden effect on reliability, even if the defects are so "minor" that they do not require reworking.

2.1.3 The Obstacles to Progress in Testing

As solderability testing interest has grown, so have the number and quality of equipment suppliers. It is now possible to purchase and set up test equipment at a reasonable cost compared to other analytical equipment and required tests. Such was not the case in earlier years and this is the reason a large data base of test results is not readily available.

Despite these improving equipment trends, the present tests are proving very difficult to use in round robin testing between independent testers. The repeatability of test results from location to location is poor [Edgington 1987; Reed 1990]. Equipment manufacturers are providing more automated test devices to help in this effort.

There exists, at any rate, a growing body of literature on solderability testing. We shall examine the available data in light of the several identifiable problems facing the solderability test community.

2.2 THE "NUMBERS" PROBLEM

2.2.1 Dip Tests

The early attempts to evaluate solderability (or more precisely, wettability, as stated above) involved dipping a fluxed test piece into molten solder and observing the result. A cam-operated dipping device adds some repeatability to the test procedure (MIL STD 202, Mthd.208; MIL STD 883, Mthd.2003.1), but the big disadvantage of these dip tests is the subjective visual evaluation which follows.

It should be emphasized that these tests are still useful for the purpose of solderability testing. However, the human visual basis of the evaluation introduces possible misinterpretations and inconsistencies that can not be tolerated in a scheme for improving quality.

Figure 2-2 shows the prescribed test device. A sample for testing is hung from the arm and lowered into the molten solder at a specified speed and then withdrawn, also at a specified speed. After solidification, the sample is examined for evidence of wetting.

The principal problem with this test centers on the inspection criteria. In a vertical dip, the solder coating may sag in places; the resulting unevenness is sometimes mistaken for dewetting. While the test was not originally meant to involve magnification (evaluation by direct unaided vision was intended), the evolution of smaller and smaller components has made magnification necessary. The reason magnification was not wanted originally is that small defects in wetting will always occur because of defects in the substrate being tested. In an actual joint, however, these defects are covered over by the solder, and do not generally affect solder joint function. Magnification shows these "defects" and biases the evaluation.

Only very careful training and experience allow the successful use of these dip tests for solderability determination. Even with this careful training, however, the test result is a "go-no go" evaluation, which is not truly compatible with recent quality improvement schemes requiring quantitative data.

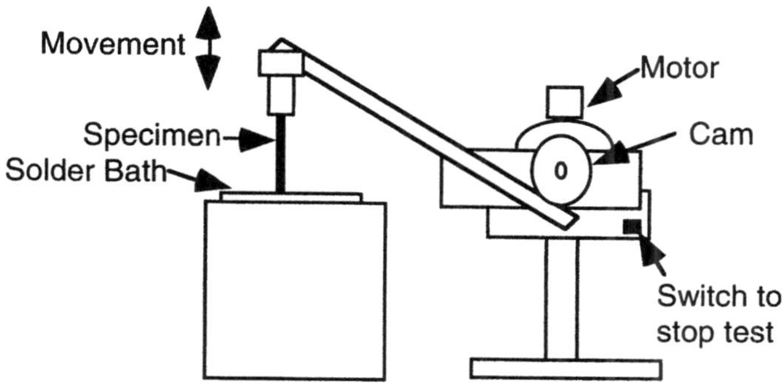

Figure 2-2 Schematic illustration of an automated vertical dip test device

2.2.2 Dip Tests Versus Six Sigma Quality

The wetting balance and microwetting balance solderability test methods will be described in detail in the following sections. However, it is first instructive simply to note that both methods provide variable data, not just the attribute data of the standard dip test. Variable data quantifies the amount of a parameter. For example, wetting balance parameters are wetting time and wetting force or wetting angle.

The theoretical wetting force relates to the interfacial surface tension between the solder and the flux. For clarity, the wetting angle is the angle that the liquid solder makes with the solid substrate at the substrate-solder-flux triple point. To understand wetting angle, refer to Figures 2-3 to 2-5.

Attribute data provides only pass or fail information. Experience with six sigma quality goals reveals that attribute data is impractical, while variable data is much more useful in practice.

For example, 50 specimens can be run on either the microwetting or wetting balance, and basic statistical information such as the mean and the standard deviation can be obtained. The distribution can be checked for normalcy, and if it does not deviate significantly from normal, it is a simple matter to determine six sigma capability. Specifically, if the mean wetting angle is 30°, the standard deviation is 5° and the upper specification limit is 60°, then the capability is (60-30)/5 = 6 sigma. Please note: statistical control must be known before capability is measured.

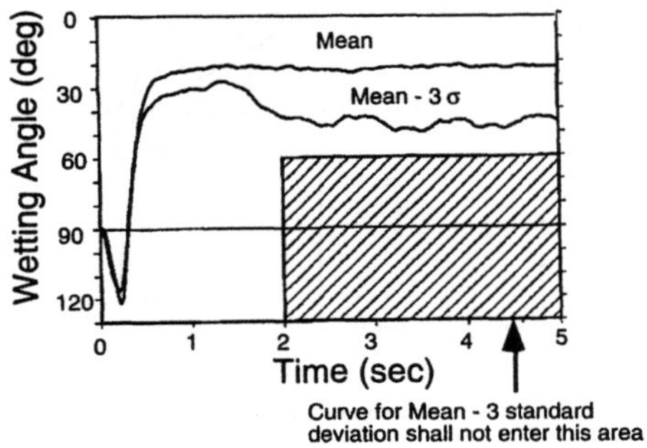

Figure 2-3 Example of a wetting balance acceptance requirement for four specimens, RMA flux. Courtesy, J. Gordon Davy.

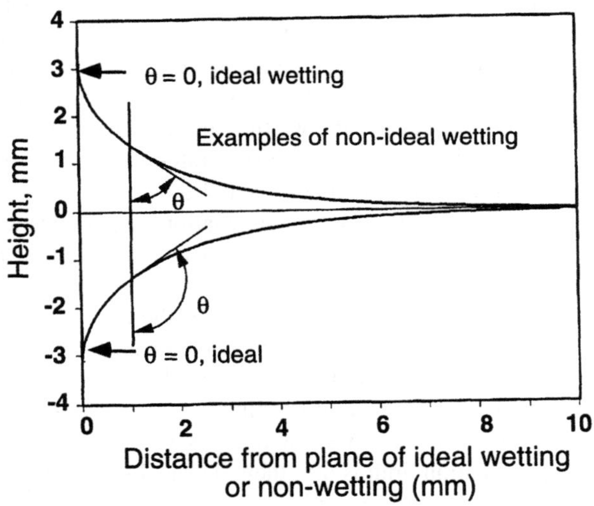

Figure 2-4 Elastica curves for liquid solder against a flat vertical suface[Klein-Wassink 1989]Courtesy, J. Gordon Davy.

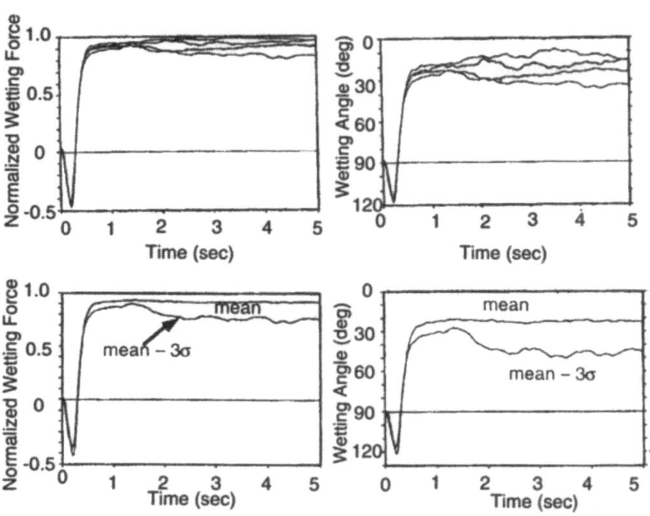

Figure 2-5 Wetting curves for 20-mil Cu-plated wire RMA flux,no steam. Courtesy, J. Gordon Davy.

Six sigma information could be obtained from attribute data. However, reasonable confidence would require thousands of specimens instead of 50. It is no wonder that the attribute data of the dip solderability tests have never been correlated to solder joint quality: they cannot be correlated in a practical sense; the work would be too costly because it would require too many specimens.

Solderability test researchers in the past, faced with this problem, relied upon "over-engineering" the test. One notable example is the selection of flux. Pure rosin fluxes were selected for use in solderability tests, in order to provide a margin of safety over the rosin mildly activated or other fluxes of the soldering process. This margin of safety was vague and not quantified. Test method designers did not apply statistics back then. So, management by relevant data was hampered.

Current test methods more closely match actual processes. The flux, dwell times, etc. of the test can be set to match the levels of the soldering process. The key requirement now is that solderability test data must be variable data. Such data can be correlated to solder joint function (e.g. electrical, mechanical, thermal) or quality. A number of researchers are presently endeavoring down this path, notably Porter [1992] and Kwoka and Mullenix [1992].

With variable data, a solderability test parameter's specification limit can be derived from a solder joint's functional or quality requirement. The six sigma solderability quality is achievable simply by ensuring a three sigma control limit which is no more than half way from the mean to the specification limit of the test parameter (e.g. wetting time or wetting angle).

2.2.3 Wetting Balance (Surface Tension Balance)

To overcome the "numbers" problem, most testers are turning to the wetting balance. Perhaps its liability is producing too much data.

The wetting balance uses a sensitive pressure transducer from which the sample under test is hung. (See the schematic in Figure 2-6.) A bath of molten solder is raised (in the most common approach) so that the sample is immersed to a specified depth. During this process of immersion and the following withdrawal, the forces measured by the transducer are continuously recorded, either on a strip recorder or on a computer work station.

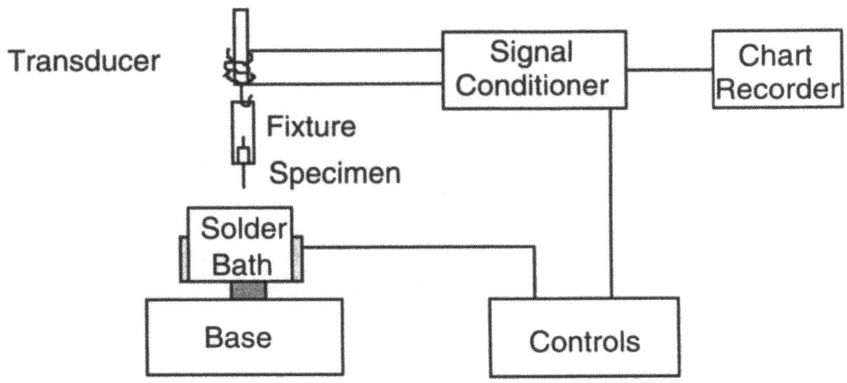

Figure 2-6 Schematic of wetting balance

Figure 2-7 shows an idealized force (wetting force on the vertical axis and time on the horizontal) curve that might result from a round component lead. The curve begins in one direction due to the buoyancy effect attempting to "reject" the lead from the solder. (Figure 2-7, point B.) When wetting begins, the force changes direction (point B of the curve). If wetting is reasonably good, the wetting force will achieve the point of balancing the buoyancy force (point C). The time from the beginning of the test (contact with the solder) to this point is called t_0. Good wetting will be shown by a continuing curve up to some maximum wetting force, F_{max}, which is achieved at some time t_{max} (point D). When the specified test time is finished, the sample is withdrawn and the curve shows a spike before finally decreasing. This area of the curve is said to provide some information on the tendency toward dewetting.

This complex curve gives many "numbers" for analysis, but sometimes it becomes difficult to decide the relative importance of each. The International Tin Research Institute has long maintained that the times are most important, provided an adequate wetting force is achieved. This concept has some physical basis, for if wetting is not achieved in a certain time (e. g. the time in the solder wave for wave soldering), then surely soldering will not occur. A typical time specification is for $t_{2/3}$: the time to two-thirds of the maximum wetting force and a typical specified value would be less than two seconds. Some feel that the wetting force is important, although the physical significance seems less obvious. Presumably the wetting force represents the "driving force" towards wetting and perhaps towards forming a good bond. A typical specification might be that the maximum wetting force exceed 250 μN/mm within 2 s of wetting time. Unfortunately, a recent study by the Ad-Hoc IPC/EIA Steam Aging Task Group suggests that adequate soldering can be achieved with parts showing a wetting force of 150 μN/mm. Some have suggested that a wetting force need only be positive for good soldering. Obviously, the significance of the wetting force needs more study.

[Kwoka and Mullenix 1990] have done a recent study of the significance of these various wetting curve values. This work is being extended to relate the measured values to subsequent soldering success [Kwoka and Mullenix 1992]. Studies of this sort are sorely needed for the wetting balance test, and for all the solderability test methods.

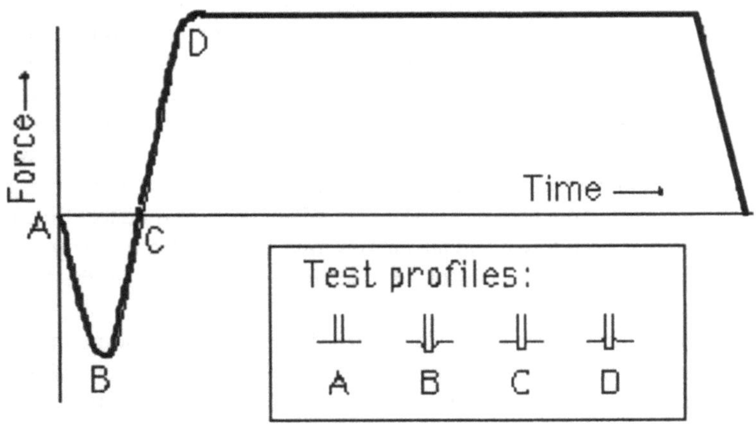

Figure 2-7 Idealized wetting curve

2.2.4 Microwetting Balance

The microwetting balance is particularly suited to small, surface-mountable components. This test is similar to the wetting balance, with the following differences. The microwetting balance has more sensitivity and spatial resolution than does the wetting balance. The microwetting balance is sensitive to aging and very sensitive to thermal demand.

A small solder globule is used, instead of the relatively large solder bath of the wetting balance method. This molten solder globule sets atop an iron pin, which in turn is press-fit into aluminum for heat transfer. It can be as small as 25 mg. Conversely, a 200 mg globule on a 4 mm iron pin can help to provide sufficient energy for the solder to heat the lead or termination and then wet up onto it. There is a basic reason why a globule must be used for very small components: the resulting data are not overwhelmed by the small length of the solderable lead. Consider a large, flat plate whose end is dipped into a solder bath. If the plate has perfect wetting, then the solder meniscus will rise approximately 3.0 mm above the solder bath. We know this because it has been calculated from the elastica curve, based on the surface tension property of the commonly used 63Sn-37Pb solder alloy.

If the component's solderable finish is smaller than 3 mm in length, clearly the wetting balance test is not sensitive enough to evaluate the small device. To avoid confounding the device's finish quality and size, a globule must be used. If the finish extends only 1 or 2 mm above the globule, the surface finish solderability can then be easily measured.

Experience has shown [Porter 1992] that, in addition to testing the wettability of small, surface mount components such as chip capacitors, the microwetting balance also provides a larger relative wetting force. This larger wetting force provides a higher resolution than one can obtain from the wetting balance. The larger wetting force is not fully understood, but it might be caused by a distortion of the little globule of liquid solder to a relatively vertical (i.e. narrow) shape as it wets the termination.

The microwetting balance is very sensitive to thermal demand. For example, leads of similar wettability, but different mass, have a similarly shaped force vs. time curve. However, the initiation of wetting takes more time for the more massive lead.

In practice, the microwetting balance is a bit more amenable to testing individual leads compared to the wetting balance. Depending on clearances, alternate leads might have to be bent back to provide room for testing.

The flux issue is particularly instructive. Non-activated R flux has three problems compared to pure rosin with 0.5% halide: 1) R flux leaves a spike on the termination when removed from the solder (due to oxide film); 2) it leaves a polymerized crust on the globule due to inadequate cleaning; and 3) it results in a 45% larger standard deviation, as a percentage of mean wetting force.

The variability is improved (i.e. reduced) when the 0.5% halide flux is substituted for non-activated R flux. Variability reduction in test methods assists efforts to achieve six sigma quality. This is because the variance of the used measurements is equal to the variance of the material plus the variance of the test method.

2.2.5 Configured Capillary

Going beyond the wetting aspects, which are measured in most other solderability tests, the configured capillary solderability test has the advantage of including the solder joint design.

The qualitative idea of this test is described by Thwaites [1982]. He shows a "V" shaped basis metal, dipped into solder, like the bottom of a slightly opened book. At the point of the "V," analogous to the location of the book binding, the solder joint is thinnest. Continuing the analogy, the joint is thickest where the open pages are farthest apart. The solder will rise higher where the joint is thin than where the joint is thick.

Wolverton and Ables [1991] tried this approach but were frustrated due to problems of quantification, so they designed a configured capillary test block of rectangular shape, consisting of two pieces. One piece is configured with machined-in grooves of varying depths from 1.0 logarithmically up to 17.8 mils. The other piece is a cover of the same area. When screwed together, the assembly forms a test block.

The test blocks are preheated, fluxed, and dipped into the solder melt. The capillary height response is plotted as a function of the natural logarithm of the solder joint thickness (Figure 2-8). A nearly perfect parabola results, with a maximum height indicating an ideal solder joint thickness for the given processing parameters combined with the given designed and manufactured solderable finish.

An alternative response of the configured capillary solderability test is solder joint voids. The test block is viewed with X-rays to determine the amount of voids as a function (again) of solder joint thickness. Figure 2-9 shows a typical result for two tested finishes. The gold over nickel finish has less voids at 3.2, 5.6, 10.0, and 17.8 mil thickness than does the tin over copper over nickel finish.

The configured capillary solderability test measures processes and designs concurrently. It compares solderable finishes, soldering processes, and/or solder joint designs (i.e. thicknesses). It provides variable data. Two responses are optimized: solder joint voids (minimized) and capillary height (maximized).

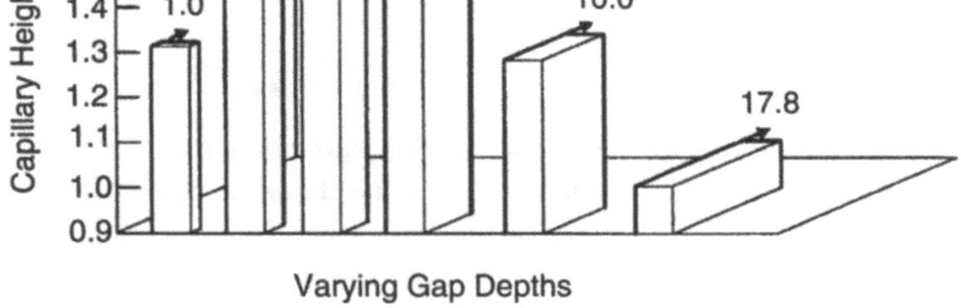

Figure 2-8 Solder capillary height vs. gap depth

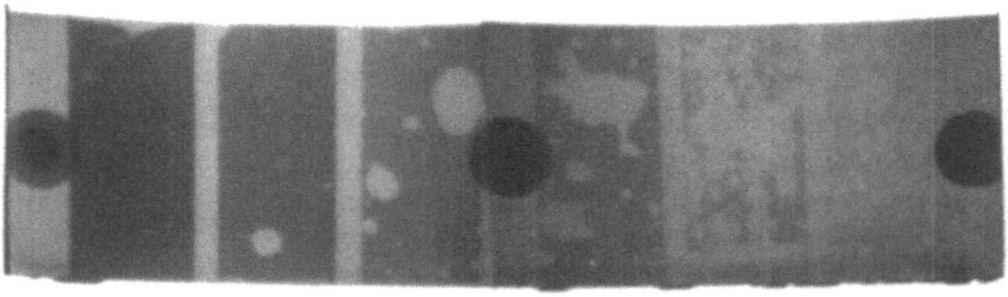

17.8 10.0 5.6 3.2 1.8 1.0
GAP DEPTHS (MILS)
(A) Ni-Au

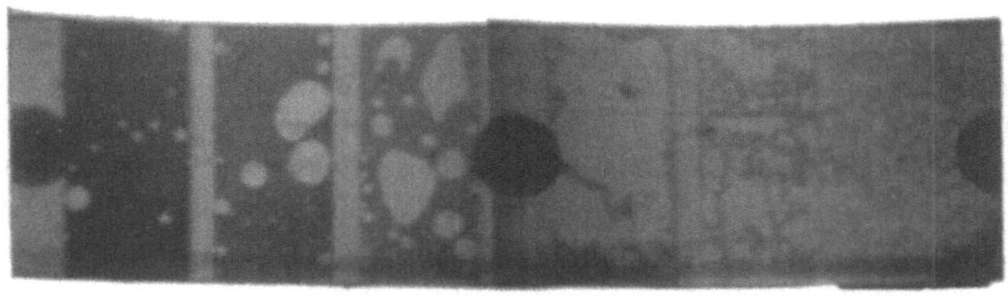

17.8 10.0 5.6 3.2 1.8 1.0
GAP DEPTHS (MILS)
(B) Ni-Cu-Sn

Figure 2-9 X-ray photographs of solder flow, other solderability test methods

Globule Test

A solderability test that is exclusively used for component leads is the globule test. With this test method, the component lead is brought down to bisect the globule as a timer is started. As wetting occurs, the globule flows over the lead and surrounds it. In doing so, the solder trips the timer, thereby recording the wetting time for the lead.

The test may not work well with very thin leads that lack the strength to bisect the globule. Also, square leads may not wet properly across the top surface. Aside from these difficulties, the globule test has proven to give reliable results when determining solderability.

PTH Globule Test

A variation of the globule test is the PTH globule test for measuring the solderability of plated through-holes (PTHs). The test involves placing the bottom of a fluxed plated through-hole into contact with the solder globule. The timer records the time required for the solder to flow up the PTH.

Other Tests

One of the oldest solderability test methods is the area-of-spread test [Thwaites 1959a, b], in which a specified volume of solder is applied to a fluxed surface. The surface is heated to melt the solder and is held at temperature for a specified time. The heat is removed so that the solder solidifies, and the evaluation is done by measuring the area of the solder globule after spreading. A larger area implies better wetting. A contact angle can be calculated using geometrical relationships. The fairly obvious limitation to area-of-spread tests is the necessity of having a large flat area in today's world of shrinking component sizes. Although the area-of-spread test is a good simple measure of wetting, as with some of the other tests, it is simply not compatible with some electronic components, especially those envisioned for future applications.

The meniscus rise test involves a vertical dip combined with a very accurate long range viewing device that allows measuring the meniscus of solder that forms during wetting (see Figure 2-7). A higher meniscus rise is indicative of better wetting. Actually, the amount of meniscus rise should correspond to the wetting force of the wetting balance test because a higher rise would imply the existence of a higher force on the sample. The equipment for this test is not considered sufficiently robust for measurements outside of a research lab; this is a major drawback to the method. The literature has relatively few references to this method.

The rotary dip test was developed by Colin Thwaites at the International Tin Research Institute in the mid-1960s [1964a, b]. An arm holds the fluxed test piece and rotates down to skim across a solder bath, exposing the surface to the bath. A visual examination of the solidified area shows the extent of wetting. The speed of the arm travel is adjusted to give a known time of exposure to the solder, and the method involves a series of tests to determine the minimum time at which full wetting is achieved. The limitations of this test are the need to conduct several tests, or at least two tests for each sample: one to assure wetting within, say, two seconds; and the other to assure no dewetting in, say, five seconds. The test does feature a relative motion of test piece and bath, simulating to some extent the physical environment of wave soldering.

2.3 THE CONSISTENCY PROBLEM

If the wetting balance alleviates the "numbers" problem, it only seems to exacerbate the consistency problem. Perhaps unknown factors are at work which throw off the wetting balance data without materially affecting its validity, but it cannot be denied that wetting balance data seem to show unacceptable variations from tester to tester.

2.3.1 Sample-to-Sample

With training, the dip tests seem to show relatively good sample-to-sample agreement. This behavior is perhaps attributable to the relative insensitivity of the test to small variations in solderability. Certainly, the longevity of dip testing suggests that it yields some useful data. There is a tendency to dismiss the dip test as an outmoded technology.

Wetting balance data from sample-to-sample can be very good as shown in work by many researchers [e. g. Lin and Harry 1982]. So, given consistent input material, the wetting balance shows good behavior. When combined with steam aging, however, the situation becomes more problematic. Steam aging is described in detail in MIL STD 202, Method 208. Briefly, parts are placed two inches above boiling water for a period of time, typically eight hours. After this aging, a vertical dip test can be run.

Control of the commonly used steam age preconditioning procedure was recently improved. (This preconditioning might simulate aging, as it metallurgically and chemically degrades component leads just prior to the solderability test.)

Weight gain experiments on foil specimens of copper and Sn-Pb reveal a steam temperature for maximum weight gain [Russell et al. 1989a, b]. This maximum weight gain correlates with maximum degradation of solderability. Figures 2-10 and 2-11 show the weight gain vs. temperature curves.

Weight gain of copper (and Sn-Pb) is maximum near a 95°C steam aging temperature, when the steam chamber is slightly open to air at one atmosphere. In contrast, when nitrogen is substituted for air, no weight gain occurs [Wolverton et al. 1989]. It is noted that the vapor pressure of water falls from 100°C to 80°C. Hence, the composition of the steam atmosphere is very much influenced by temperature just below 100°C.

Component solderability test specifications have responded to the data by specifying control of steam temperature when such preconditioning is used. MIL STD 202, Method 208 and ANSI J STD 002 require steam temperature control at 93°C ± 3°C, corrected for geographic elevation.

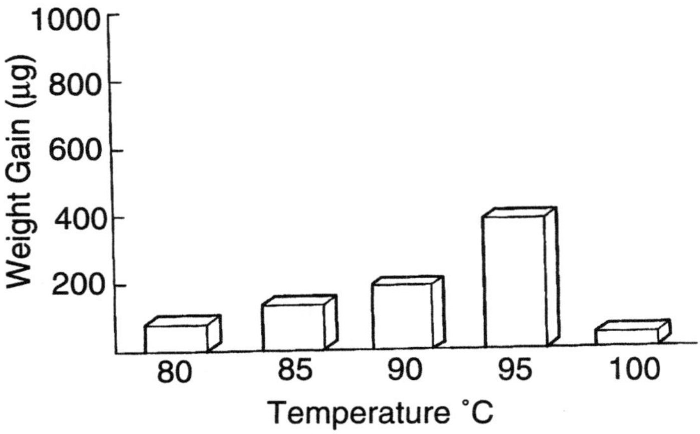

Figure 2-10 Weight gain as a function of temperature for Sn-Pb coupons under steam aging conditions

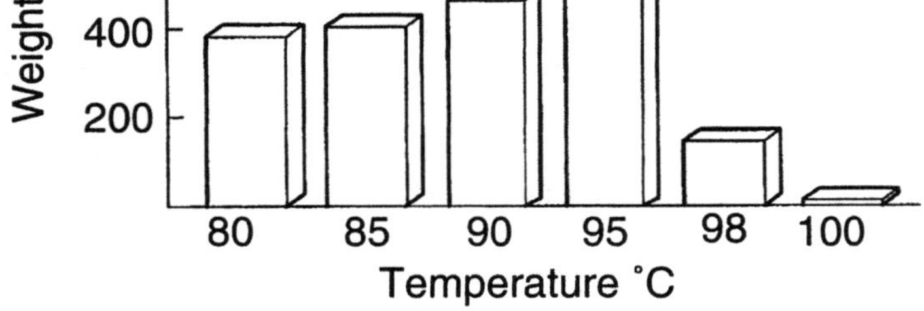

Figure 2-11 Weight gain of copper coupons as a function of temperature under steam aging conditions

2.3.2 Operator-to-Operator and Location-to-Location

There seem to be relatively few studies on these variations for dip tests. Some years ago an industry study examined an early version of the IPC timed rise test for plated through holes. It found an alarming inconsistency among five testers, but this may have been caused by inadequacies in the test method, rather than a condemnation of dip tests in general. Still, it might be comforting to see some studies done in this area, especially since dip tests remain widely used in industry.

The variations shown by wetting balance tests have been well publicized, as mentioned above. The IPC effort, in particular, involved several of the acknowledged experts in wetting balance testing, and they were unable to get satisfactory consistency, even when testing very simple samples. Although the cause of these variations is unknown, the new generation of test machines is making the test more automated, so that perhaps machine-based variations will be minimized.

It should be remembered that the wetting balance test is fairly gentle with the surface under test. In actual soldering, especially wave soldering, there is considerable physical force (akin to erosion) acting on the surface to help break up and break through surface oxides and films. The wetting balance, therefore, may unduly emphasize the surface condition of the sample, a notoriously difficult area to control.

2.4 THE REQUIREMENTS PROBLEM

Assuming that the wetting balance test is to be preferred over dip testing because of the additional data it provides and that the consistency problem can be overcome, there is still doubt as to the necessary test results to assure adequate solderability. As alluded to several times above, there are a lack of published data on the correlation of solderability test results to actual soldering success. This correlation is the whole reason for solderability testing.

Making this correlation is no simple task. There may be several process steps before assembly that could alter the correlation, so a great number of test samples would have to be incorporated in the study and followed very carefully. Often the facility producing the component is well separated from the facility doing the assembly, making the data trail very difficult. These difficulties must be overcome, however, because of their bearing on the future of solderability testing.

A part of this overall problem may be that solderability testing is necessarily done on the individual parts. It may be that in filling the capillary space that makes up the solder joint between two surfaces, good solderability from one of the surfaces may help the solder to wet and spread through the space, even when the other surface possesses somewhat poor solderability. If, in fact, a good surface can make up for a poor one, then acceptable solder joints can result even when a surface of poor solderability is involved. Such a condition seems plausible and would hamper attempts to achieve a sensible correlation of solderability and defects.

An example of such an effect is the "weak knee" problem in printed wiring boards. In this condition, reflowed boards, due to surface tension effects, have a very thin solderable coating layer at the rim of the plated through-holes. If such a board is subjected to a float test for solderability, it will fail the test criteria for a high-reliability product because solder will not flow up the hole and onto the surrounding pad. When actual soldering is done, these boards can perform satisfactorily, and it seems likely that the presence of a lead in the hole can help to pull the solder up and over the non-wetting area to the edge of the hole. It is very important, therefore, that studies trying to correlate solderability test results with actual soldering characterize the solderability of all surfaces involved.

2.5 THE AGING PROBLEM

Perhaps the most interesting and frustrating aspect of solderability testing is that a good result from a solderability test today will in no way guarantee a good result next month, let alone next year. Even with the use of just-in-time manufacturing techniques, there is still a need for shelf life. Spare parts must be stored and kept solderable for some time. For the military users, long-term storage of spare parts is a necessity, so shelf life is very important.

Figure 2-12 shows the results of some International Tin Research Institute studies into the capability of steam aging to simulate actual storage [Ackroyd 1976, 1977]. First, note the left hand set of curves which show the wetting times as measured by the globule test for wires coated with various thicknesses of Sn-Pb and after a given amount of steam aging. Analogous results have been obtained for normal storage, except the storage times are much longer than the steam aging times. For the reasonable thicknesses (> 1 μm), it can be seen that the wetting time stays at a low value for some time before very quickly deteriorating to a high wetting time. The time delay before deterioration seems proportional to the coating thickness. This is the basis for the above assertion that solderability tests by themselves are no measure of shelf life.

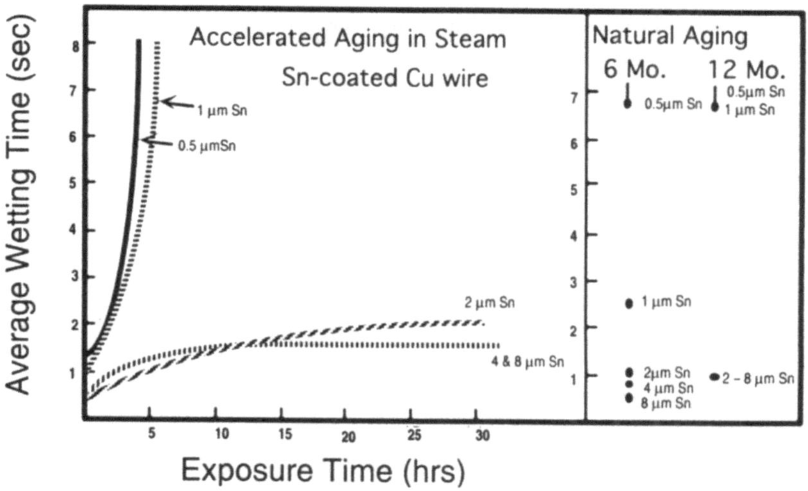

Figure 2-12 Plot of wetting time as a function of aging exposure time

The deterioration of solderability is the subject of Chapter 6 of this book, but it is worthwhile to state here that for normal storage conditions in the electronics industry, oxidation of the solderable finish surface (usually Sn-Pb) is not a major factor in solderability shelf life. The oxides which form on tin or Sn-Pb surfaces are thin, self-limiting, and friable; thus, so long as there is fusible material under them (residual tin or Sn-Pb), the oxide layer dissolves and/or breaks up under the action of the flux and the melting of the coating underneath, and it floats away, if not fully dissolved [Klein-Wassink 1989]. This view is consistent with the observed rapid deterioration of solderability after some incubation period. A simple oxidation attack of the surface should produce, in contrast, a gradual decrease in solderability.

The major reason for loss of solderability with such surfaces, then, is growth of intermetallic compounds which consume tin from the solderable coating [Ackroyd and MacKay 1977a, b; Davis et al. 1982; Davis et al. 1983a, b; Thwaites 1981; Warwick and Muckett 1983a, b]. As this tin is depleted, the coating becomes more Pb-rich and less able to melt during the soldering process. So long as tin is present to a level above its solubility limit in the lead phase, there will be a phase present that melts at 183°C. Once the tin content falls below that level, the melting point increases to 327°C. Now the coating will no longer melt under typical soldering conditions, and therefore the oxide is not so easily removed.

So, to simulate storage and thereby evaluate shelf life, these degradation processes need to be accelerated; this is most easily done by elevated temperature exposure. A convenient method of exposure is to use "steam aging." This method has been the source of much controversy in recent years and it is perhaps based on the changes that have taken place in solderability testing since the inception of steam aging.

Figure 2-12 shows the effect of steam aging on coated wires as measured by changes in the globule test wetting time. The right side of the figure shows the effect of "natural" storage on the wetting time. It can be seen that 16–20 hours of steam aging give about the same extent of degradation on these wires as does one year of storage at ambient. This study was the basis for the original IPC recommendation on steam aging. Subsequent work has suggested that perhaps 8 hours of aging is sufficient to screen out potential solderability shelf life problems. From Figure 2-12 one can see that 8 hours is indeed likely to sort out good performance from bad.

For more information on surface oxidation and its consequences, see Chapter 6 of this book.

2.6 CONCLUSIONS

We have seen some of the major problem areas facing solderability testing today. The need to obtain values for solderability is being addressed by the wider use of the wetting balance and the promise of alternative tests still being developed. The consistency problem is being investigated by the Institute for Interconnecting and Packaging of Electronic Circuits (IPC) Wetting Balance Task Group and others, and is being addressed by the test equipment makers who hope that more automation will produce more consistent results.

The requirements problem is a difficult one because of the complications of following specific solderability-tested components through to final soldering. Full characterization of the assembly is needed to separate the effects of differing solderabilities for the components of a joint. And, finally, the aging problem has overcome a hurdle with industry standardization on 8 hours of steam aging. Despite the apparent problems, steam aging seems to be a viable method for thermally stressing components to evaluate shelf life tendencies.

One of the most promising aspects of solderability testing has been the cooperation of industry groups, namely, the IPC and the Electronic Industries Association (EIA), in working on the steam aging problem. Improved communication between these groups and the involvement of other industry, government, academic, and national laboratory personnel, provide a more optimistic view that a solution to these problems is possible.

2.7 APPENDIX

2.7.1 A Descriptive and Practical Definition of Wettability and Solderability

Many definitions have been offered for the term solderability and a few can be found for wettability. The definition of solderability found in *Solders and Soldering* by H. H. Manko reads:

> "The properties of a base metal to be wet by molten solder under specified conditions of time, temperature and environment";

and that found in *Capillary Joining–Brazing and Soft-Soldering* by C. J. Thwaites is:

> "The property of a surface which allows it to be wetted by molten soft solder which can be measured by prescribed test methods."

In *Soldering in Electronics,* R. J. Klein-Wassink provides an excellent description of solderability, but a succinct, practical definition is lacking. Other definitions, very similar to those quoted above, can be found in various engineering specifications for soldering and in sales brochures from solder producers. In addition to the ambiguity (e.g. the words properties and property) and confusion of terms (such as wet and wetted) these definitions of solderability are descriptive and useful to the layperson and student but have no practical value to the expert interested in manufacturing process yield and the contemporary interpretation of quality.

Precise descriptions of the terms wettability and solderability will require careful thought and wordsmithing; that is beyond the scope of this communication. Rather, it is our purpose to suggest definitions which are quantitative and useful, and to relate these definitions to solder process yield. To most who have had experience in the engineering and science of soldering, the terms wettability and solderability are related but distinct attributes. Wettability is thought of as a complex combination of solder, flux, and substrate properties which can be characterized by careful laboratory measurements of, for example, contact angle or wetting force. Solderability is reflective of a solder process margin or "window" associated with solder assembly of parts. As such, the idea of solderability contains more variables and tolerances and is thus distinguished from the idea of wettability, although the terms are no doubt related.

Two examples of the distinction between wettability and solderability are provided here. Suppose there are N_t bare surface-mount devices which are earmarked for pretinning and soldering to an assembly board. After pretinning and inspection, N_w devices survive to be soldered. After soldering and inspection, N_{sw} devices are considered both wetted (pretinned, in this case) and soldered adequately. Wettability can be defined as $W = N_w/N_t$, solderability can be defined as $S = N_{sw}/N_w$, and yield can be defined as $Y = N_{sw}/N_t = S \times W$.

A similar scenario can be imagined for N_t bare resistors which are inserted into plated through-holes and soldered without pretinning. The wettability assessment is made again by inspection (e. g. of contact angle) with a similar number passing. The solderability assessment can be made by some functional measurement or by some other destructive/nondestructive test. The definitions of wettability, solderability, and yield remain as described above. Consequently, we offer the following definitions.

2.7.2 Wettability

We define a wetting "measure," w, to be an actual measurement such as contact angle, meniscus rise, or capillary penetration depth. If this measurement is made repeatedly on a given large number of parts, a parent distribution of w values is generated and is associated with that part and the technology used. Hence, we define wettability as the probability a given value, w, is greater than or equal to an arbitrary value, w_0, which is judged as acceptable. Stated in a quantitative way:

$$\text{WETTABILTY} = W = P(w \geq w_0).$$

2.7.3 Solderability

We define a soldering "measure," s, to be an actual measurement such as a nondestructive push/pull force, a resistance value, or visual inspection (1 or 0). If performed repeatedly in the manner described above for wettability, a parent distribution of s values is generated. However, the definition of solderability is slightly different. We define solderability as the conditional probability a given value, s, is greater than or equal to an arbitrary value, s_0, which is judged as acceptable given that the part is wetted. Stated in a quantitative way:

$$\text{SOLDERABILITY} = S = P(s \geq s_0/w \geq w_0).$$

These definitions are capable of generalization to the case of multiple inspection criteria. Suppose that inspection requires that $w_1 \geq w_{10}$ and $w_2 \geq w_{20}$ up to $w_n \geq w_{n0}$. Wettability is then

$$W = P\,(w_1 \geq w_{10} \ll w_2 \geq w_{20} \ll \cdots\, w_n \geq w_{n0})$$

and a similar definition can be offered for solderability.

2.7.4 Solder Process Yield

We define solder process yield as the probability a given part is judged both wettable and solderable; this is given quantitatively by the simple expression:

SOLDER PROCESS YIELD = S x W.

Quantitative and practical definitions of wettability and solderability should be required to

- recognize wettability and solderability as dependent but distinct attributes;
- be valid regardless of the wetting and soldering tests or technologies used;
- include the concepts of inspection and yield;
- be compatible with any inspection philosophy such as the Taguchi method;
- be compatible with multiple inspection criteria;
- allow forecasting of solder process yield;
- enable solder process control;
- provide a metric for continuous improvement feedback; and
- provide a metric for benchmarking.

It is our opinion that the definitions given of wettability and solderability comply with these requirements.

(The inequalities in these definitions may be reversed depending on the exact measure chosen.)

REFERENCES

(Solderability testing is often addressed in conferences, and the proceedings of these meetings are not always widely available. An extensive list has been compiled to assist the reader in getting further information on the subject and is available from the authors.)

Ackroyd, M. L. 1976. Proc. INTERNEPCON UK. Brighton, England. pp. 214-35.

—. 1977. *Publication No. 531*. Uxbridge, England: International Tin Research Institute.

Ackroyd, M. L., and C. A. MacKay. 1977a. Circuit World. 3(2):6-12

—. 1977b. *Publication No. 529*. Uxbridge, England: International Tin Research Institute.

Davis, P. E., M. E. Warwick, and P. J. Kay. 1982. *Plating and Surface Finishing*. 69:72-6.

Davis, P. E., M. E. Warwick, and S. J. Muckett. 1983a. *Plating and Surface Finishing*. 70:49-53.

—. 1983b. *Publication No. 634*. Uxbridge, England: International Tin Research Institute.

DeVore, J. A. 1978. *2nd Annual Tech. Seminar Proc*. China Lake, CA.

—. 1984a. *Circuits Manuf*. 24(6):62-4, 70.

—. 1984b. *J Metals*. 36(7):51-3.

—. 1984c. *Proc. 8th Annual Seminar*. China Lake, CA. pp. 123-29.

—. 1985. *IPC Tech. Rev*. 26(2):17-20.

—. 1990. IPC Meeting. San Diego, CA.

Edgington, R. J., et al. 1987. *Connection Technology*.

Fazekas, D. A. 1987. *11th Annual Electronics Manuf. Seminar Proc*. China Lake, CA. NWC TP 6789. pp. 75-85.

Klein-Wassink, R. J. 1989. *Soldering in Electronics*. Ayr, Scotland: Electrochemical Publications, Ltd.

Kwoka, M. A., and P. D. Mullenix. 1990. *14th Annual Electronics Manuf. Seminar*

Proc. China Lake: Naval Weapons Center. NWC TP 7066 and EMPF TR 0010. pp. 269-93.

—. 1992. *16th Annual Electronics Manuf. Seminar Proc*. China Lake, CA. NWC TP 7163. pp. 237-63.

Lin, K. M., and T. R. Harry. 1982. *Western Electric Engr*. 26(2):11-9.

Porter, J. 1992. *16th Annual Electronics Manuf*. Seminar Proc. China Lake, CA. NWC TP 7163. pp. 7-19.

Reed, J. 1990. IPC Fall Meeting. San Diego, CA.

Russell, W. R., B. M. Waller, P. S. Barry, and W. M. Wolverton. 1989a. *13th Annual Electronics Manuf. Seminar Proc*. China Lake, CA. NWC TP 6986 and EMPF TP 0007. pp. 31-53.

—. 1989b. *Sold. and Surf. Mt. Tech*. No. 3. pp. 25-9.

Thwaites, C. J. 1959a. *Trans. Inst. Metal Finishing*. 36(6):203 1 9.

—. 1959b. *Publication No. 304*. Uxbridge, England: International Tin Research Institute.

—. 1964a. *Electrical Manufacture*. 8(5):18-19, 22-3.

—. 1964b. *Publication No. 334*. Uxbridqe, England: International

—. 1981. *Brazing and Soldering*. No. 1. pp. 15-18.

—. 1982. *Capillary Joining--Brazing and Soft-Soldering*. John Wiley and Sons, Ltd.

U. S. Department of Defense Specifications. MIL STD 202, Method 208: Solderability. MIL STD 883, Method 2003.1: Solderability for Microelectronics.

Warwick, M. E., and S. J. Muckett. 1983a. *Circuit World.* 9(4): 5

—. 1983b. *Publication No. 632*. Uxbridge, England: International Tin Research Institute.

Wolverton, W. M., W. R. Russell, B. Waller, and P. Barry. 1989. *Microelectronic Packaging Technology*. ASM.

Wolverton, W. M., and B. Ables. 1991. *15th Annual Electronics Manufacturing Seminar Proc*. China Lake, CA. NWC TP 7110. pp. 7 25.

3

Fluxes and Flux Action

Colin A. MacKay

3.1 INTRODUCTION

Because the manufacture of fluxes is a proprietary formulation business, contents and compositions of the numerous "magic mixes" are necessarily unknown. It is their very proprietary nature that allows the range and variety of fluxes to exist. In this environment, attempts by user companies to determine even which product is best for their application or assembly process seldom reach the level of material differentiation. They are usually simple performance studies evaluated against specific functional performance requirements on a pass-fail or level-of-defect basis [Cassidy and Lin 1981a, b; Lambert 1978]. Attempts to identify important individual flux components and to assess the relative performance of material variants remain largely the perogative of the manufacturers. Soldering flux development is usually achieved by working either within individual internal perceptions of requirements or, more commonly, working with individuals within user companies to solve individual problems related to a specific product, design, and process. By its very nature, such work involves minimal resource and fast response time variants, usually meaning substitution of one component by a simple generic variation of the same class of material. This means that a group of products from a particular manufacturer continues by using the same basic materials types; the products are improved by modifications to formulations, usually within the same generic group of chemicals and based on the same concepts as to performance mechanisms. One recent exception to this development mode is the SA (Synthetic Activated) group of fluxes [Kenyon and Emig 1987]. Here, a major cleaning chemicals manufacturer intervened to formulate a flux system whose residues specifically required solvent cleaning chemicals. This resulted in a product with highly cleanable residue (provided a solvent was used) and the first major new flux formulation since the introduction of rosin-based fluxes for electronic applications. The drive for this initiative was fairly obvious.

However, study and identification of the causes of problems such as inadequate solder wetting, burned or high levels of residue, entrapped residues in the solder, inadequate soldering production rates etc., and the by-products that caused them have over the course of twenty to thirty years produced a body of information from which inferences and deductions give evidence of what materials must have caused the original problem. Review of the literature about fluxes indicates that.

3.2 FLUX HISTORY

Originally the idea of a flux as a soldering aid developed from the metal coatings industry, where the essential nature of a clean surface prior to tin or zinc coating was well known. Whether iron or copper, sheet metal was first immersed in an acid pickle solution before dipping in the molten metal bath. When soldering of two components was required, it was obvious that similar cleanliness requirements applied. Consequently, it became soldering practice to use the same chemicals for soldering as had been used for hot dipping, but applied in small amounts onto the joint area.

When hot dipping, the molten metal bath surface from oxidation was protected with a "pot cover" of mixed fused salts that floated on the molten metal surface. This same property was also required of a solder flux to protect the cleaned metal of the joint components between the time that the surfaces were cleaned and the molten solder wetted the joint interface. It is not surprising that an analogous mixture of the materials used for engineering soldering applications was formulated for electrical soldering. The liquid acid component cleaned the joint surface, and as the temperature continued to rise and the liquid evaporated, remnant salts fused *in-situ* to provide the pot cover. Wetting of the joint metals by the solder with subsequent capillary penetration completed the soldering operation. Copious washing to remove the fused salt residues was at that time no disadvantage.

Under these conditions it was easy to show that as the acid concentration dropped, the time taken to clean the surface increased; consequently, it took longer for the soldering operation, capillary rise, and penetration to initiate and complete. Soldering in the fledgling electrical industry used the same technology, but it soon became apparent that washing to remove the residues had to be more discrete, localized, and thorough since corrosion by water and salt residues represented a far more serious hazard in electrical circuitry than did corrosion in a heavy over-engineered mechanical assembly. The natural precautions of fluxing only a small area around the joint and carefully washing the residues from that area carried over into the earliest electronics assemblies where heavy wiring harnesses of individual wires and large components were involved. With the advent of the printed wiring board and the associated wave soldering operation, residues were distributed over the whole surface of the electronics assembly, and their presence became a serious consideration.

It was at this time that rosin-based fluxes were developed and rapidly dominated the market as electronics fluxes. The assumptions about how rosin fluxes worked and what controlled their efficacy were simply carried over from the earlier hot dipping ideas. As the complexity of the formulations of fluxes developed, so did their proprietary nature. Black magic pays when there aren't many witch doctors, so there was not a great effort at enlightenment as to how these improved fluxes worked. Because there was so little information available about how soldering fluxes worked mechanistically, classifications based upon such functional parameters as their action in removing tarnishes and oxides, enhancing wetting and improving spreading, were ignored in favor of classifications based on post-soldering effects such as the effects of residues [ISO Doc. Flux Residues; IPC J-STD-004, Reqs. for Soldering fluxes; Mil Spec F14256 Rev. F], the effects of residues upon testability, corrosion, surface insulations, and interactions that produced visual blemishes. It was in this way that tests and activity classifications became increasingly concerned with the activity of residues and not with the effectiveness of a flux in promoting the soldering process. Little concern and less explanation was available about how these complex formulations worked or what controlled the magnitude of their effectiveness.

3.3 FLUX REQUIREMENTS

A flux is required to clean and prepare the surface of a substrate or component so that it can be wetted by molten solder to produce a truly metallurgically bonded join between the joint components. Having prepared the interface surfaces, at some temperature below the melting point of the solder alloy, the flux then protects the surfaces from re-oxidation until the solder melts and wets the surfaces. To ensure that joints are sound, completely filled, and with the minimum of voids, temperatures at least 35°C and commonly 50°C above the solder melting point are used. It is an additional function of the flux to protect the molten solder from oxidation with the atmosphere during the time it is first molten, up until it passes through its maximum temperature, and until it re-solidifies and cools down to ambient. In part this is achieved by the flux dynamically reacting with any substrate or compound oxides as they form, but is realized mainly by formulating a component in the flux which, after the solvent has evaporated and the solids of the flux are molten, forms a continuous film over the surfaces that inhibits access of oxygen to the hot metal surfaces either of components or the molten solder. Fluxes also serve to reduce the surface tension of the solder/substrate/flux triple point interface at the advancing solder boundary, as indicated in a later section.

When solder cream is used, the flux vehicle in which the solder powder is suspended has several other requirements. The flux, in addition to cleaning and protecting substrate and component, must also react with or disrupt the natural oxide film that covers each solder particle. Only by fulfilling this function can the individual solder particles coalesce to form a continuous metal interface filling. Particle-to-particle coalescence creates a homogeneous metal film, producing a homogeneous advancing surface film interface that uses a combination of surface tension force and density difference arising from the density difference between molten solder (~8.5 gm/cc for a 60%Sn–40%Pb alloy) and rosin (0.9 gm/cc), to exert a homogeneous hydrostatic force. This then sweeps the flux droplets from within the solder joint volume to the outside of the joint and eliminates the entrapment that produces porosity.

The difference in density between solder and flux vehicle causes another flux requirement for solder creams. This is a capability to support the very heavy metal particles in a low density flux matrix so that the composite material does not rapidly separate into a layer of metal particles and a layer of organic flux material. A successful formulation is required to provide settling times (for a density difference of about 7.6 gm/cc) for up to about six months.

The success with which this latter ability is achieved creates yet a further problem, that of corrosive reaction within the oxide-reducing interaction of the chemical formulation within the flux. Designed specifically to react with oxidation films at modestly elevated temperatures, the flux vehicles are also required to remain in contact with the oxide films at or near room temperature around each solder particle without corrosive interaction or degradation of the fluxing performance for the whole storage lifetime. Although the equilibrium reaction temperature of a formulation may be a hundred or so degrees above ambient, any small ambient temperature reaction, particularly if it results in solid precipitated by-products, can over time produce a significant amount of product. This results in reduced fluxing effectiveness, increased viscosity, and eventually a hard crust of basic lead carbonate with tin oxide forming over the surface of the solder cream. The wide range of temperature ambients between winter and summer aggravates this problem.

These are all requirements of the pre-soldering performance of the flux. There are also some requirements of the flux after it has passed through the soldering process. The residue of the original flux, or any combination of the original material with any temperature or chemically modified by-products, is required to remain inert and non-corrosive to the assembly and components after soldering, irrespective of either oxidative or moisture reactions. The residual post-soldering film is also required to allow electrical tests without adversely affecting the test equipment and to be chemically compatible with any post-solder coating treatment that can be part of the completed assembly processing.

3.4 ROSIN-BASED FLUXES

3.4.1 Rosin Flux Formulations

Although commercial liquid fluxes are formulated as solutions in alcohols so that the dispersed ions do have some mobility at room temperature, there is very little evidence of room temperature reaction on an oxidized copper surface within normal fluxing times. The main purpose of the solvent is either to allow the application of a thin uniform film of the flux or, with further formulation, to allow a specific process application method (for example, foaming), as in the foam fluxing method. Preheating of the fluxed substrate, while serving to reduce thermal shock as the substrate contacts the molten solder surface, also serves to remove the solvent of the flux system and create a thin uniform film of molten rosin together with whatever activators have been added.

The rosin starts to soften at between 50°C and 70°C and is fully fluid at about 120°C. In this condition it is weakly acidic and, without any augmentation, will clean lightly oxidized copper. However, printed wiring boards and components do not come in only one grade of oxidation. Storage conditions and controls produce a range of differently oxidized surfaces, and the soldering assembly process is expected to easily cope with all of these and produce 99.99+% yields.

Another aspect of the soldering process is the production rate. Line speeds that allow economical production rates have to be achieved, so that although a plain rosin flux might be able to fully clean a surface eventually, the time required to do this could be unacceptably long. In order to improve both aspects of a plain rosin flux, ionic additives (originally chlorides), were formulated on the basis of the same idea about fluxing mechanisms as applied for inorganic acid fluxes. Therefore, the flux series designated RMA (rosin mildly activated) contains less activator than does the RA (rosin activated) series, and so on up to RSA (rosin super activated)[IPC J-STD-004].

The problem with unbridled application of the most activated materials was that the residues remaining after soldering were corrosive to varying degrees. This led in turn to specification of materials according to the amount and activity of the residue and the inclusion of tests such as the copper mirror test of residue extracts [Mil Spec F14256 Rev. F; QQS 571E] within the activity specification for electronics fluxes.

Technological developments within the formulation industry also served to make soldering performance/composition correlation more difficult. By preparing formulations with two-part activators which reacted to form the active constituent *in-situ*, the temperature at which reaction with the substrate began was raised from room temperature to above that of the reaction temperature of the two parts of the formulation. Activators in which the active halogen is hydrogen-bonded to a second molecule (as a hydrohalide molecule, hydrogen-bonded as an amine hydrochloride) are also used. The temperature required to break the bond and release the halogen compound represented the minimum activity temperature. Yet higher activity temperatures could be realized with activator compounds which incorporated halogens or some other active group ionically or covalently, so that the more stable bonding of the active elements required higher dissociation temperatures. Implied in these developments was the idea that because they required higher temperatures to initiate activity, the residues from such activator molecules would somehow be less active after soldering. This ignored the fact that often the compound used as carrier was volatile and was not available for recombination after soldering, so that one half of the initial compound was no longer available for recombination after soldering. Such an example would be hydrazine chloride, which on decomposition releases hydrohalide that reacts with the tarnish layer. The hydrazine breakdown products are volatile; after soldering, the radical needed to reform the chloride is no longer present and the halogen remains to participate in any corrosion mechanisms. Only in the case of covalently bonded compounds was there evidence that residues were less active.

3.4.2 Flux Contents

The principal constituents of liquid electronics fluxes are isopropyl alcohol as the solvent and gum rosin as the vehicle. To extend or enhance the action of a simple rosin flux, activators are added. These are organic acid, halogenated compounds, and amines, or mixes of these materials. Originally considered as being required to increase reaction rates for the reduction and dissolution of oxides and tarnish, more recent solderability measurements indicate that their role is more to extend oxide reduction capacity than reaction rates [Audette and MacKay 1981, 1982]. Additional chemicals such as wetting agents or foaming agents are often added to facilitate special purpose fluxes such as those applied as foam suspensions or as sprays. Further special requirements, like elevated temperature activity or prolonged tackiness, can necessitate small additions of higher boiling point solvents.

When a solder cream is formulated, the flux vehicle is produced as a gel or cream using a more viscous solvent such as butyl carbitol. In order to maintain the dense solder metal particles, oleo resinous materials like thixatrol or thixcin are commonly added [MacKay 1980a, b; 1989; Johnson and Kevra 1989] to generate a gel whose cellular structure helps support the metal. Viscosity control is achieved by addition of dimer and trimer derivatives of the abietic acid, the basic constituent of rosin.

SA fluxes are formulated without rosin. Instead, a high molecular weight carbowax-type compound with an iso-alkyl acid phosphate material such as the octyl or butyl compound is used. Some of the most common materials found in fluxes are shown in Table 3-1.

Table 3-1 Some Materials Found in Liquid and Cream Fluxes

Flux Form	Purpose	Chemical Type	Flux
Liquid Flux	Solvent	Isopropyl Alcohol	R, RMA, RA, RSA
		Glycol	R, RMA, RA, RSA
		Glycerol	Low residue, Water soluble
	Unactivated Vehicle (Low Melting Point Solders)	Water White Rosin	R, RMA, RA, RS
	Unactivated Vehicle (High Melting Point Solders)	Tall Oil Rosin	R, RMA, RA, RSA
	Activators	Organic Halogen Acid Adducts	RMA, RA
		Halogenated Organics	RMA, RA
		Di-, Tri-alkyl amines	RMA, RA
		Ammonium, Hydrozinium etc., Halides	RA, RSA
		Carboxylic (Aryl and Alkyl) Acids	RMA (non halogen), RA, RSA

		Carboxylic (Aryl and Alkyl) Acids Chlorhydrates	RMA, RA, RSA
		Anilene Chlorhydrates	RMA, RA, RSA
		Halo Pyridenes	RMA, RA, RSA
		Halo Hydrate Hydrozino salts	RMA, RA, RSA
Cream	Solvent	Butyl Carbitol	R, RMA, RA, RSA
	Unactivated Vehicle (Low Melting Point Solders)	White Rosin	R, RMA, RA, RSA
	Unactivated Vehicle (High Melting Point Solders)	Tall Oil Rosin	R, RMA, RA, RSA
	Activators	Organic Halogen Acid Adducts	RMA, RA
		Halogenated Organics	RMA, RA
		Di-, Tri-alkyl amines	RMA, RA
		Ammonium, Hydrozinium etc., Halides	RA, RSA
		Carboxylic (Aryl and Alkyl) Acids	RMA (non halogen), RA, RSA
		Carboxylic (Aryl and Alkyl) Acid Chlorhydrates	RMA, RA, RSA
		Anilene Chlorhydrates	RMA, RA, RSA
		Halo Pyridenes	RMA, RA, RSA
		Halo Hydrate Hydrazino salts	RMA, RA, RSA

	Amine modified Abietic Acid, i.e. Polyrad 1110	R, RMA, RA, RSA Aqueous Cleaning
Suspension Agents	Thixatrol	R, RMA, RA, RSA
	Thixcin	R, RMA, RA, RSA
	Klucel	R, RMA, RA, RSA
Viscosity Control	Dymerex	R, RMA, RA, RSA
	Trimerex	R, RMA, RA, RSA
	Pentolyn	R, RMA, RA, RSA
	Stabelite Esters	R, RMA, RA, RSA

3.4.3 Chemistry of Rosin

The principal constituent of an electronic flux is rosin. This is a natural product derived from the resinous excretion of trees. It is a glassy mix of abietic acid with some isomorphic compounds of abietic acids and numerous hydrogenated modifications of that acid. The basic chemical structure is shown in Figure 3-1.

Owing to its solid state and glassy structure, the acid action of rosin is not apparent until it melts. Only then do the ions of the material have sufficient mobility to enter into reactions with substrate surfaces. The material exhibits ionic character only in the fused state; that is, it has a pK rather than a pH number [Glasstone and Lewin 1960].

Because of the numerous reactions between the various constituent parts of rosin and the critical relationship this must have upon the identity of the residues which can form during the soldering operation, it is appropriate to understand the origins and composition of the starting materials and refined rosin used in fluxes. As a natural product, the proportions of the various constituents will vary from genus to genus and also for different variations within a botanical group [Harris 1952; Feiser and Feiser n.d.]. For example, in Europe, the material commonly used is derived from a specific type of Portuguese pine; in America, a North Carolina variant is specified. In addition to the gross differences in product constitution found between types of trees, trees of the same type grown in different locations, even material from the same woods in different years, can show significant variations. This obviously affects product uniformity and repeatability.

The oleoresinous secretion collected from scarred or "barked" pine trees of the genus Pinus Palustrus or, less often, Pinus Caribaea, is the most common, but not the only source of the raw materials from which American rosin is produced. This produces gum rosin. Wood rosin is extracted from aged stumps and tall oil rosin is distilled from the rosin extract from pulped whole trees. The composition of the rosin varies in each case, in part reflecting the different proportions of sapwood and heartwood in each source, the rosin constitution in each being different.

The oleoresinous exudate material contains turpentine and a range of resin acids. Its essential composition is shown in Table 3-2 [Harris 1948; Harris and Sanderson 1948a, b; Harris and Sparks 1948; Palkin and Harris 1933].

All of these resin acids are closely related; being diterpene acids of general formula $C_{19}H_{29}COOH$, they are two related groups based upon either abietic or pimaric acids. Their close stereochemical relationship is illustrated by their structures, as shown in Figure 3-1.

As can be seen from the ring numbers illustrated in the abietic acid structure, all have methyl groups at positions 1 and 12; while the abietic acids have propyl or isopropyl groups at 7, the pimaric types have methyl or vinyl groups in this position. The differences within the hydroaromatic nucleus of the structures are essentially the positions of the double bonds and the presence (as in abietic and neoabietic acids) or absence (as in levopimaric, dehydroabietic acids) of bond coordination. All these acids are extremely labile, and readily isomerize or dehydrogenate from one constituent to the others. Thus, the refining of oleoresinous exudate produces a modified rosin composition which is the gum rosin of composition shown in Table 3-3 [Campbell n.d.; Harris and Sanderson 1948b].

Table 3-2 Composition of Oleoresinous Exudate

Rosin Acid	% Composition
Levopimaric	30–35
Neoabietic	15–20
Abietic	15–20
Iso-dextro-Pimaric	8
Dextro-Pimaric	8
Dehydroabietic	4
Dihyroabietic	4

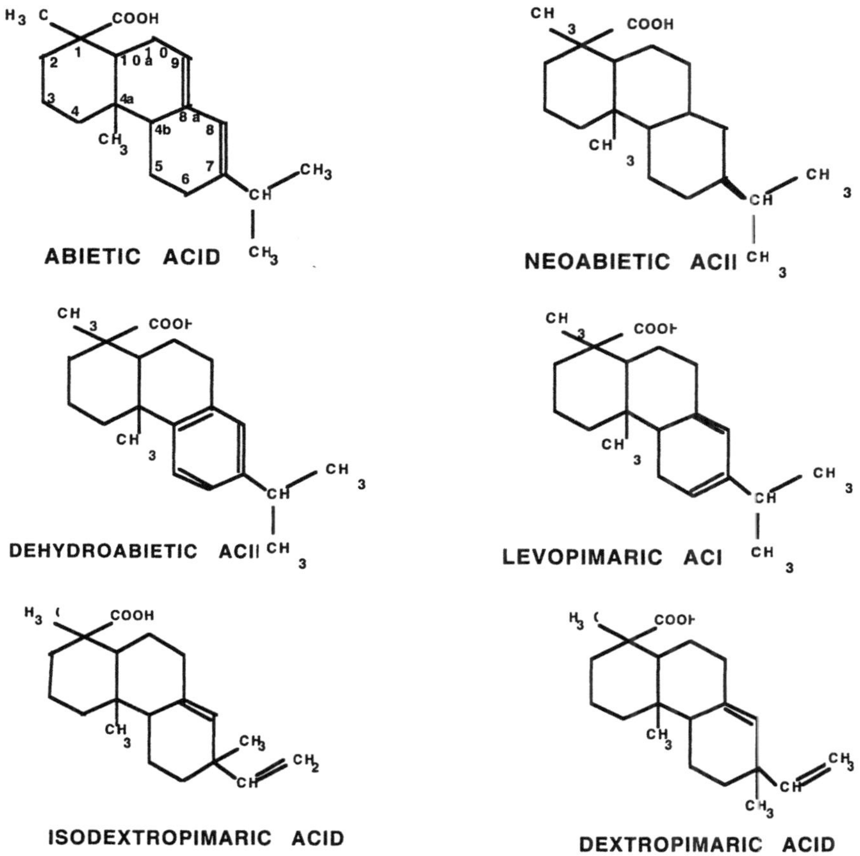

Figure 3-1 Stereochemical formulae of some resin acids

Table 3-3 Composition of Gum Rosin

Rosin Acid	% Composition
Levopimaric	Trace
Neoabietic	10–20
Abietic	30–40
Iso-dextro-pimaric	8
Dextro-pimaric	8
Dehydroabietic	5
Dihyroabietic	16
Tetra-abietic	16

The principal casualty of the distillation stage at 150°–170°C is levopimaric acid, which is isomerized and disproportionated. Two double bonds are rearranged to the more stable abietic acid configuration, which raises the abietic acid composition to 30–40%. A disproportionation reaction with hydrogen exchange converts two molecules of levopimaric acid to dehyrdoabietic acid and dihyroabietic acid; another similar reaction involving levopimaric acid with abietic acid yields tetra-abietic acid and some dihydroabietic acid.

So, while the major proportion of gum rosin is abietic acid, it is by no means the only constituent; and to treat the identification of the results of fluxing reactions as solely the result of abietic acid reactions is a gross oversimplification.

The reactions mentioned above are the result of the distillation at 150°–170°C. Heating regimes up to 300°C, as in soldering, provide a far wider range of possibilities. Reaction among the abietic acid types via dehydrogenation at 250°C-270°C lead to further production of dehydroabietic acid, while at 300°C hydrogenation of neoabietic acid initiates a reversible reaction to abietic acid. By comparison the pimaric acid types appear relatively stable.

Reactions of this type must be responsible for many residues, together with a number of metal salts of these acids. Investigations to identify these under more controlled conditions than the full complex mix have not been undertaken since the extensive structural derivations by Ruzicka et al. [1922, 1933, 1940; see also Feiser and Feiser n.d.; Feiser and Campbell 1938] and the refinement studies by Harris et al. [1933, 1948: all refs.]. The complexities of these reactions when halogens are added to the mix have had no investigation either.

3.4.4 Synthetic Resin/Rosins

Polyvinyl acetate and pentaerythritol tetrabenzoate and bile acids [UK Patents 1550648 and 1577519] have been used as synthetic resins, particularly in low solids formulations. Substituted rosins such as ethoxylated versions have also been used with specific properties such as water solubility being obtained by amine substitution at acid sites.

3.4.5 Mechanistic Inference

As mentioned previously, the reasons for using fluxes are well known. First, they clean the surface to be soldered by removing the oxidation and/or other chemical contamination. Second, they form a protective blanket over the clean surface to prevent re-oxidation until the temperature becomes high enough for soldering. The assumed mechanism by which fluxes act, and hence by which improvements have been based, is one of concentration dependence of the acidic constituents.

Assessment of a flux's activity and assignment to a designated activity level among the classes acceptable for electronics applications depend upon a combination of criteria that often reduce soldering performance to a secondary consideration. Some classifications are by percentage of activator, others distinguish fluxes by the activator used, while still others combine water extract resistivity and flux corrosiveness on a thin copper film deposited onto glass as the definitive classifier. It has been suggested that in part the reason that a relatively poor correlation between wetting balance performance and flux formulations has been synergistic effects between the components of increasingly complex formulations.

3.4.6 Wetting Balance Performance of Rosin Fluxes

How well do rosin fluxes of different types perform and what can be deduced from their performance? Wetting time tests with the standard Surface Tension Balance procedures [Mil Spec 883C; MacKay 1970; Wallis 1974], using a range of different flux formulations with copper treated to produce a range of different surface oxidation levels [Audette and MacKay 1981, 1982], have produced interesting results (Figure 3-2).

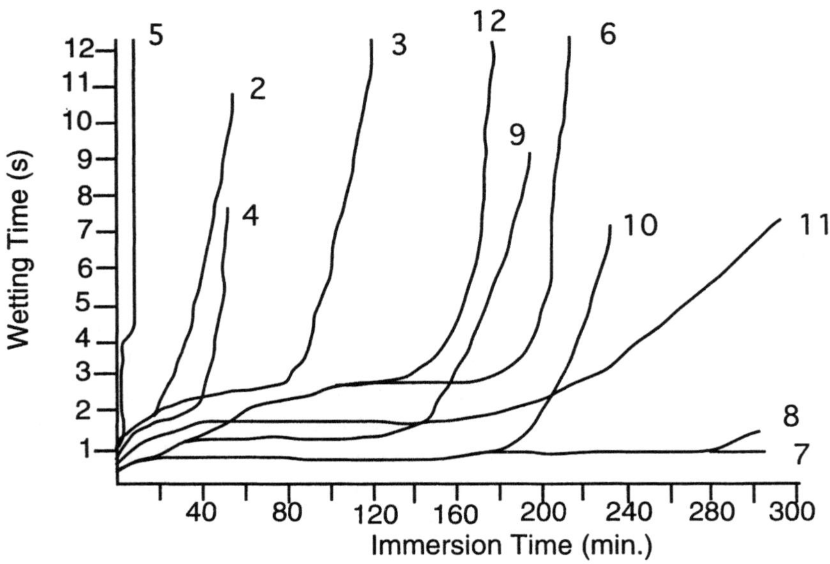

Figure 3-2 Wetting times of progressively oxidized copper for a range of fluxes as defined in Table 3-4

The curves of Figure 3-2 essentially exhibit the same form, consisting of an initial sharp increase in wetting time followed by a plateau that remains nearly horizontal as oxidation levels, represented here by immersion times in an oxidizing solution, increase. Further increases produced an abrupt rise in wetting time immediately prior to the surfaces becoming non-wettable. In this region the results diverge considerably.

Certain features and the shape of the wetting time/immersion time curves become significant. Assuming that since all fluxes have low viscosities, the same solvent, and a fairly narrow range of solids content, they deposit an equal film over the test coupon so that the same volume of active constituent is involved in all cases. In this way a controlled quantity of rosin plus activator was available in each case to react with the oxide layer. Then each sample becomes representative of the bulk flux from which it was taken, and flux results then become comparable.

The fluxes tested were all of the rosin type with grades RMA and RA. They were also from a single commercial supplier so that the numbers quoted are for that generic product series containing essentially the same types of chemicals. Details of the application and performance of these fluxes are given in Table 3-4.

Plain rosin non-activated flux (flux ref. no. 1) was also examined, but since the maximum immersion time oxidation treatment that this flux was able to solder was only 1 minute, this would not fit on the scale of Figure 3-2, although the general shape of the curve was the same.

All fluxes exhibited essentially the same features (as shown in Figure 3-2): an initial zone in which the wetting time increased sharply, followed by a level plateau, and then a second sharply increasing region which rose to an infinite wetting time, i.e. non-wetting. In some cases, the plateau was vestigial, as for 1 (R) and 5 (RMA); while for others, such as 7 (RA), it was extremely extended. In general, the lowest and longest plateaux were for the most active, i.e. RA, fluxes. Plateau levels also tended to mirror activity levels, occurring at the shortest wetting times for the RA fluxes and at the longest wetting times for the RMA and R grades. Some RA fluxes, however, exhibited the same plateau wetting times as the RMA materials (compare, for example, 6 (RA) and 12 (RA) with 3 (RMA)). It was also clear that another major effect of the activators, particularly the complex or augmented systems such as 8 and 9 (RAs) and 3 (RMA), was not to accelerate fluxing but to extend the capacity of the material so that it will react far greater amounts of tarnish at the same rate.

3.4.7 An Active-Constituent, Concentration-Dependent Mechanism

What is actually happening along the wetting time vs. oxidation curves shown in Figure 3-2 is that the oxide is steadily being removed until a clean copper surface is exposed. When the metal is clean, the solder metal reacts with it and wets the clean copper. The basic wetting time for clean copper is indicated by the intercept of the wetting curves with the wetting time axis when oxidation immersion time is zero. Experimental results indicate a base wetting time of approximately 0.5 s. However, further experiments indicate that the time shown here was limited by the rise time of the test equipment and that using equipment with a rise time response of ~ 0.002 s gave a wetting time of between 0.1 and 0.15 s for clean copper with 60% tin–40% lead solder. Any increase in wetting time above this minimum value would then represent the time taken for the test flux to remove the oxidation.

In fact, most of the wetting time recorded was the time taken to remove the oxide film; thus, with a near linear relationship between immersion time and oxide thickness, the ratio of the oxidation time to the wetting time at any point on the curve would be a measure of the reaction rate for dissolution of the oxide in the flux. Figure 3-3 shows the results of plotting these ratios for all the wetting time curves in Figure 3-2.

Table 3-4 Properties of Fluxes with Test Results in Figure 3-2

Flux Ref. No.	Classification	Properties
1.	R	40% pure water white rosin in alcohol. Not activated, used for electronic soldering where residues are left on boards.
2.	RMA	Rosin in alcohol, 37% solids activated with a non-ionic agent. Used for electronic soldering by both foam dip and wave fluxing. Residue removal necessary only in most critical applications.
3.	RMA	Most active rosin in alcohol flux in its class. 37% solids activated to higher level than (2.) with different non-ionic activating agent. For electronic soldering; also for dip and wave applications. Residue removal as above.
4.	RMA	Rosin in chlorinated solvent. Medium activity flux as (2.) Activated with low level of ionic activator (37% solids).
5.	RMA	Special low solids content rosin in alcohol activated with non-ionic agent (25% solids).
6.	RA	Rosin in alcohol, organic acid activated flux. Completely halide free. Reduced residue corrosivity (15% solids).
7.	RA	Rosin in alcohol, high solids content (57%) with an ionic activating agent. Does not degrade during extended soldering times.
8.	RA	Rosin in alcohol, fully activated with an ionic activator. Residues generally must be removed (25% solids).
9.	RA	Rosin in alcohol, low solids (13%) with an ionic activator. Easy cleaning flux.
10.	RA	35% rosin in glycol ether solvent with an ionic activator. Low volatility, low maintenance flux.
11.	RA	Rosin in alcohol (25% solids), moderate activity. Residues efficiently removed by chlorinated solvents.
12.	RA	Rosin in alcohol spirit, special low wicking formulation with an ionic activating agent (25% solids).

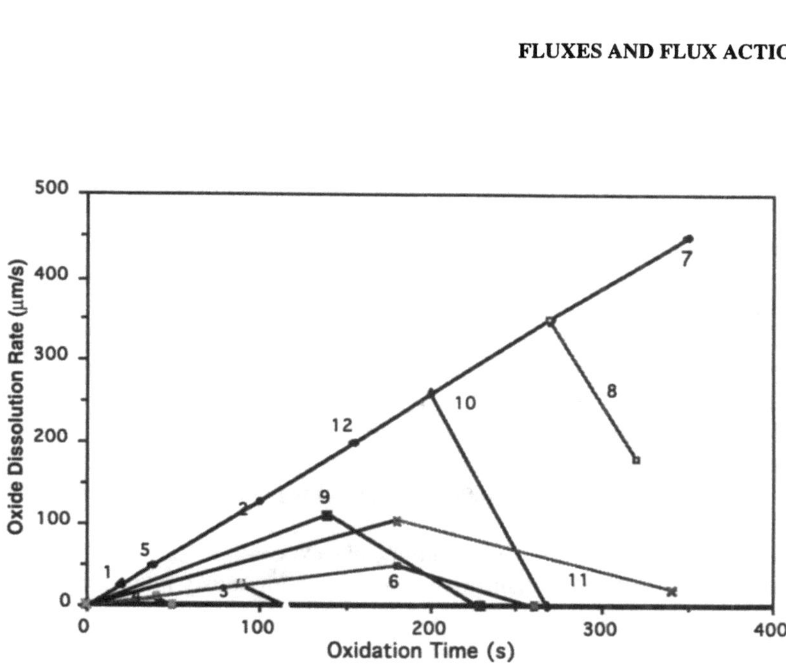

Figure 3-3 Oxide dissolution rate from solderability values in Figure 3-2

When these ratios were measured for all points on the curves, it was found that they all fell on a single straight line (as shown in Figure 3-3), which shows a positive slope and increasing reaction rate with oxide film thickness.

If flux action were dependent upon reactant concentration in the flux, either rosin itself or the activator or both, as has previously been supposed, then from classical considerations the reaction rate of the flux with an oxide film would be simply dependent upon the concentration of the active species in the solution. In physical chemical terms, for a system that depends only upon a single reactant, i.e. a first order reaction, then the form of the concentration with time curve would be expressed as [Glasstone and Lewin 1960]:

$$c_t = c_i \, e^{-kt} . \tag{3-1}$$

With more than one component in the system the equation is of the same form, with the starting conditions modified to

$$dx = k(A - x)(B - x) \, dt . \tag{3-2}$$

Both curves are of the same form, as shown in Figure 3-4. Even when a solid film dissolving into a solution is one of the constituents, the concentration of the dissolved solid remains constant until all of the solid has dissolved. Only then does the concentration of that component begin to decrease, in which case the first part of the reaction behaves like a first order reaction, while the last part behaves like a second order reaction. In either case, the general form of the relationship between concentration and time is a negative exponential, as before—a negative exponential as in Figure 3-4, which does not yield a reaction rate curve of the form shown in Figure 3-3.

Nor are these effects altered by having only a restricted flux volume; this would accelerate the rate of decrease of concentration so that the reaction rate drops more sharply. Higher order reactions would have the same general negative exponential form.

Whatever the precise order of reaction, the expected wetting time curve, a concentration-dependent mechanism, would be a continuously increasing wetting time as oxide thickness increased. This is clearly not what is observed during flux action in Figure 3-3.

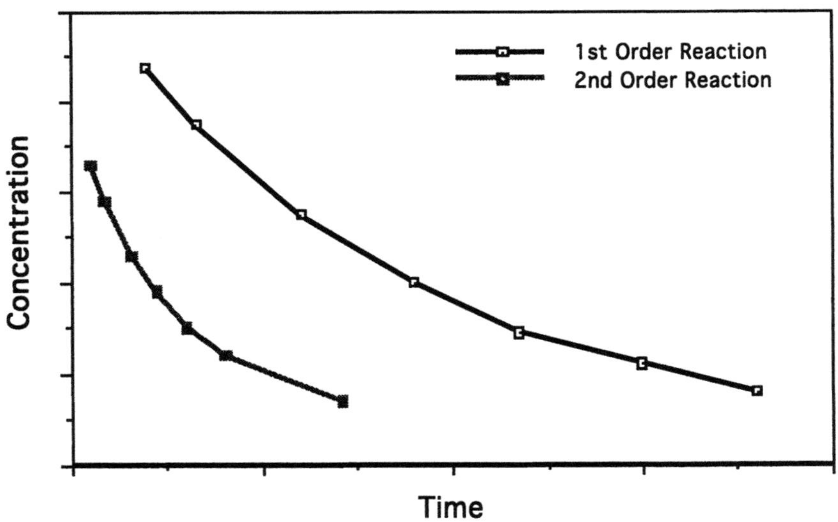

Figure 3-4 Form of first and second order reactions

3.4.8 Autocatalysis and Chain Reaction Equations

As a general rule the rate of a catalytic reaction is proportional to the concentration of the catalyst present. An example of this would be the catalysis of ether by iodine vapor where the equation for the rate of decrease in the amount of ether in the presence of iodine could be written [Glasstone and Lewin 1960]:

$$\frac{-d\,[\text{ether}]}{dt} \qquad [I_2]\ k_1\ [c_{I2}]\ k_2\ [c_{\text{ether}}] \tag{3-3}$$

leading ultimately to a function

$$\frac{x}{P_o} = \frac{(e^{at} - 1)}{(1 + be^{at})} \tag{3-4}$$

which is a steadily increasing function with time as in Figure 3-4 and not one that decreases exponentially.

Alternatively, a chain reaction in which an intermediate product of the reaction reacted with one of the constituents to create yet more of the intermediate would give a function [Glasstone and Lewin 1960]:

$$\frac{d\,[AB]}{dt} = \frac{K_a\ [c_{A2}]\ [c_{B2}]}{K_b + [c_{ab}]\ [c_{B2}]} \tag{3-5}$$

This also represents a continuously increasing function.

From a comparison of the form of the reaction rate curve which is observed in the surface tension balance experiments (Figure 3-4), it seems clear that the mechanisms that would support the fluxing process, as indicated by Figures 3-2 and 3-3, would be either autocatalysis or chain reactions. In the first case the reaction could be catalyzed by the same ions, either copper or hydrogen, which were being generated during oxide removal. A combination involving the halogen ions from the activator system would also be possible. Such driven catalytic reactions are not unknown.

A chain reaction system could be such that a secondary reaction involving copper ions, possibly even involving transitions from valence state Cu^{++} to Cu^{+}, assumes the rate determining function. Combinations of these could happen, with the possibility that an exothermic reaction could further accelerate reaction rates.

The possibility that the increase in temperature arising from an exothermic reaction increased reaction rates to the extent indicated, would tend to be discounted, since calorimetric results do not show large exothermic peaks.

3.4.9 Implications of Alternative Mechanisms

The implication of the differences in fluxing mechanism from the simple concentration-dependent mechanism assumed for many years can be wide–reaching. Perhaps injections of less corrosive compounds than currently used activators, but which contain the necessary ions, could replace the corrosive residue inorganic halides currently used; such replacements might be inert materials. A source of copper ions, for example, could show a significant activity increase. In this way very highly activated systems that have no reactive residues could be formulated. In a similar vein, questions involving the use of other ionic species, possibly other metal systems, certainly deserve consideration and study.

An example of where this type of study could lead is the problem currently looming in the field of solder creams for surface mount applications: where finer and finer detail requires finer solder powders for satisfactory screen printing performance. This in turn means higher oxide percentages in the powder; requiring, in turn, more flux activity to allow solder reflow during the soldering operation. Finer printed wiring board detail is usually associated with smaller printed surface board/component joint gaps, making it more and more difficult to clean residues from increasingly more activated solder creams. This seems to be a classic vicious cycle, which only a more enlightened approach to studies on how flux systems work can break.

3.4.10 Effect of Impurities in Solder with Rosin Flux

As with fused salt fluxes, the general effect of impurities in solders is to decrease the area-of-spread results, as in Figure 3-5, for example. The effects on surface tension balance results were mixed and depended upon the phase diagram of the ternary systems that the impurity elements formed with the tin/lead alloy. For cadmium, whose results for both wetting rate and equilibrium wetting force are shown in Figure 3-6, wetting times increase and the equilibrium wetting force decreases. When antimony was the impurity (Figure 3-7), there was no significant change in wetting time, while the equilibrium wetting force showed a significant decrease.

In this situation the area of spread would have shown a decrease. By comparison, when bismuth was present (Figure 3-8), wetting times actually decreased while the wetting force increased. This represented a smaller area of spread which was reached more quickly.

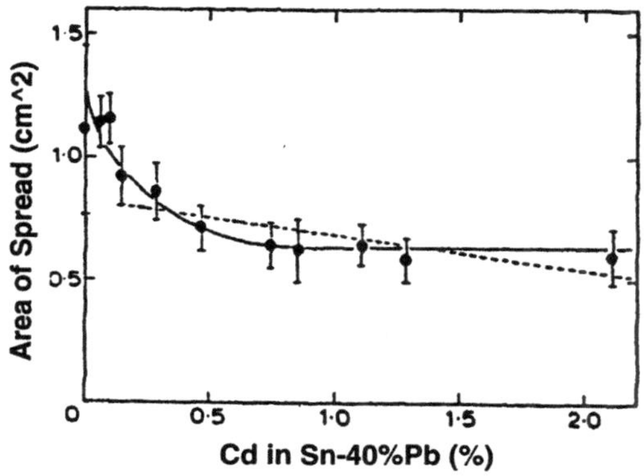

Figure 3-5 Effect of cadmium on area of spread of 60Sn-40Pb solder on copper base metal (non-activated rosin flux)

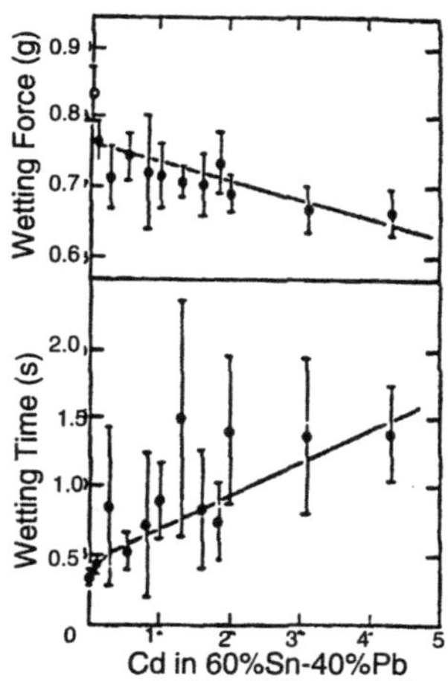

Figure 3-6 Effect of cadmium on wetting properties of 60Sn-40Pb solder on copper base metal (non-activated rosin flux)

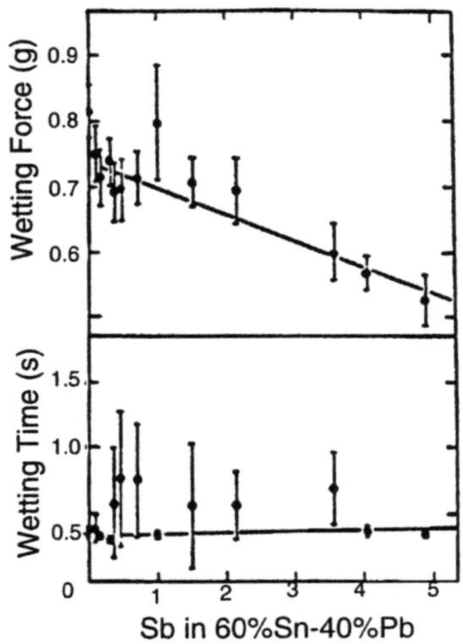

Figure 3-7 Effect of antimony on wetting properties of 60Sn–40Pb solder on copper base metal (non-activated rosin flux)

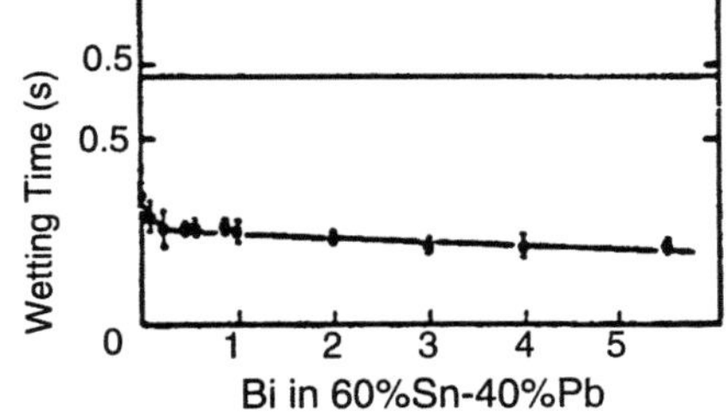

Figure 3-8 Effect of bismuth on wetting properties of 60Sn–40Pb solder on copper base metal (non-activated rosin flux)

3.4.11 Effects of Fluxes on Surface Tension

While the principal role of fluxes is essentially one of providing pure clean interfaces so that the molten solder can readily wet the substrate/component surfaces, the simple equilibrium configuration shown in Figure 3-9 invokes the interaction of several different interfacial tension forces, depending upon the environment of the interface. From this it is obvious that changes induced in the surface tension of the solder must have a significant effect. A reduction in surface tension of the metal surface affects the shape of the metal surface at the flux/solder interface and also the degree to which it will spread. Thus, the degree of penetration of a solder into a capillary gap and the level and rate that it will wet and climb into a hole will be greatly influenced by the flux/solder system.

The surface tension of tin, lead, and solders generally decreases with temperature [Pokrovskii and Galanina 1949; Semenchenko et al. 1953; Pokrovskii and Sosnina 1950; Hagness 1921; Draht and Sauerwald 1926] and shows a continuous smooth exponential function from the high surface tension component (tin), to the low (lead). They do not show discontinuities or inflections at special metallurgical compositions, such as phase changes or eutectic points, across the full phase diagram composition range [Semenchenko et al. 1953; Hagness 1921; Bircumshaw 1928a, b, c; Matuyama 1927; Pokrovskii and Saidov 1955; Howie and Hondros 1982; Klyachko and Kunin 1949; Greenway 1949].

Table 3-5 lists surface tensions for a variety of elements found in solders from which the form of the curves [Semenchenko 1961] for binary systems can be inferred.

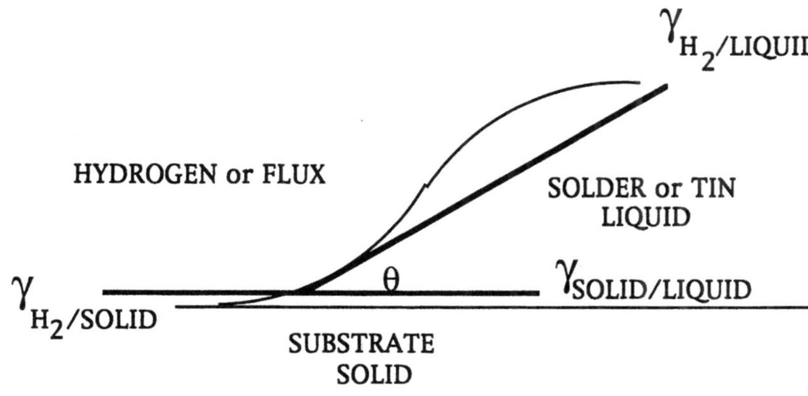

Figure 3-9 Wetting force equilibrium

Table 3-5 Surface Tension of a Variety of Elements Found in Solders

Element	Temp. (°C)	Surface Tension (Dynes/cm^2)	
		Measured	Computed for 400°C
Tin	418	617.1	523.9
Lead	425	447.3	441.6
Cadmium	444	628.8	624.8
Bismuth	450	365.9	366.3
Antimony	432	244	---
Zinc	420	773.0	755.7

The effect of atmosphere is to reduce surface tension (Matuyama 1927; Semenchenko 1961). When air is replaced by hydrogen, nitrogen, or argon, while maintaining the smooth exponential relationship with composition the degree of spreading and wetting increases.

In a rosin-type flux, the smooth curve was depressed to lower surface tensions such that at 400°C the surface tension of tin was reduced to 500mJm-2 and lead to ~380 mJm-2. Surface tension was further reduced when a $ZnCl_2$ flux was used [Matuyama 1927]. Table 3-6 shows the surface tension values for a variety of atmospheres.

It seems generally considered [Greenway 1949; Semenchenko 1961] that it is difficult to separate any effect of flux reduction of surface tension for a highly ionizable material such as $ZnCl_2$ from possible surface contamination of the molten solder surface by the active flux ingredients.

Table 3-6 Effect of Atmosphere above Metal on Surface Tensions of Sn and Pb

Element	Atmosphere and Surface Tension (Dynes/cm^2)				
	Air	H_2	Vac	Organic Flux	$ZnCl_2$
Tin	617	540	545	500	300
Lead	450	449		380	260

3.4.12 Flux/Oxide Reactions for CuO

The effects of oxides on soldering have long been known, and several investigators have studied the influence of oxidation level on soldering performance [Drummond et al. 1981; Cassidy and Lin 1981a,b; Henry and Girouard 1990]. Work investigating a laboratory test used to determine an activity number for fluxes studied the interaction of fluxes with oxides [Roche and MacKay 1989]. The test employs the reaction of fixed volume of flux to reduce an oxide in an excess of the flux. The total metal ion in the separated liquid fraction, which represents the total amount of the specific oxide which that volume of flux can reduce, was used as the flux efficacy indicator. In the initial tests the oxide selected was cupric oxide, CuO.

The possible effect of nitrogen in the atmosphere or, by corollary, the effect of oxygen, was investigated along with the effect of the degree of temperature control. Extension of the technique to other oxides was also studied and the results for the two possible copper oxides were compared. Five different commercial fluxes were examined: one was a non-activated rosin, one was described as a "No Clean" flux, two were RMA, and the other was RSA. The fluxes were from three different manufacturers. Two additional metal oxides were used, Cu_2O and SnO_2.

The method used closely parallels that described earlier [Roche and MacKay 1989], with a weighed quantity of flux being added to an excess of metal oxide. Samples were reacted for times ranging from 0–1500 min. at 50°C, 100°C, and 150°C. The reaction was quenched by addition of isopropyl alcohol (IPA) and the residue separated by filtration. After evaporation to dryness the resulting residue from the filtrate was digested with sulfuric acid, nitric acid, and hydrogen peroxide until all residue was in solution. After dilution with water the solution was analyzed for copper by inductively coupled plasma source atomic absorption (ICP).

Figure 3-10 shows the results for three different activity levels of rosin-based flux from the same manufacturer. Preliminary results [Roche and MacKay 1989] indicated a plateau-like form of the relationship for concentration with time for the reaction. More recent results (Hiatt et al. 1991), Figure 3-10, did not reproduce this but indicated a peak shape with a long trailing decrease for the weight of reacted copper per gram of flux. Peak values and reaction rates for all three fluxes are shown in Table 3-7. Reaction rates were determined from the initial slopes of the graphs in Figure 3-11, which are for the initial reaction times. As can be seen in Figure 3-10, reaction rates appear near linear up to the peak reaction time.

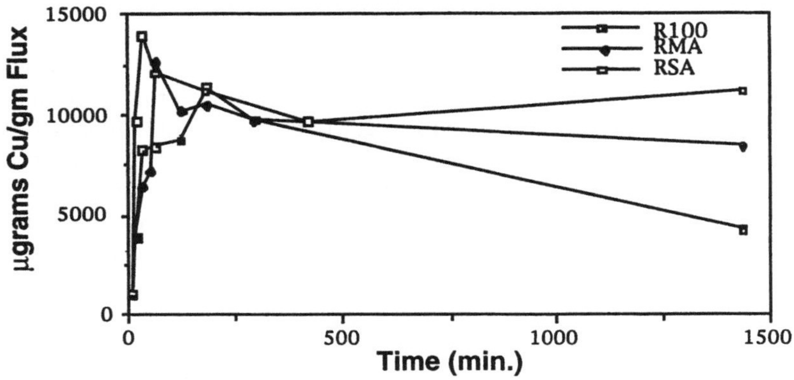

Figure 3-10 Copper concentration for three different fluxes reacted with CuO at 100°C for three different flux classifications

Table 3-7 Fluxing Rates for R, RMA, and RSA Fluxes at Three Temperatures with CuO

Temperature °C	Fluxing Rate (µgs Copper/g Flux/min)		
	R	RMA	RSA
50°C	0.035	0.037	0.035
100°C	0.140	0.207	0.207
150°C	0.143	0.251	0.148

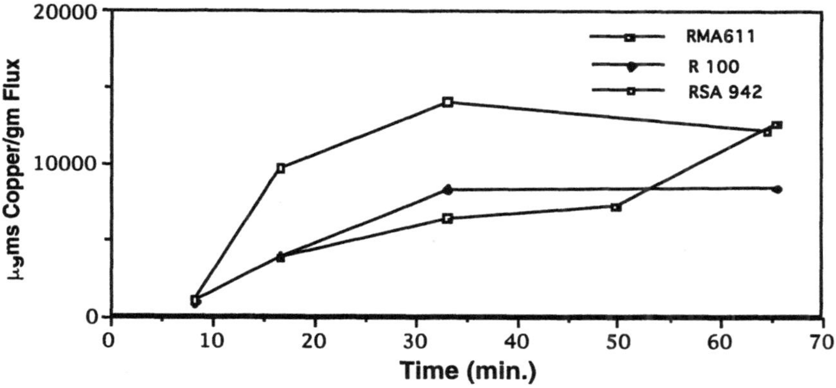

Figure 3-11 Initial copper concentrations as a function of reaction time of CuO with three different fluxes

3.4.13 Effect of Atmosphere

A major concern about these results was the drop in copper concentration with extended reaction times. It was considered that this was due to prolonged exposure of the solution to the air in the head space of the reaction tube, which initiated reoxidation of the reacted soluble copper, causing precipitation of the freshly oxidized copper; hence, removing copper from the solution. In order to test this, experiments with the air replaced by nitrogen were performed. The results of these are shown in Figures 3-12 and 3-13.

As can be seen, these experiments were inconclusive in that an increase in copper level was observed in the RSA solution while in the RMA solution the copper level was reduced.

3.4.14 Reaction Temperature Effects

The effect of reaction temperature was next examined using temperatures at 50°C, 100°C, and 150°C. Figures 3-14, 3-15, and 3-16 show these results, which are also tabulated in Table 3-7, while Table 3-8 shows the activation energies calculated from these values.

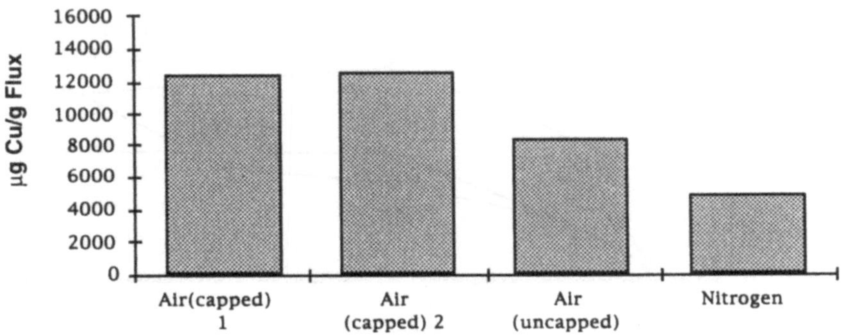

Figure 3-12 Effect of headspace atmosphere over extracted solution from RMA flux reaction

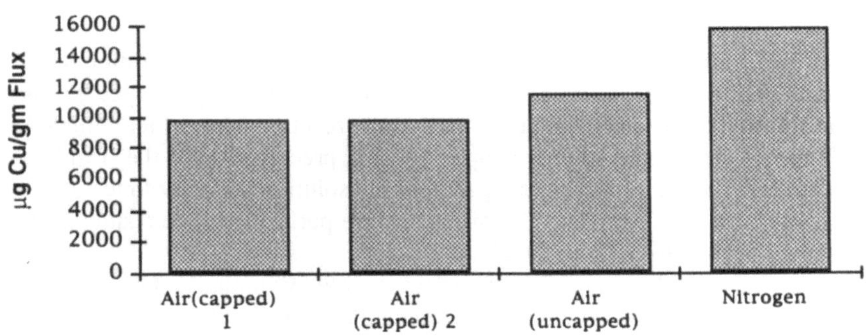

Figure 3-13 Effect of headspace atmosphere over RSA flux

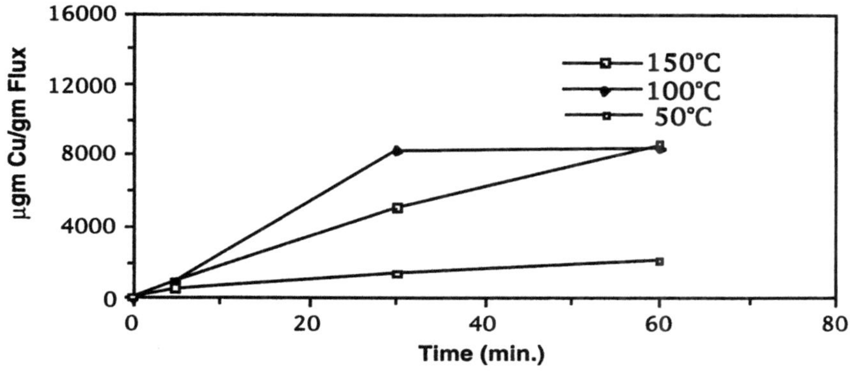

Figure 3-14 Effect of temperature on flux activity for non-activated rosin flux with CuO

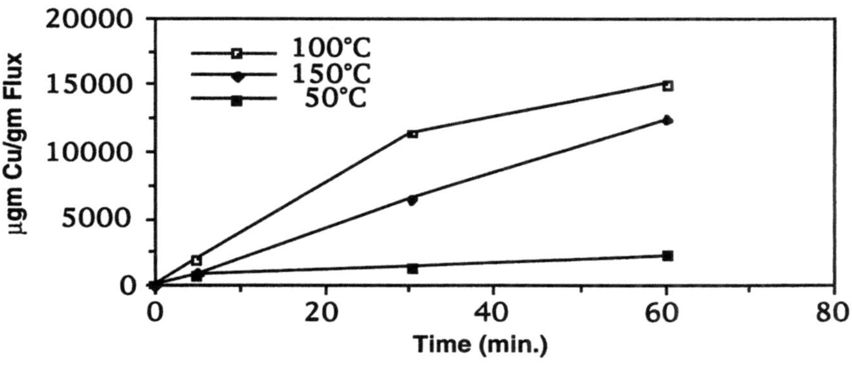

Figure 3-15 Effect of temperature on flux activity for RMA flux with CuO

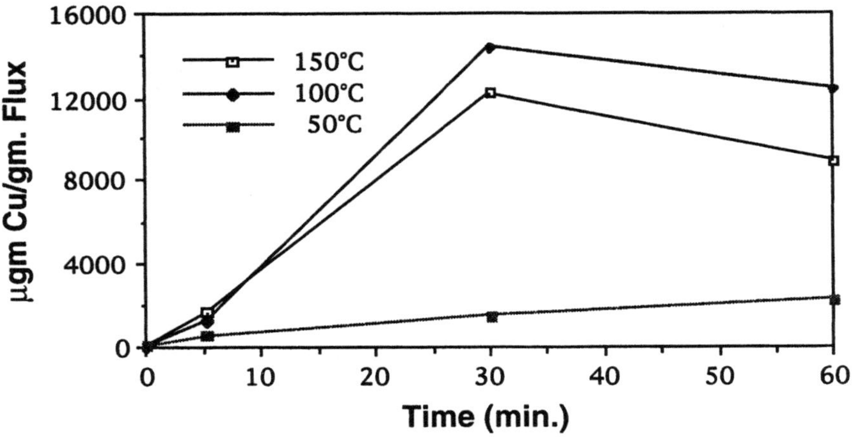

Figure 3-16 Effect of temperature on flux activity for RSA flux with CuO

Table 3-8 Activation Energies Calculated from Fluxing Rate Values

Flux Type	Activation Energy kJ/Mol
R	28.6
RMA	25.8
SRA	17.1

These results in Table 3-8 would indicate that the effect of increased activation of a flux is to reduce the energy required to initiate reaction with copper oxide. Fluxes from another generic group, i.e. a different manufacturer using different formulations and a different group of chemicals, were tested; the results for two of these are shown in Figure 3-17.

Comparison with Figure 3-11 shows that, while this RMA flux exhibits the same shape as the other materials examined, the peak height is nearly three times that of the RSA material provided by the other manufacturer. The other material illustrated on this figure, a low activity flux designed more for its cover coat abilities rather than for any cleaning activity and specifically requiring no cleaning, shows very little ability to reduce oxide. This figure, in conjunction with Figure 3-11, emphasizes the fact that fluxes from a low activity group, as defined by criteria other than their soldering capability, can have markedly different behaviors with respect to their abilities to reduce oxide.

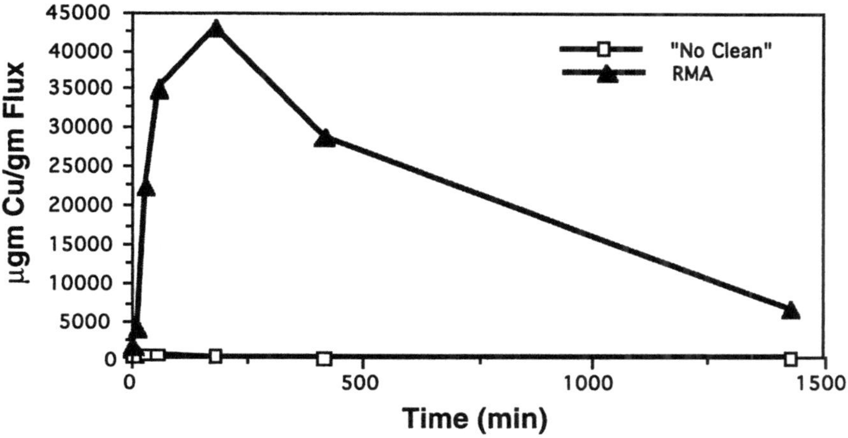

Figure 3-17 Results for flux activity with products from another generic group

3.4.15 Use of Other Oxides

Similar tests have been performed with other oxides. SnO and Cu_2O have both been reacted, with the results for SnO shown in Figure 3-18. As might be expected considering the stability of this oxide and its relative chemical inertness, very little oxide was reduced to metal in solution. Considering the low level of copper produced by the non-clean flux illustrated in Figure 3-17, it was unexpected that any tin oxide could be reduced.

Early results suggested that the selection of cupric oxide as the model oxide was, upon reflection, not perfect because other studies indicated that it was reduced in a two-part mechanism. When the original RMA material was compared for both CuO and Cu_2O, results as shown in Figure 3-19 were obtained. As can be seen, the sensitivity of the test would be increased nearly 50% if Cu_2O were used because of the increased amount of copper which was reduced.

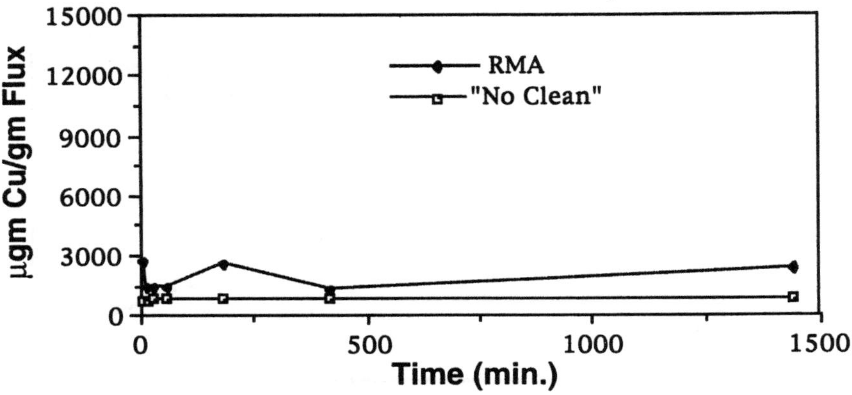

Figure 3-18 Tin concentrations for RMA and "No Clean" fluxes refluxed with SnO

3.4.16 A Flux Activity Number

It does seem feasible to grade fluxes in this manner, which is more nearly the action they perform during a soldering operation and could lead to a quite different ranking than that obtained from the existing flux activity designations used by flux manufacturers.

These results would indicate that most of the chemical action of a flux is to reduce surface oxides and that solder oxides are relatively unaffected. More work is needed if soldering performance specific to a particular lead material is to be evaluated in this way; mixes of oxides such as iron and nickel oxides would need be studied in the case of Kovar leaded devices, for example.

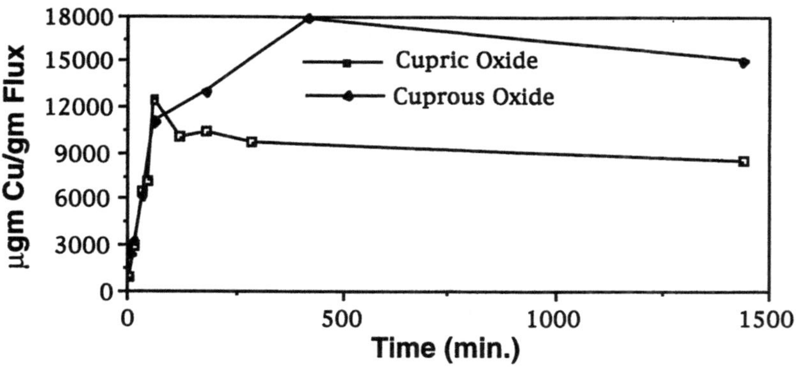

Figure 3-19 Copper concentration in RMA flux reacted with cuprous and cupric oxides

3.4.17 Thermodynamics of Flux Action

Thermodynamic studies of the processes taking place during flux action have been reported using Differential Scanning Calorimetry (DSC) [Sherman and MacKay 1989], Thermogravimetric Analysis (TGA) (Sherman and MacKay 1989), and Fourier Transform Infrared Analysis [Momose and Tamia 1970; Snyder 1987].

When combinations of DSC and TGA results were examined, a peak in the DSC scan and in the DTA scan could be assigned to a weight loss mechanism. Peaks which did not correspond in both types of scan were then clearly due to exotherms arising from reaction mechanisms. In this way results from a progressive series of mixes of the components used in a typical flux formulation could be compounded until a complete model of a formulation was assembled and certain thermodynamic features of the fluxing/soldering process could be seen.

3.4.18 Abietic Acid, Decomposition, and Interactions

Abietic acid begins to decompose at about 150°C, with most of the reaction taking place near 275°C, as shown by the sudden, sharp weight decrease in Figure 3-20. By about 475°C the material fully decomposes, yielding a residue weighing only about 3% of the original sample weight.

Twin endothermic peaks, the first beginning at 49°C, the second at 52°C, and both persisting to 148°C (which did not correspond to any weight loss temperatures indicated by the TGA scan) represent the complex melting behavior of abietic acid. The twin peaks at about 250°C correspond to the position of a maximum rate of weight loss in the TGA and represent the decomposition of abietic acid into various molecular fragments. Two well-resolved endotherms at 380°C and at 425°C indicate reaction between fragments of the molecule. Possibly this could be evidence of polymerization to dimeric or trimeric products.

When a sample with a vent hole in the lid was scanned in the DSC module (Figure 3-21), the twin melting peaks were once again apparent. A similar rise was observed at 149°C in the scan for the sealed sample, but instead of decaying gradually, this peak persisted until suddenly decaying at 275°C. This sudden decay coincides exactly with the complete decomposition of the abietic acid in the TGA scans. This supports the contention that the volatiles interact amongst themselves.

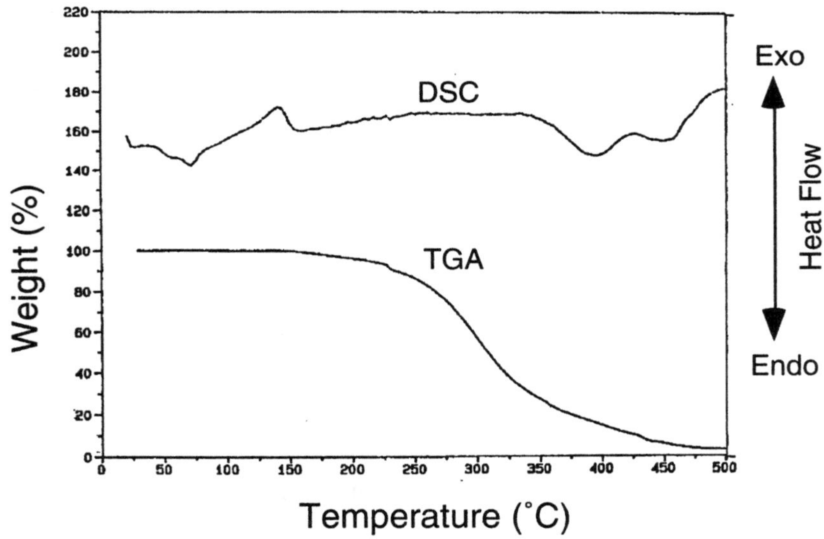

Figure 3-20 Thermogravimetric and Differential Scanning Calorimetry (DSC) scans for abietic acid

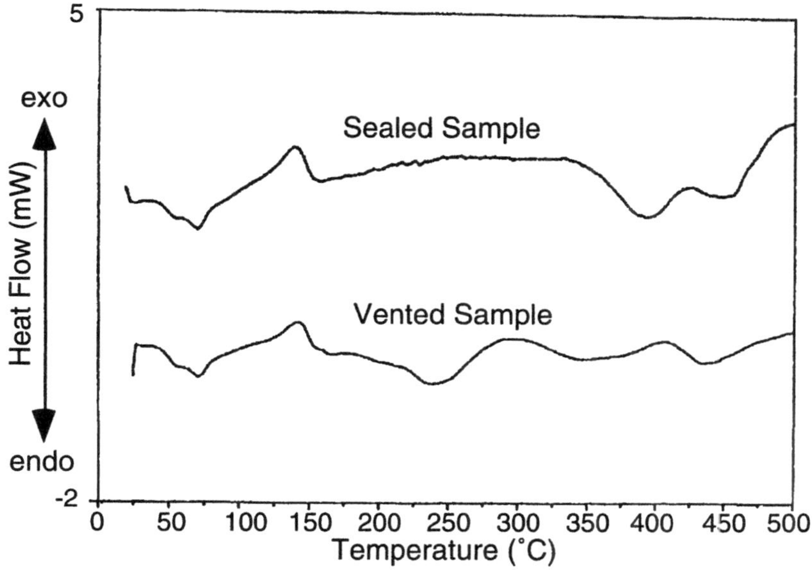

Figure 3-21 DSC scans of abietic acid with sealed and vented pans (sealed samples). Scans produced further interaction peaks in the DSC scan, representing some other interactions occurring among the volatiles or with the volatiles and the resulting residues.

3.4.19 Triethanolamine-Hydrochloride (TEA-HCl) and Decomposition

The TEA-HCI weight loss scans showed a complex arrest starting at 51°C, with possible side reactions starting at 100°C, and ending at 185°C. This arrest represents a weight loss of about 17.5%, which agrees relatively well with the theoretical percentage of hydrochloric acid (19%) in TEA-HCI. The slightly low value could be due to adsorbed moisture in the deliquescent TEA-HCI, making the apparent molecular weight higher than 185.7. The first small arrest between 51°C and 100°C could be the evolution of water vapor. The major arrest at 275°C represented either the boiling or decomposition (or both) of the triethanolamine which had formed immediately following hydrochloric acid evolution.

The DSC scan for this material (Figure 3-22) showed an endotherm starting at room temperature, probably associated with adsorbed water vapor, followed by a second, which occurred between 75°C and 90°C. This latter peak was probably due to the evolution of any water of crystallization. The large peak at 177°C, while indicative of the evolution of hydrogen bonded HCI molecules, was due to the melting of triethanolamine which was formed spontaneously at above its melting point when the HCI molecule broke free. The dissociation temperature of the adduct was 177°C, while the melting point of the parent compound was only 25°C.

The boiling point of triethanolamine, 277°C, was shown in the endotherm starting at about 250°C. Superimposed fine structure on this peak indicated that this material also decomposed at these high temperatures.

3.4.20 Materials Combinations: Building a Solder Cream

Abietic acid + IPA

In combination, the abietic acid and isopropyl alcohol system represented the solvation of the basic rosin vehicle in a typical flux solvent. These components are considered inert and unreactive with one another.

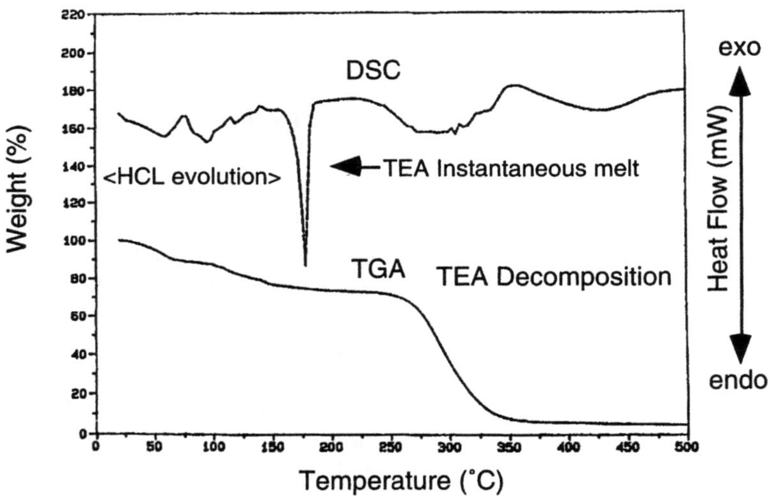

Figure 3-22 TGA and DSC scans of triethanolamine-hydrochloride

The TGA scans for a 45% colophony solution, which simulates common R grade flux, showed an intermediate arrest, indicating that the system was not simply one of classic evaporation. The presence of the arrest was evidence of some type of bonding between the abietic acid and the solvent, resulting in a small amount of thermal energy being required to disrupt the association. Once this bonding was disrupted, the solvent was released and a sudden weight loss was registered. The precise temperature at which this occurred was 150°C. The final residue in all cases was only 3–5% of the original sample, consistent with the residue from abietic acid.

The associated DSC scans for this system, illustrated in Figure 3-23, indicate a large composite endothermic reaction peak consisting of at least two components. The first initiated at a temperature corresponding to the first weight loss arrest in the TGA scan and represented the evaporation of solvent. The second, which merged progressively with the first, occurred in the middle of the weight loss arrest and represented the melting curve of abietic acid. This was followed by an extended, more shallow peak corresponding with the decomposition of abietic acid. The melting peak also showed a small satellite at the two lowest heating rates, indicating reactions occurring at ~120°C.

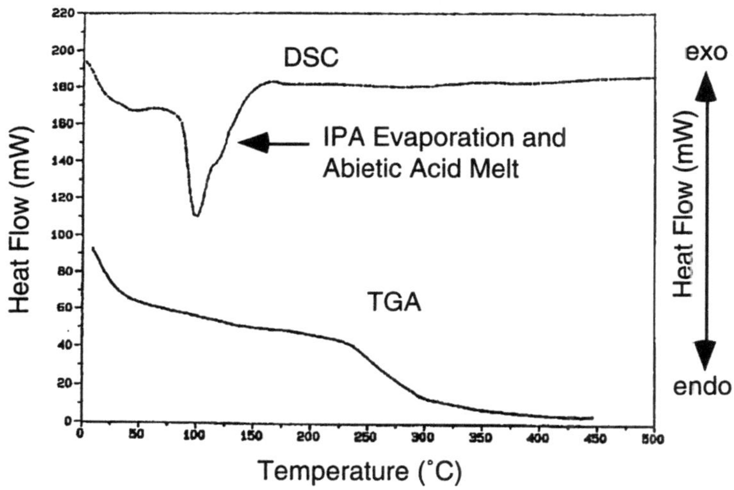

Figure 3-23 DSC and TGA scans of abietic acid and isopropyl alcohol

Abietic Acid + IPA + Activator

The next stage in the study involved adding a typical activator, in this case 0.3% triethanolamine-hydrochloride. The TGA scans were almost identical in comparison to the trace with no activator. The slight subsidiary arrest observed corresponded with the total evaporation of the IPA solvent. The only effect of the activator addition appeared to be that the decomposition temperature of the abietic acid was delayed by 5°–10°C. The only other evidence of the small addition of activator was a noticeable change in the slope of the weight loss at ~150°C, which would correspond to the weight loss due to evolution of HCl, as shown in the TEA-HCl scan.

On the DSC scans, illustrated in Figure 3-24, the IPA evaporation peak merged into the abietic acid melt peak. The effect of the TEA-HCl addition appeared as a satellite peak starting at 130°C, in agreement with the temperature assigned to the evolution of HCl on the pure TEA-HCl curve and the associated large melting peak of tetraethanolamine, which formed spontaneously.

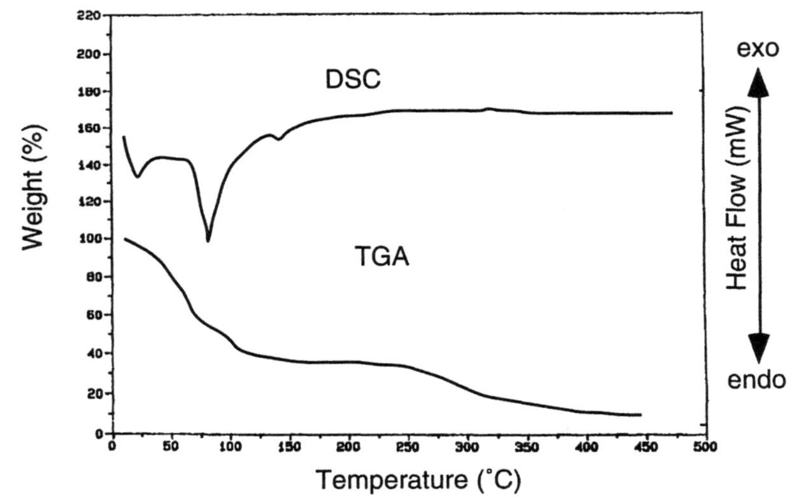

Figure 3-24 DSC and TGA scans of abietic acid plus isopropyl alcohol and triethanolamine-hydrochloride

Abietic Acid + IPA + Activator + Copper Oxide

Cupric oxide was then added to the mixture as a representation of the oxide normally present on a copper surface of an assembly. It was added in excess in an effort to maximize any reaction occurring. A major uncontrolled weight loss problem was found to be associated with splattering of the slurry which formed as the solvent evaporated, and the mix boiled explosively. Once this had been overcome by using a vented sample pan lid, continuous curves were obtained, typically showing three arrests. The first ended with a 68% weight loss, corresponding to complete evaporation of the IPA solvent. The second corresponded to the decomposition of the abietic acid and the third, occurring at ~420°C, was associated with the presence of cupric oxide in the sample.

Calculations using values from the traces with and without CuO indicated that probably the CuO had not been reduced to Cu metal in that the weight loss was insufficient to account for the total loss of oxygen during this experiment.

DSC scans of this formulation with vented samples (see Figure 3-25) showed the same trends as both abietic acid + IPA and abietic acid + IPA + TEA-HCl traces. The twin peaks of the IPA evaporation and the abietic acid melting were the same as for the above two systems, with the melting peak being larger than the evaporation. The abietic acid decomposition endotherms were also similar to those observed previously. One additional large feature was the exotherm at 400°–450°C associated with the presence of cupric oxide. There was also a small additional satellite peak at ~140°C.

Sealed DSC samples, while essentially exhibiting the same features as the vented samples, did reveal some small differences. In these, the IPA evaporation endotherms were larger than the abietic acid melting endotherms. The exotherm associated with cupric oxide was not so pronounced and occurred at slightly lower temperatures. These two series of DSC scans also indicated that there were differences in the reactions when the volatile components of the reactions were allowed to escape and when they were contained.

The presence of the peaks associated specifically with the cupric oxide, although absent from the raw cupric oxide scans, indicate a reaction taking place. The exothermic nature of these peaks suggests a re-oxidation process, although the weight loss seen in the TGA scans seemed to show continued reduction. When the pans were opened after performing the scans, the black cupric oxide had changed to a coppery red color. Microscopic examination of this showed gray/black films over parts of the residue, which might confirm the reoxidation theory.

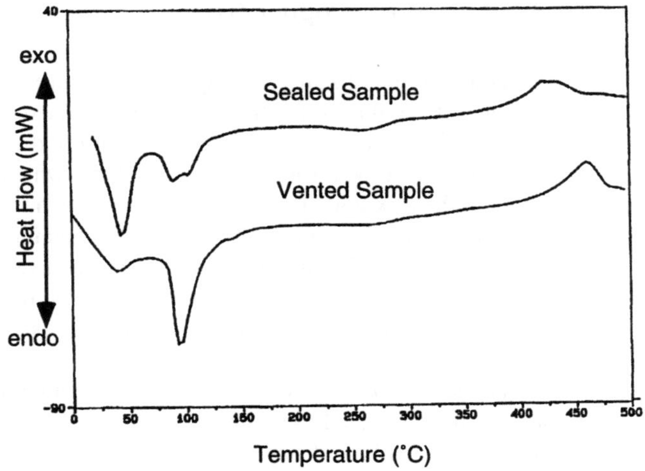

Figure 3-25 Comparison of DSC scans for abietic acid plus isopropyl alcohol plus triethanolamine-hydrochloride plus cupric oxide in sealed and vented pans

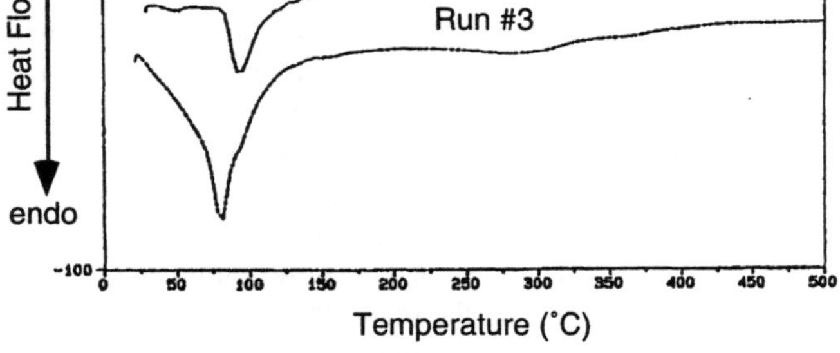

Figure 3-26 DSC scans showing progressive reduction of cupric oxide with repeated applications of flux

In an additional set of experiments, the residues from a DSC scan were repeatedly refluxed and rerun. The successive scans (Figure 3-26) show the gradual reduction of the peak at 400°C and its eventual disappearance. Visual examination of the residue after the final scan showed a sintered coppery mass indicative of complete reduction of the oxide. From these observations it appears that the reaction progressed in two steps. Cupric oxide was first reduced to cuprous oxide, which is reddish in color, and then was further reduced to copper metal.

Abietic Acid + IPA + Activator + Solder Powder

As an intermediate for the full simulation of the complete soldering process, small additions of 62%Sn-36%Pb-2%Ag alloy solder powder (~325 mesh) with a melting point at 179°C were added to the flux formulation without any cupric oxide being present. The TGA traces of this system showed a minor arrest occurring at a slightly higher weight percent than the weight loss corresponding to loss of all the solvent and a major arrest corresponding to the decomposition of the abietic acid when TEA was present. The weight loss percentage at this point corresponded exactly with that expected for decomposition of the abietic acid when allowance was made for the solder powder.

The main features of the DSC scans, both vented and sealed, were the abietic acid melting and decomposition peaks together with the solder melting peak. In the traces for the two lowest heating rates, vestigial peaks at ~130°C were observed, indicating the evolution of HCl from the small amount of activator present. The interaction of the flux with oxide on the solder powder surface did not show up as a thermodynamic peak in these traces.

Abietic Acid + IPA + Activator + CuO + Solder Powder

When vented samples of these full soldering process simulation samples were scanned in the DSC, the major features were again the melting and decomposition of abietic acid and the melting peak of solder. A peak at ~400°C associated with CuO was also present. At the lowest heating rate there was a side peak at ~125°C, as seen in Figure 3-27.

The scans from samples with sealed pans were more complex than those from vented pan samples. A complex of secondary peaks were observed between ~120°C and 150°C; also, the single exothermic peak from CuO was split into two, presumably due to an effect of reaction with the constrained vapors.

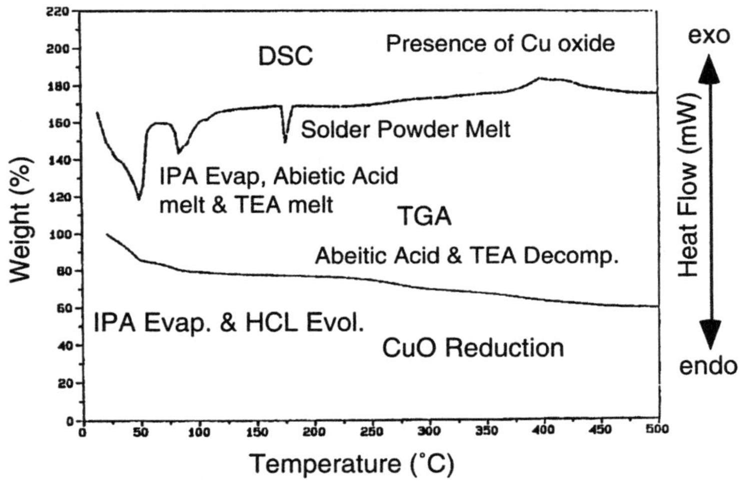

Figure 3-27 DSC and TGA scans of a simulated soldering process using abietic acid plus isopropyl alcohol plus triethanolamine-hydrochloride plus cupric oxide plus solder

It was clear from comparison of the results of the vented and sealed DSC experiments that the exact sample conditions have an effect on the reactions taking place. In a real soldering situation, the assembly to be joined would not be contained and volatiles would be able to escape. However, the volume of flux used would be considerably more than used in these experiments; therefore, it would be entirely possible that with some of the flux acting as a blanket, conditions similar to those existing in our sealed tests could occur at the flux/metal interface.

The testing using CuO as the simulated surface oxide was clearly too severe for comparison with practical soldering reactions, but it showed some interesting features of the fluxing mechanism, in particular a stepwise process for reducing higher oxides. Exo-electron studies by Momose and Tamai [1970] suggest that the lower oxide, Cu_2O, would have been reduced at ~180°C so that Cu_2O would have been a better choice of surface oxide simulation. These results indicate that CuO had a somewhat more complex behavior, possibly associated with reoxidation taking place at ~400°C.

Abietic Acid + IPA + Activator + Copper Powder + Solder Powder

The results of experiments using copper powder are indicated in Figure 3-28. In these, the peak at ~400°C associated with the CuO was absent. Two shallow extended peaks, corresponding to the formation of Cu_6Sn_5 plus Cu_3Sn, formed at temperatures consistent with the formation temperatures of these compounds, as shown in the phase diagram for this system.

The study was able to show many aspects of the sequence of events taking place during a complete soldering operation. However, it had not been able to show all the sequential details of the reactions involved. In part this was due to the magnitude of the thermodynamic energies involved. The large energies involved in change-of-state reactions, such as melting and evaporation, tended to mask the smaller energies associated with chemical reactions such as the formation of copper abiate. The glassy nature of the molten rosin which produced the melting peak over such a broad temperature range was obviously important since it tended to mask any sign of the other reactions taking place in that temperature range in the mixtures studied.

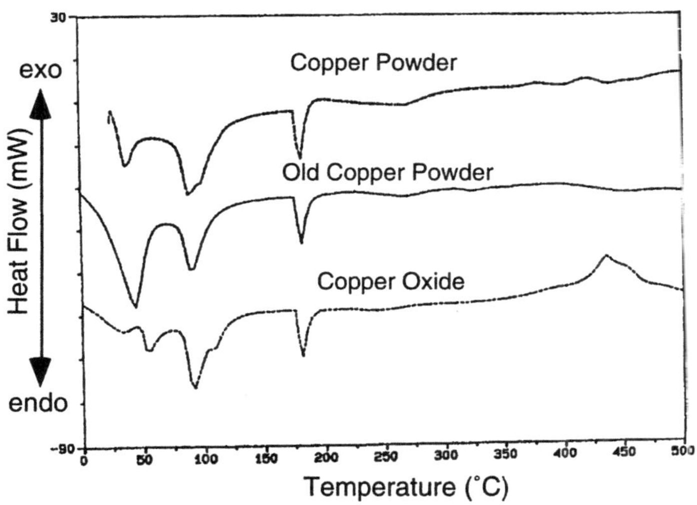

Figure 3-28 Comparison of DSC scans with fresh and aged copper powder with copper

A formulation with only 0.3% activator addition, while it corresponded to the practical situation of an RMA grade flux, did not really allow observation of activator reactions in the flux systems. A higher level of activation, although not necessarily a practical composition, could have been more informative.

3.4.21 Fourier Transform Infrared Analysis

Diffuse Reflectance Fourier Transform Infrared Analysis (DRIFT) has been used to study the surface reactions of rosin with a variety of metal oxides common to soldering systems [Snyder 1987]. By measuring the diffuse scattering from the surface of a sample and comparing it to the scattering from a reference surface (usually KBr), the absorbance characteristics of the sample can be measured. In the case of flux/oxide powder interactions, the diffuse reflectance spectra showed changes arising from reactions within the flux film on the surface of the oxide powder. Copper, tin, and lead oxides were studied, together with salts of copper and lead. In the case of cupric and stannous oxide (SnO) the absorption band at 1600 cm^{-1}, arising from the acid group $[COO]^{-1}$, was monitored. From these spectra the presence of copper and tin carboxylate compounds was detected at the lowest test temperatures used, 260°C. With stannic oxide (SnO_2), no reaction was observed, even up to the highest temperatures used (350°C). This is not surprising considering the stability of this last compound. The same generalized conclusions did not hold for the lead compounds: Pb_3O_4, a highly stable oxide, showed evidence of reaction at the lowest temperature (260°C), but with the absorption band shifted to 1520 cm^{-1}. Other lead compounds reacted differently, in that lead chloride did not react below 300°C, while lead sulphate was unreactive even at 350°C. Lead sulphate is highly stable, while lead chloride has a similar free energy to stannous oxide.

These studies show that the reactivity with the flux and the solubility of the oxide, or the reaction product with the flux, were important factors and that these reactions in the case of lead compounds were complicated.

3.4.22 Flux Residues

With so many materials available on a hot reactive interface during the soldering process, it is not surprising that a variety of by-product residues are created by interactions between these chemicals. Most of the products have some effect on the performance (reliability), corrosion, or testability. This is what led to the requirement that all of these be removed from the assemblies. Logically, the progression from the heavy use of chlorinated fluorocarbons (CFCs) for removal of rosin flux residues to a reduced use for light rosin flux residues to no use if rosin were eliminated from the flux formulation is reasonable, but flux residue levels are not the sole criterion for flux selection. A flux must work for a solder process to be effective.

3.4.23 Rosin Flux Formulations

Rosin flux formulations are principally mixes of solvent with natural rosins at between 35% and 46% concentrations. Other additions seldom total more than 15%. Typically the activators, and halogenated and amine mixtures tend to range from about 0.3–1% for RMA and to 1.5% for RA. RSA grades have the maximum halogen activator augmented with up to 3% organic acids, singly or in combinations. Thickeners are usually found in the range 7–10%, while viscosity control agents are added from ~5–15% depending upon the ultimate viscosity requirement. Nonhalogen formulations utilize only organic acid activators from about the 2% level.

3.4.24 Low Solids Fluxes

Concern for environmental destruction of the earth's ozone layer centers on the specific interaction of chlorinated fluorocarbon (CFC) materials with ozone itself. Since the quantity of flux residue that must be cleaned influences the amount of solvent needed, a system that deposits less flux on each board assembly has proportionately less impact upon the amount of the cleaning material used and, correspondingly, a smaller contribution to the overall effect.

The major parameter in a soldering flux that contributes towards the flux residual on a board is the percentage of solids in the flux. Solids are mainly the rosin and the activator. Lower levels of rosin will produce less residual. Table 3-9 shows how solids levels have dropped in recent years. Figures 3-29 and 3-30 show how these levels have been progressively reduced in both USA [Ostrander et al. n.d.; Sinclair et al. 1985; Rubin 1982, 1990; Rubin and Warwick 1990; Langan n.d.] and Japan [Okano 1991].

Table 3-9 Solids Contents Reductions with Introduction of Low Solids Fluxes

Date and Flux Form	Flux Content		Flux Class
	Rosin %	Activator %	
1985 (Liquid Flux)	35–20	0.3–1.5	RMA–RA
1985 (Cream)	55–45	0.3–1.5	RMA–RA
1989 (Low Solids)	8	0.5	RMA
1991 (Low Solids)	0 –2	0.5 (Halide)	RMA
		~1 (No Halide)	RMA

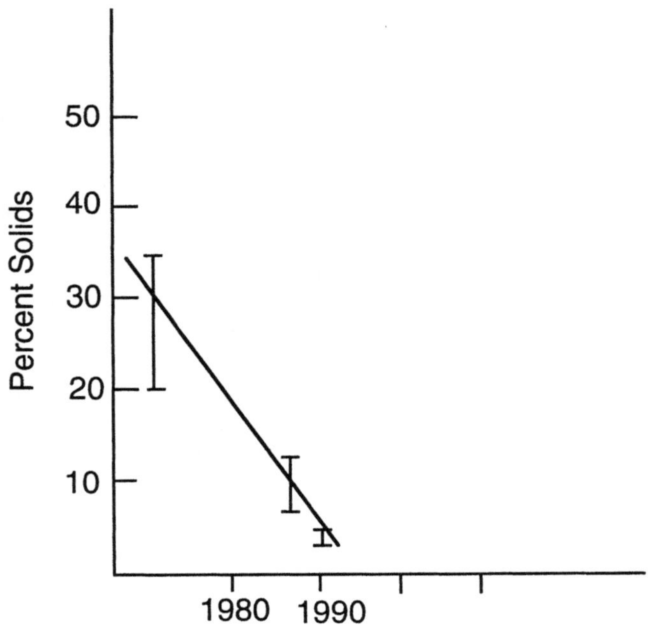

Figure 3-29 Percentage of solids in fluxes as a function of year.

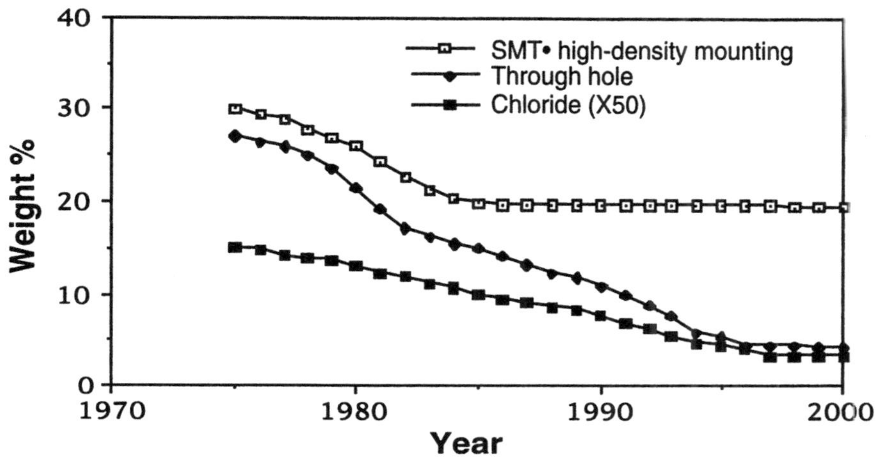

Figure 3-30 Trend for percentage of solids in fluxes in Japan

These changes have not been without compromises. When molten, natural rosin has a tarnish reduction capability of its own. As rosin contents of fluxes have been reduced, overall reduction capability has been correspondingly reduced. This has meant that either more activator or some other acidic material has to be formulated in low residue fluxes [Rubin 1982, 1990; Rubin and Warwick 1990; Kawakatsu and Yokoi 1990] or much more careful component and board storage is necessary to achieve satisfactory soldering [Kocka 1990] with these materials. Another problem is how to compensate for the much thinner fused rosin protective film that inhibits reoxidation of an interface between the time that the tarnish on a surface is removed and it is wetted and covered by solder.

Diffusion of oxygen through a fluid film is dependent upon the nature of the material, its permeability to oxygen and its stability towards oxidation, the film thickness, and the temperature [Kawakatsu and Yokoi 1990]. As the weight percentage of the vehicle is reduced, the residual solids film after evaporation of the solvent has become thinner, so that this protection role has become less effective. To correct for this, either a more potent substitute activator or acid must be added to the formulation, or a less oxidizing atmosphere cover must be arranged [Sinclair et al. 1985; Okano 1991; Kawakatsu and Yokoi 1990; Ohki 1991].

The additional function of protection of the molten solder surface until the temperature falls again to below the melting point of the solder also exists, but this is far less significant because after wetting, the molten solder protects the underlying substrate.

A major safeguard arising from the use of rosin is that both before and after melting to be an acid vehicle, it exists as an inert moisture-resistant lacquer. As a solid, it is capable of encapsulating highly reactive compounds and rendering them inert even though they might be highly ionic. As the flux vehicle loading is reduced, the activator proportion becomes progressively higher compared to the vehicle, and the thinner molten film present after soldering becomes less effective in encapsulating active residues. Chemical modifications of natural rosin or use of synthetic resin has reduced the pK effectiveness of the vehicle because the -COOH group in the resin acids is affected greatly since it is the most active site of the acid and thus the most readily modified. In most synthetic resins there are no acid sites and consequently no contributions toward oxides reduction. A possible major benefit of substituting a synthetic for natural rosin is that doing so produces residues that are more soluble in solvents and require less solvent to achieve cleaning. By suitable selection of side groups, such as incorporation of amine or soap groups in the rosin acid skeleton, a level of specificity such as water solubility can be imparted to a residue. This means, however, that the film might become more moisture-absorbent, necessitating less delay between assembly and cleaning.

Another side effect of the reduced amount of rosin is a reduced ability to correct solder wave exit side defects in the wave soldering process [Ostrander et al. n.d.]. These are such defects as bridging between conductors, icicle formation, and webbing. Surface webbing, the formation of a layer-network film of solder across an insulator surface, usually relates more to solder mask roughness and cure quality than to flux performance. Formation of these other defects is related to the action of a thin residual oxide film arising from either oxide in the solder wave, or on the substrate that is carried completely through the solder wave, to be present at the solder meniscus as it breaks away from the wiring board surface on the wave solder exit side. There it interferes with the clean separation of the soldered board from the solder and drags the solder film adhering to the oxide either across the gap between conductor tracks to give bridges or out from the plane of the circuit board to form icicles on leads and terminations.

Reduction of solids content in solder cream fluxes is a significantly more difficult problem than reducing solids content in liquid fluxes. First, the solids in a solder cream are not just rosin but a complex of other organic materials formulated to support the highly dense solder particles and to control viscosity and screen printing performance. Some of these actions are realized by adding inert material additions such as modified silica gels which, while helping support the dense solder particles, can be mechanically removed after any cleaning process or can remain as inert surface dirt. The general trend, however, is to incorporate materials that will react to yield volatile products and water vapor in a special atmosphere [Sinclair et al. 1985]. These then leave reduced amounts of residues.

Two material classes are emerging in liquid systems: those with halogen activators and those without. Figures 3-31 and 3-32 summarize some results of the relative soldering performance of the two classes [Rubin 1990]. The halogens are more effective at lower levels than the organic acid activators. From these it seems that organic acid additions are less effective overall than halogens but have a higher rate of reduction in wetting times as the formulation percentage increases from 2–6%. Halogen additions have a larger overall effect, but this is less well defined as to the percentage in the composition. The effect seems less marked for the spreading factor than for an effect on rates of wetting. The results for halogen activators could be a manifestation of the proposed catalysis effect of activators.

3.5 GASEOUS FLUXES

3.5.1 Nitrogen Atmospheres

The use of nitrogen alone as a protective atmosphere utilizes the principle of dilution. Oxidation will be reduced if the composition of the atmosphere reduces the amount of oxygen available to oxidize a surface. In order to affect the development of oxide films on a metal surface, the oxygen content in an atmosphere must be kept very low. About 1.5–2.5 ppm of O_2 can produce a 1μm thick oxide layer or between 0.015 ppm and 0.025 ppm of O_2 will form a 1Å thick film. Therefore, to be fully effective, an inert N_2 atmosphere would need to have very low O_2 levels, below 0.01 ppm. However, results using nitrogen with O_2 contents >10 ppm have been used and demonstrate improved performance in aiding soldering performances of low solids content flux [Trovato 1990; Schouten 1989; Pickering et al. n.d.; Klein-Wassink 1989; Mehta et al. 1991; Fodor and Lensch 1990; Brammer 1991]. The dynamics of the oxidation process are critically important if reduced oxidation rates and reduced defects such as solder voids and bridging are to be achieved. It has also been shown [Kawakatsu and Yokoi 1990] that use of nitrogen atmospheres can reduce wetting times by as much as 40%, while defects were reduced from 37% down to 0%. For solder pastes the situation was more complex [Hanaway et al. n.d.] and depended upon the formulation used, both as to the magnitude of the performance improvement and as to the O_2 level at which it was initiated.

Another important effect of nitrogen atmospheres is the reduction of difficult-to-remove residues due to a reduction in oxidation of the solder cream vehicle.

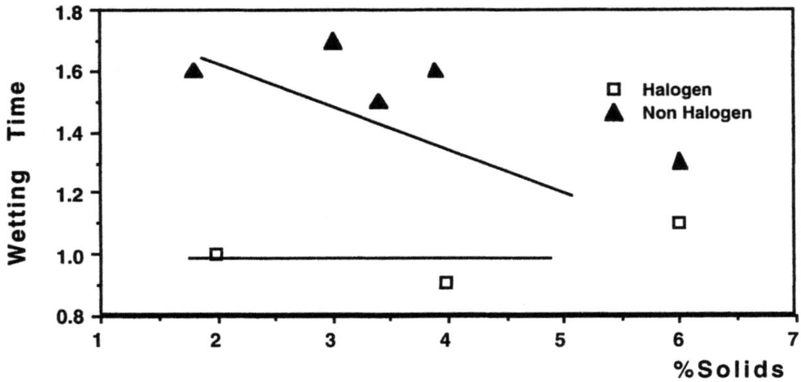

Figure 3-31 Wetting times vs. solids percentage for halogen and non-halogen activated fluxes

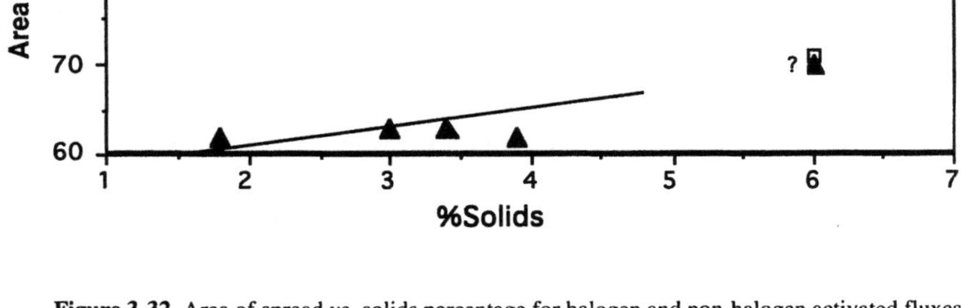

Figure 3-32 Area of spread vs. solids percentage for halogen and non-halogen activated fluxes

3.5.2 N_2/ H_2 Atmospheres (including Ar(H_2))

Addition of hydrogen to a protective atmosphere affects the dynamics of oxidation but cannot ultimately stop it. Figure 3-33 shows the temperature at which various oxides are reduced to metal. All of these temperatures are far higher than the temperature achieved during the soldering process. Any improvement must therefore be realized by modifying the reaction rates of equations of the type:

$$2M^{n+} + nO^{2-} \leftrightarrow M_2O_n \tag{3-6}$$

to change the reaction rate controlling the formation of the oxide leading to reduced amounts of oxide. The ability to create any significant reaction:

$$M_2O_n + n/2H_2 \leftrightarrow 2M + nH_2O \tag{3-7}$$

must be very limited since at soldering temperatures the oxide reducing reaction rate is very small. Modification of the equation to the ternary:

$$2M^{n+} + (n+m/2)O^{2-} + mH^{+} \leftrightarrow M_2O_n + m/2H_2O \tag{3-8}$$

with addition of hydrogen does mean that more oxygen is required to form an equivalent amount of oxide, so that in a nitrogen blanket, less oxide would be formed in the presence of hydrogen.

If the inert gas used is capable of creating a plasma (for example in Ar), then Ar^+ reactions such as $Ar^+ + H_2 \leftrightarrow ArH + H^0$ enhance the proportion of monatomic hydrogen occurring, which in turn enhances the reduction of oxides [Pickering et al. n.d.].

Thermodynamic calculations for a soldering process at 210°C (Ivankovits and Jacobs 1990) predict differences in the performance of an air-oven reflow process compared to a hydrogen-containing-nitrogen atmosphere according to the equation:

$$73.9\ Sn + 26.1\ Pb + 10\ O_2 + 40\ N_2 \leftrightarrow 61.1\ Sn + 26.1\ Pb + 5.6\ SnO + 7.2\ SnO_2 \tag{3-9}$$

compared to the equation with a 3%H_2 in N_2 atmosphere:

$$\begin{aligned} &61.1\ Sn + 26.1\ Pb\ 7.2\ SnO_2 + 5.6\ Sn + 10\ 0\ H_2 \leftrightarrow \\ &6.41\ Sn + 26.1\ Pb + 9.8\ SnO_2 + 0.48\ H_2O \end{aligned} \tag{3-10}$$

The effect of these atmospheres on the residual from solder cream vehicle was shown for both normal and low residue formulations.

3.5.3 N_2 Reactive Mixtures

By injecting acid vapors into the atmosphere, far more highly reactive gaseous constituents than hydrogen can be introduced. The most commonly used additives are formic acid and adipic acid [Ivankovits and Jacobs 1990; Hartmann 1991; Fodor 1990]. Formic acid reacts with metal oxides above ~150°C [Trovato 1990; Fodor and Lensch 1990] to produce formate salts which begin to decompose above ~200°C. Reactions of the type:

$$\begin{aligned} \text{Metal(Me)oxide} + 2\,HCOOH &\leftrightarrow Me(COOH)_2 + H_2O \\ Me(COOH)_2 &\leftrightarrow Me + 2\,CO_2 + H_2 \end{aligned} \tag{3-11}$$

and experiments using CF_2Cl_2, CF_4, and SF_6 have also been performed [Moskowitz et al. 1986], but residues from these gases were not volatile and tended to corrode in the presence of moisture.

3.5.4 Analysis of Residues

Analytical identification of various isolated residues found on PWBs have been reported and serve as an indicator of the roles various constituents have performed during the soldering cycle. Solubility comparisons of pure compounds in various solvents (Archer et al. n.d.) help show the degree of modification or heat damage that the starting materials have experienced. More heavily modified compounds or materials created by interactions of chemical components with each other and the substrates or electrical component surfaces are often apparent after more complex analytical detective work [Cabelka and Archer 1985; Archer and Cabelka n.d.]. Results from such analyses can indicate the necessity for considerable reassessment of conventional assumptions about how flux components interact. Reaction products from the major flux components often mask identification of by-products from the minor flux constituents, so products formed by reactions of activators, for example, are heavily masked by oxidized or polymerized rosin residues.

As mentioned above, results of solvent comparisons can be very meaningful; however, use of powerful solvents capable of attacking basic PWB materials raises a question about whether the residues are being dissolved or simply undermined.

Using a combination of techniques, Lovering [1984] was able to identify rosin-acid type peaks in the Infrared adsorption spectra of a particular residue which combined with a shift in the spectral position characteristic of the formation of a metal salt to indicate the presence of a metal abiate. Scanning electron microscope analysis of the residue showed the metal to be tin (with some lead, copper, and bromine). Molecular weight measurements gave a value consistent with the major residue containing two abietic acid radicals with one tin atom, clearly suggesting that the Tin II abiate salt was formed.

Similarly, Dunaway [1989] identified an organic salt containing tin and chlorine. Controlled equipment experiments also related the onset and severity of residue formation with the magnitude of the preheat conditions and contact with water prior to saponification. Immediate contact with saponifier, a compound that forms soaps with rosin residues so that they become water soluble, led to rapid formation of a soap salt which was readily soluble and was rinsed away.

Ultraviolet spectroscopy has been fairly widely used to identify the rosin acid radical [Cabelka and Archer 1985; Archer and Cabelka n.d.; Lovering 1984; Dunaway 1989; Kaier 1983; Bonner 1980; Archer et al. 1987; Klima and Magid 1987] in the natural rosin products, and this has worked well for examination of residues. Archer, Cabelka, and Nalazek [1987] used the technique to establish a calibration curve using the characteristic absorption peak at 241 nm for abietic acid, and used this to determine the amount of residue on PWBs. In doing so they also established that unmodified abietic acid was a major part of post-soldering residues. Molecular weight determinations before and after heating gum rosin showed little change in the weight fraction profile (Figure 3-34) with no indication of any increased molecular weight fraction or tail, which did not show any evidence for polymerization reactions taking place in heated rosin.

Klima et al. [1987] and Dunn [1990] showed the same results using high pressure liquid chromatography (HPLC) as the analytical tool. This has the advantage of a simplified single peak spectrum per constituent for each resin acid compound compared to a far more complex multi-peak ultraviolet spectrum. In this way a far more sensitive discrimination between the natural acids present in water white gum was possible. The same sensitivity was shown for the residue extracts, and this too showed the presence of unmodified resin acids in the residues.

3.6 Inorganic Fluxes

Although inorganic fluxes are not normally allowed for high reliability products, some fluxes that require cleaning do have compounds of this type in their formulation.

3.6.1 Mechanistic Studies

On the basis of the historical development of fluxes, the purpose and action of flux were the following: to clean the substrate of all contaminating films, and then to protect the clean surface from reoxidation until the coating metal, or solder, melts; and by reacting with it, wet the surface. According to this description of the process, any resulting response to a solderability test describes the effectiveness of the cleaning (and/or the protection). Implicit in this would be that once the surface was fully clean, the solderability would be optimum and could not be improved.

To test this, Turkdogan and Zador [1961] used a reducing atmosphere furnace to reduce the oxide film on a degreased and cleaned steel strip. After flooding the vessel with a protective, non-reducing atmosphere, a weighed quantity of tin was placed on the strip and the area of spread measured. Optimization of the reducing atmosphere and experiment parameters showed that the spread area reached a maximum and thereafter stayed at this value, which represented the ultimate spreading value possible with fully cleaned steel. When a weighed quantity of inorganic flux salts was applied with the tin or solder under the same optimum cleaning conditions, then the spread area increased to being greater than the value that had been obtained for clean steel. This indicated that flux in this form provided some additional aid to the wetting operation beyond what was possible from simply applying tin to a clean metal surface.

The explanation originally offered was on the basis of surface tension modification and a classical triangle of forces equilibrium involving the three surface tension components (see Figure 3-9), $\gamma_{(H_2/solid)}$, $\gamma_{(solid/liquid)}$, and $\gamma_{(H_2/liquid)}$:

$$\gamma_{(H_2/solid)} = \gamma_{(solid/liquid)} + \gamma_{(H_2/liquid}(\cos\theta)$$

where θ is the contact angle measured up from the solid interface to the surface of the liquid. The area of spread for the fluxed system was greater than for the hydrogen atmosphere system because $\gamma_{(flux/liquid)}$ was smaller than $\gamma_{(H2/liquid)}$.

3.7 EFFECT OF SOLDER IMPURITIES ON SOLDERABILITY

Later workers [Onishi et al. 1972, 1973, 1974, 1975] who used the area-of-spread test attributed the performance improvement they found to the reaction of flux salts with molten solder, created the reaction an alloy of the metal of the flux salt with the tin or solder, and suggested that the alloy produced spread more readily than the tin or solder. This, however, ignored results of other workers [Ackroyd et al. 1975] who had studied the effects of deliberate additions of a range of impurities to 63%Sn-37%Pb and 60%Sn-40%Pb alloys and had shown that in all cases this resulted in a decrease in solderability. Figures 3-35, 3-36, and 3-37 show the effects on area-of-spread tests of low levels of antimony, copper, and zinc impurity on the solderability of brass with 60%Sn–40%Pb when a zinc ammonium chloride flux was used. Only for the case of copper with brass as the substrate was any increase in spread area observed. This was attributed to reaction of the copper with the zinc dissolved from the brass surface to form small particles of copper-zinc compound, which was observed in the microstructure of the joint, and not the formation of a modified solder alloy. The copper actually removed impurity arising from dissolution of the base material. In all other cases for the brass (as shown), for copper, and for steel substrates, the area of spread was decreased. Similar effects were also observed for wetting balance tests.

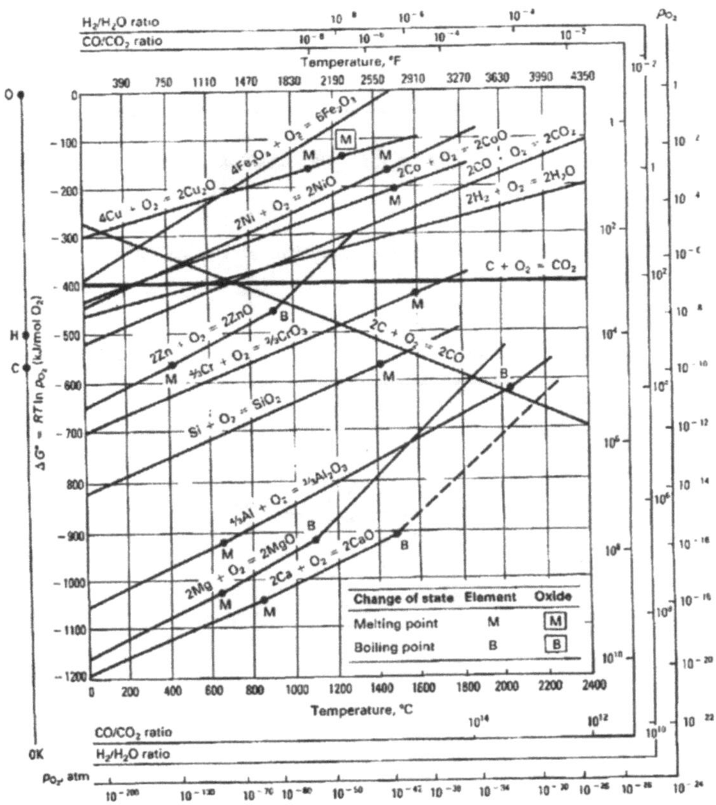

Figure 3-33 Ellingham diagram of oxide reduction temperatures as a function of gas partial pressure.

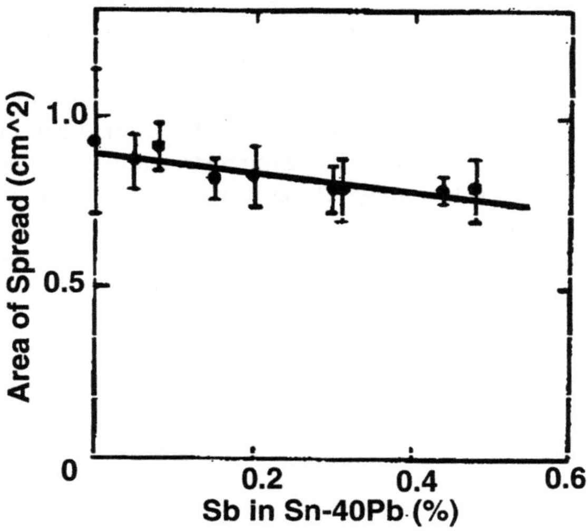

Figure 3-34 Effect of antimony on area of spread of 60Sn-40Pb solder on brass base metal (zinc ammonium chloride flux)

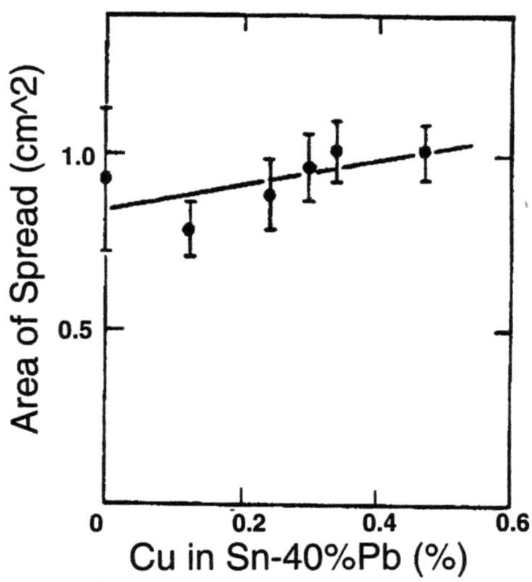

Figure 3-35 Effect of copper on area of spread of 60Sn-40Pb solder on brass base metal (zinc ammonium chloride flux)

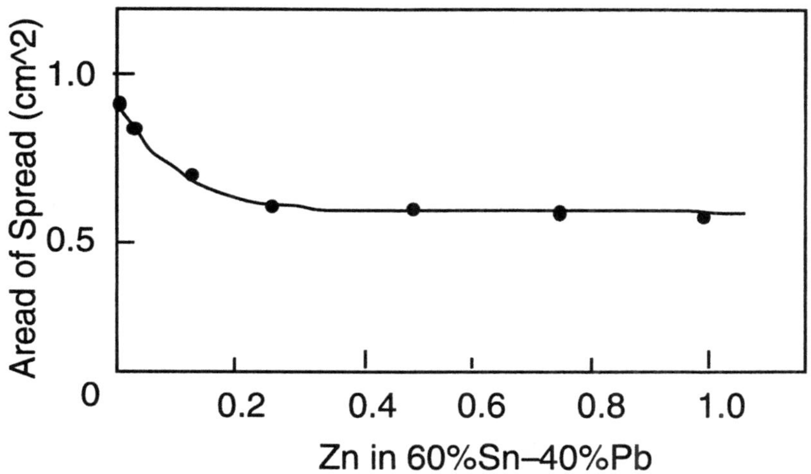

Figure 3-36 Effect of zinc on area of spread of 60Sn-40Pb solder on brass base metal (zinc ammonium chloride flux)

Table 3-10 Effect of Impurity on Area-of-Spread Values on Various Substrates Fluxed with Zinc Ammonium Chloride

Impurity	Substrate	% Impurity to Produce 25% Reduction in Area of Spread
Antimony	Brass	0.9
	Steel	no effect
Arsenic	Brass	no effect
	Steel	0.2
Bismuth	Brass	5.1
	Steel	2.8
Cadmium	Steel	1.9
Copper	Steel	0.5
Zinc	Brass	0.04

Table 3-10 shows a synopsis of these results for all impurities when fused salt fluxes were used. The percent column gives the impurity level percentage that caused a 25% reduction in the area-of-spread value. Decreased areas were observed both in systems which reacted with the solder to produce intermetallic compounds with the impurity, as with copper, and in systems which dissolved the impurity to form a solution in the melt. Only 0.4% copper was required to initiate the formation of Cu_6Sn_5 intermetallic in 60%Sn–40%Pb alloy, in which it formed a fine suspension of needles in the compound. This caused the viscosity of the melt to increase with a corresponding inability to flow and spread. Silver reacted in a similar manner, forming tiny platelets of Ag_3Sn. The effect of this compound was less marked than for copper. Antimony compound does not form until the composition reaches about 8%, so at the low levels used in these experiments it appeared to affect the spreading properties only as a fully dissolved element.

3.7.1 An Electrochemical Mechanism

From the results mentioned earlier [Onishi et al. 1972, 1973, 1974, 1975], it is clear that a large range of inorganic salts could be formulated into fused salt type fluxes and would demonstrate very significant improvements in solder flow characteristics. The most probable mechanism for the fluxing effect with inorganic fluxes as put forwards by Latin [1938] and supported by both Bailey and Watkins [1951], and Turkdogan and Zador [1961], was electrochemical. All of these studies detected an electrochemical potential between either tin or solder and the substrate material in a cell with the fused salt as electrolyte. Experiments were conducted [MacKay 1977] with a cell of the type shown in Figure 3-38 using fused salt mixes of ammonium chloride with stannous chloride and ferric chloride, and with tin or solder as one electrode and iron or copper as the other. The grey film referred to by Latin [1938] and Bailey and Watkins [1951] was analyzed by scanning electron microscopy and was found to be crystallites of Cu_6Sn_5 on the substrates, when either formulation was used at a temperature above the melting point of the solder or tin. When lower temperatures were used, a thin film of solid tin was observed. With steel as the substrate, a thin film of $FeSn_2$ was found. This would indicate that the flux acts as an electrolyte of a fused salt electrochemical cell.

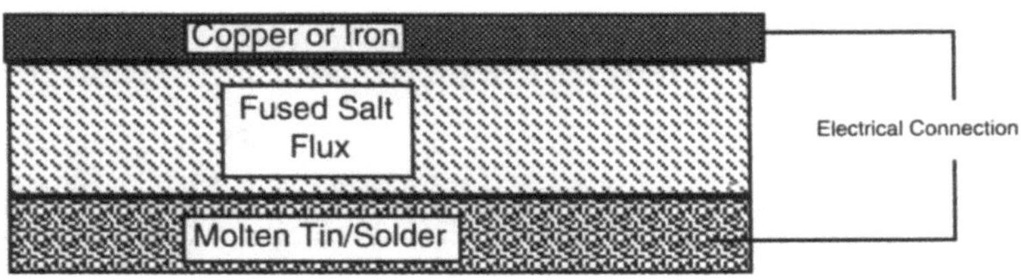

Figure 3-37 Test cell demonstrating electrochemical effect with fused salt fluxes

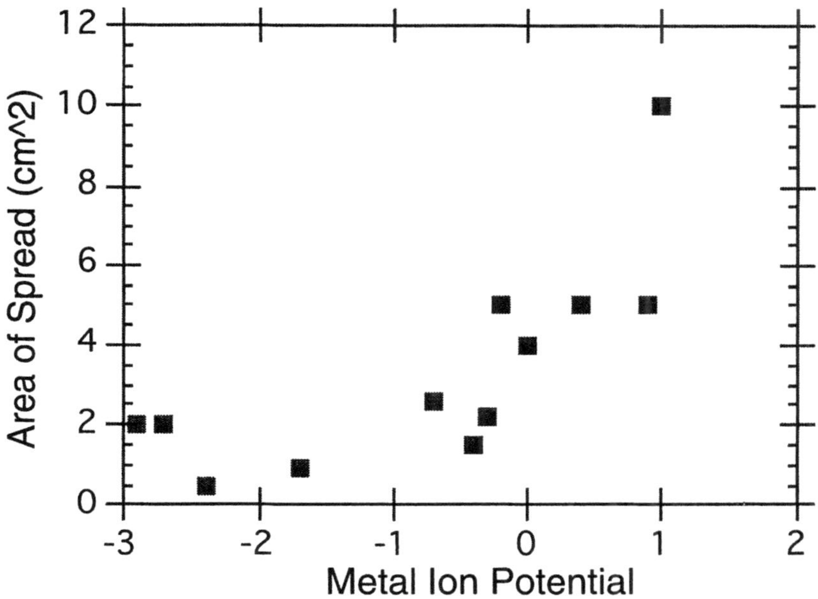

Figure 3-38 Effect of metal salt ion electron potential on area-of-spread values for nitrate compounds

The positive effects on solderability of a tin coating layer are well known: the deposition of tin had a clear beneficial effect since the substrate surface was replaced by one which was more solderable. The reason for other materials having beneficial action could be attributed to the extreme reactivity of freshly reacted materials. Nascent hydrogen, freshly evolved, is a well known example of this phenomenon, and freshly deposited metals also show this property. In a soldering situation, any freshly deposited metal would be laid down immediately before the advancing solder front at the wetting interface, ideally situated for establishment of a metal–metal interaction layer, and essential for surface wetting.

When the work with a wide range of inorganic salts was examined, some large effects were observed. Although the explanation given was faulty, the effects were real. Scrutiny of these results in the light of the materials properties of the compounds used showed a clear correlation between the metal ion potentials and the free energies of the compounds, with the observed wetting responses. For all the systems tested (chloride, sulphate, and nitrate compounds), those with metal ion potentials below about –0.75 volts had little, if any, effect on area of spread, which was small. Metal ion potentials above this showed a progressively larger effect (see Figure 3-39). Similarly, compounds with high free-energies were found to be less effective than those with low values, with the lower energies showing progressively larger area of spread. The free energies of chlorides and nitrates, at which an effect began to be seen, were similar: at about 100 Kcals/mol, as shown in Figure 3-40. Sulfates in general had high free-energy values and were effective at values below about 250 Kcals/mol. Other properties included in the analysis were the valence of the metal, the compound melting point, and the atomic weight. All these showed no obvious correlation.

It seems, therefore, that for spreading capability of solder with inorganic salt fluxes, the prime controlling factors are the combination of the electropotential of the metal ion and the ability of the compound to dissociate in a fused state. This agrees with an electrochemical mechanism being involved in the deposition of a metal ion onto the substrate in front of the advancing solder.

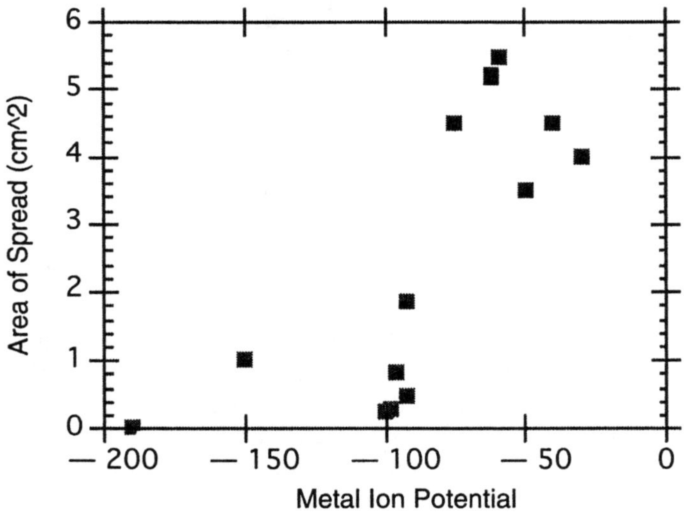

Figure 3-39 Effect of metal salt ion electron potential on area-of-spread values for chloride compounds

3.8 SOLDERABILITY TESTS

The most important requirement of any attempt to measure flux effectiveness is a quantitative test which would allow discrimination between different formulations. Development of tests that emphasize various soldering process features–solderability tests–offer these forms of discrimination. Many tests have been reported in the literature, most of them based upon some aspect of the soldering process. These are reviewed below in relation to what is being measured in each test and with some comments on their suitability for flux testing and flux mechanism studies.

3.8.1 Visual Assessment

Relies upon inspector experience and memory. It offers no quantitative evaluation but with experienced personnel offers a rapid comparison of a result with previously agreed-upon visual standards.

3.8.2 Area-of-Spread Test

The area-of-spread test is a quantitative test that involves the spread, when melted, of a weighed cylindrical tablet of solder, fluxed and heated on a square coupon of substrate. The area is usually determined by planimetric methods. This test does not have a time element, so it cannot indicate anything about wetting rates. The test is allowed to continue until the spreading action has stopped. It indicates the combination of the maximum substrate/solder interaction with the minimum solder/air or solder/flux surface tension.

3.8.3 Edge Dip and Capillary Rise Tests

Both the edge dip and capillary rise tests involve the edgewise dipping of either a single flat piece of substrate material or a piece bent so that it creates a capillary gap. Assessment is usually qualitative, the edge dip test in particular being used to indicate solderability quality aspects, i.e. non-wetting, dewetting, pinholes, etc. The capillary rise variant could be used quantitatively if the meniscus rise height were measured. These tests, like the previous one, measure an equilibrium situation, stressing to a different extent the same interfacial forces.

3.8.4 Globule Test

The globule test is a very geometry-specific test designed for solderability assessment of wires (preferably round wires). A weighed cylindrical tablet of solder is heated on an anvil. When the solder is molten, a wire is lowered into the solder bead. The time from initial contact until the bead coalesces around the wire is a measure of wetting time. This test also includes an element of applied pressure to the wetting interface resulting from the hydrostatic pressure of immersion, and is slightly biaxial in that the bead may wet along the wire as well as around it. This is a rate measuring test.

3.8.5 Rotary Dip Test

Rotary dip testing is designed to measure the wetting time of flat sheet material in a system that emulates the wave soldering process. It copies the relative motion of solder and substrate by sweeping the test coupon tangentially across the surface of a solder bath (as opposed to the solder wave sweeping the wiring board surface). A contact timer indicates the wetting time for each test, and a series of samples is required for a series of different rotation speeds of the sample arm. For a determination of optimum rotary dip solderability, the minimum solder contact time that gives complete coverage of the test coupon is determined. This test involves an element of squeegee action, a complex combination of applied pressure and enhanced thermal contact behavior, that aids and speeds the wetting process. This is a dynamic test.

3.8.6 Surface Tension Balance Test

The surface tension balance, or wetting balance test, measures the vertical force acting on a test coupon from the time wetting is initiated until it is completed or the force profile up until the time exceeds some preset maximum. Both dynamic and equilibrium parameters are measured by this test, in that the time to reach any specified position in the wetting process can be determined. Also, provided the wetting interface reaches equilibrium within the duration of the test, the wetting force at the equilibrium wetting condition can likewise be determined. The endpoints commonly chosen are: the time for the wetting force to equal the upthrust force for a particular set of immersion depth and immersion rate conditions, the time to initiate wetting, and the time for wetting to be completed. Wetting rates at any position in the process can be calculated from the profile, and the final wetting force is measured directly.

3.8.7 Flux Action from Solderability Measurements

A quantitative test is required in order to make an effective measurement of flux action, and the type of test used determines which aspect of flux action is being quantified. The area-of-spread test determines factors independent of time that relate to the advance of the wetting interface across a surface. The surface tension balance test determines rate factors that relate to the speed of wetting. Both the rotary dip and the globule tests are more complex tests, and are probably not good tests for determining flux mechanisms because the results tend to be test-parameter dependent. The globule test is influenced by the size of the wire and the size of the solder pellet; the rotary dip test is influenced by the rotation speed of the test arm.

3.8.8 Status of Flux Development and Flux Action Studies

One major concern of flux users is to satisfy the legislative requirement for elimination of CFC usage. Flux formulations therefore are tending towards very much reduced levels, possibly the total absence, of rosin content in their formulations. Liquid fluxes are already being formulated with no rosin.

Gaseous atmospheres containing organic activators are being extensively studied as blanket materials in modified soldering machines. These must invoke a rethink of the mechanisms involved, and a comprehensive assurance that the byproducts produced are truly inactive and do not react with normal environment constituents.

While these approaches offer an attractive solution to the problems of through-hole component soldering, they do not address the requirements for surface mount assembly-requirements being pursued by most electronics equipment design and assembly. A start has been made to formulate solder cream vehicles which will interact with soldering atmospheres to yield gaseous products that reduce the amount of residues. The complexity of the vehicle itself, with its many components, makes this a daunting task. More sophisticated gaseous chemistry approaches must be sought, along with new approaches to formulation of the solder cream vehicle.

New approaches to the soldering process, such as controlled solder jets or applications of ink jet printing techniques may well create new ways of assembling electronics. Very likely, these will bring their own challenges to the control and removal of the oxide and tarnish films whose presence is so critical to the final assembly process.

REFERENCES

Ackroyd, M. L., C. A. MacKay, and C. J. Thwaites, 1975. *metals Technol.* Feb. 1975. pp. 73-85.

Archer, W. L., and T. D. Cabelka. 1988. *Proc. ISHM Conf.*

Archer, W. L., T. D. Cabelka, and J. J. Nalazek. 1987. *Ass. Eng.* Dec. 1987. pp. 20-23.

Archer, W. L., T. D. Cabelka, and J. A. Tromba. 1986. *Proc. ISHM Conf.* Oct. 1986.

Audette, D. E., and C. A. MacKay. 1981. *Proc. Inst. Met. Fin. Conf.* London, England. March 1981.

—. 1982. *Circuits Manufacturing* May 1982.

Bailey, G. L. J., and H. C. Watkins. 1951. *J. Inst. Metals.* 19:57-76.

Bircumshaw, L. L. 1926a. *Phil. Mag.* 2:341.

—. 1926b. *Phil. Mag.* 3:1286.

—. 1926c. *Phil. Mag.* 12:1931.

Bonner, K. 1980. IPC Ann. Mtg. April 1980. IPC TP 323.

Brammer, D. 1991. *Circuits Ass.* June 1991. pp. 41-5.

Cabelka, T. D., and W. L. Archer. 1985. *Proc. ISHM Conf.* Oct. 1985. pp. 520-528.

Campbell, W. P. *Hercules Powder Company Reports.*

Cassidy, M. P., and K. M. Lin. 1981a. *Proc. NEPCON West.* Anaheim, CA. Feb. 1981.

—. 1981b. *Electronic Packaging and Production* Nov. 1981.

Draht, G., and F. Sauerwald. 1926. *Zeit. fur Anorg. und Allegem. Chem.* 162:103.

Drummond, C. D., W. M. Wachel, and D. L. Yenowine. 1981. Soldering Conf. Naval Weapons Center. Jan. 1981.

Dunaway, J. B. 1989. *Circuits World* 15(4).

Dunn, M. H. 1990, *Surface Mount Technol.* March 1990. pp.40-1.

Feiser, and W. P. Campbell. 1938. *J. Amer. Chem. Soc.* 60:159.

Feiser, and Feiser. *Biological Resins.*

Fleck, E. E., and S. Palkin. 1942. *Ind. Eng. Chem. Anal. Eng.* 14:146.

Fodor, P. 1990. *Proc. NEPCON East.* pp.179-87.

Fodor, P., and P. J. Lensch. 1990 *E. P. and P.* April 1990. pp 64-66.

Glasstone, S., and D. Lewin. 1960. *Elements of Physical Chemistry*. Princeton, NJ: Van Norstrand Reinhold.

Greenway, H. K. 1949. *J. Inst. Met.* 74:133.

Hagness, Th. R. 1921. *J. Amer Chem. Soc. 43:1621.*

Hanaway, E., L. Hagerty, and R. Crothers. *E. P. and P.* 31(11):118-20.

Harris, G. C. 1948. *J. Amer Chem. Soc.* 70:3674.

—. 1952. *Wood Resins.* Amer. Chem. Soc. Reinhold Press.

Harris, G. C., and T. F. Sanderson. 1948a. *J. Amer Chem. Soc.* 70(1):334.

—. 1948b. *J. Amer Chem. Soc.* 70:2079, 2081, 3870.

Harris, G. C., and J. Sparks. 1948. *J. Amer Chem. Soc.* 70:3674.

Hartman, H. J. 1991. *Circuits Ass.* Jan 1991. pp.60-65.

Henry, J. J., and N. S. Girouard. 1990. *Proc. Int'l. Conf. on Flux Technology.*

Pittsburgh, PA.:Mellon Inst. Sept. 1990. Also: IPC TP 886.

Hiatt, B., G. Jordhamo, P. J. Singh, and C. A. MacKay. 1991. *Proc. ASM Microelectronics Packaging, Materials, and Processes Conf.* Montreal, Canada.

Howle, F. A., and E. D. Hondros. 1982. *J. Mat. Sci.* 17:1434.

Ivankovits, J. C., and S. W. Jacobs. 1990. *Proc. NEPCON East* pp. 629-48.

Johnson, C. J., and J. Kevra. 1989. *Solder Paste Technology: Principles and Applications* Blue Ridge Summit, PA: TAB Books.

Kaier, R. J. 1983. IPC Ann. Mtg. April 1983. IPB TP 468.

Kawakatsu, I., and K. Yokoi. 1990 *Proc. 8th IEMT Int'l Electronics Manufacturing Conf.* May 1990.

Kenyon, W. G., and D. A. Emig. 1987. *Proc. China Lake Conf.* China Lake, CA. Wilmington, DE: E I Dupont de Nemours Co. NWC TP 6789; IPC 719.

Klein-Wassink, R. J. 1989. Soldering in Electronics Ayr, Scotland: Electrochem Publishers.

Klima, R. F., and H. Magid. 1987. *Proc. NEPCON West* pp. 736-46.

Klyachko, and Kunin. 1949. *Dokl. Akad. Nauk.* SSSR. 64:85.

Kocka, D. C. 1990. *Electronics Packaging and Prod.*

Lambert, L. 1978. *Proc. NEPCON West* Anaheim, CA. Feb. 1978. p. 75.

Langan, J. P. *Proc. 3rd Int'l SAMPE Electronics Conf.*

Latin, A. 1938. *Trans. Faraday Soc.* 34:1384-395.

Lovering, D. G. 1984. *Brazing and Soldering* No. 7 (Autumn).

Mackay, C. A. 1977. *Tin Research Inst. Report.*

—. 1980a. *Electronic Packaging and Production.* Feb. 1980. pp. 116-26.

—. 1980b. *Proc. INTERNEPCON Int'l. Electronic Packaging Conf.* Brighton, England. Oct. 1980.

—. 1989. *Proc. ASM Microelectronic Pkg. Tech. Conf.* Philadelphia, PA. April 1989. pp. 405-718.

MacKay, D. 1970. *Proc. INTERNEPCON Conf.* Brighton, England. p. 11.

Matuyama. 1927. *SCi Reps.* Tokyo Univ. 16:555.

Mehta, A., S. Adams, J. Keegan, and J. Savage. 1991. *Proc. NEPCON East* pp. 481-90.

Momose, Y., and Y. Tamia. 1970. *J.I.M.* 93.

Morris, J. R., and N. Bandyopadhyay. 1990. *Print. Circ. Ass.* 4(2):26-31.

Moskowitz, P. A., H. L. Yen, and S. K. Ray. 1986. *J. Vac. Sci. Technol.* A4. p. 838.

Nakajima, M. 1991. *J.E.E.* 28:N292.

Ohki, K. 1991. *JEEE.* April 1991.

Okano, T. 1991. *JEEE.* Oct. 1991.

Onishi, J., I. Okamoto, and A. Omori. 1972. *Trans. Japan Welding Inst.* 1(1):23-7.

—. 1973a. *Trans. Japan Welding Inst.* 2(1):113-19.

—. 1973b. *Trans. Japan Welding Inst.* 2(2):97-102; 103-10.

—. 1974. *Trans. Japan Welding Inst.* 3(1):99-103; 105-09.

—. 1975. *Trans. Japan Welding Inst.* 4(1):79-84;85-90.

Ostrander, M., K-L Wun, and J. Baker.

Palkin, S., and G. C. Harris. 1933. *J. Amer. Chem. Soc.* 55:3677.

Pickering, K., P. Southworth, C. Wort, A. Parsons, and D. J. Pedder.

Pokrovskii, and Galanina. 1949. *Zh. Fiz. Khim.* 23:324.

Pokrovskii, and Saidov. 1955. *Zh. fiz. Khim.* 29:1601.

Pokrovskii, and Sosnina. 1950. *Dokl. Akad. Nauk.* SSSR. 86:1105.

Roche, N. G., and C. A. MacKay. 1989. *Proc. ASM Microelectronics Packaging, Materials, and Processes Conf.* Philadelphia, PA. April 1989.

Rubin, W. 1982. *Welding Journal.* Oct. 1982. p. 39.

—. 1990. *Electronic Production* May 1990.

Rubin, W., and M. Warwick. 1990. *Surface Mount Technology.* Oct. 1990.

Ruzicka, and J. Meyer. 1922. *Helv. Chim. Acta* 5:315.

Ruzicka, and Sternbach. 1940. *Helv. Chim. Acta* 23:333.

Ruzicka and Waldmann. 1933. *Helv. Chim. Acta* 16:842.

Schouten, G. 1989. *Circuits Man.* Sept. 1989. pp. 50-53.

Semenchenko. 1961. *Surface Phenomena in Metal Alloys* Trans. from the original Russian, *Poverkhnostyye Yavleniya v Metallaakh i Splavakh* Oxford: Pergamon Press Ltd.

Semenchenko, Pokrovskii, and Lazarev. 1953. *Dokl. Akad. Nauk.* SSSR. 89:102.

Sherman, K., and C. A. MacKay. 1989. *Proc. ASM Microelectronics Pkg. Tech. Conf.* Philadelphia, PA. April 1989.

Sinclair J. D., et al. 1985. *J. IEEE.*

Snyder, R. W. 1987. *Applied Spectroscopy.* 14(3):460-63.

Trovato, R. 1990. *Circuits Man.* April 1990. pp. 66-7.

Turkdogan, E. T., and S. Zador. 1961. *J. Iron and Steel Inst.* March 1961. pp. 233-39.

Wallis, D. R. 1974. *Metallurgist* Jan. 1974. p. 15.

4

Reactive Wetting and Intermetallic Formation

William J. Boettinger, Carol A. Handwerker, and Ursula R. Kattner

4.1 INTRODUCTION

Despite its apparent simplicity, the spreading of molten solder on copper or on solder-coated copper to form a solder joint involves many complex physical processes. Poor solderability of electronic components, while infrequent in absolute numbers, can cause significant manufacturing difficulties when acceptable performance requires less than one failure in 10^6 joints. An assessment of solderability involves consideration of the entire soldering process including the details of the soldering equipment, the design of the joint geometry, and the wettability of the surfaces to be joined. Wettability involves consideration of the intrinsic rate and extent that solder can spread on a particular surface and is the primary focus of this chapter. Thus we need only consider macroscopically simple geometries of wetting; we will focus on the fundamental physical processes involved in solder wetting and spreading. The ultimate goal is to determine those processes that are most important in order to provide a scientific basis for the analysis of engineering problems in solder technology.

Our discussion begins with consideration of the spreading of a molten solder droplet on a horizontal copper substrate. Figure 4-1 shows a sketch with an indication of various processes that may occur during the spreading of a solder droplet on a Cu substrate. The initiation of solder spreading clearly depends on heat flow and on the melting of the solder. Subsequent solder spreading involves motion of the fluid within the droplet. In a reference frame fixed in the substrate, fluid streamlines terminate on the free boundary because it is moving. The flow patterns can be driven by capillarity (reduction of surface energy), buoyancy (density), and liquid surface tension gradients (Marangoni effects) at rates limited by the viscosity of the liquid solder and the kinetics of the moving contact line.

For spreading of a non-reacting fluid on a defect-free substrate, a thin liquid layer ahead of the moving contact line with a thickness of ~10 nm, known as a "precursor foot," plays a major role in the contact line kinetics [de Gennes 1985]. This precursor foot exists because a disjoining pressure occurs between the liquid-vapor and liquid-substrate interfaces that extrudes a thin layer of fluid away from the macroscopic contact line. Solder droplets sometimes are observed with "halos" [Klein-Wassink 1989] leading the macroscopic spreading contact line, but the relationship between halos and the precursor foot has not been established.

Small amounts of volume diffusion occur during solder spreading. Diffusion of copper into the liquid solder and diffusion of tin into the solid copper (albeit in a very thin, perhaps undetectable layer) provide a mechanism for dissolution of a thin layer of the copper substrate. Formation of intermetallic compounds during the diffusion process also occurs. An example of our current lack of understanding about reactive wetting is the uncertainty regarding the location of the intermetallic formation with respect to the position of the moving contact line. In Figure 4-1 the intermetallics are assumed to form slightly behind the position of the moving contact line, exposing a small region of direct liquid/copper contact. This unproven assumption is based on details of the grain structure observed within the intermetallic layers, as described in a later section of this chapter. As the contact line slows relative to the lateral spreading rate of the intermetallics, the region of direct liquid/copper contact should disappear. These chemical interactions by interdiffusion provide a reduction in the free energy of the system and should play a role in the overall dynamics. Exactly how this is coupled to the motion is unclear at the present time. Attempts to understand this coupling using studies of solder spreading that involve diffusion on a time scale of minutes to hours [Bailey and Watkins 1951,1952; Turdogan and Zador 1961] are not particularly relevant for the dynamics of solder spreading as related to technical soldering problems.

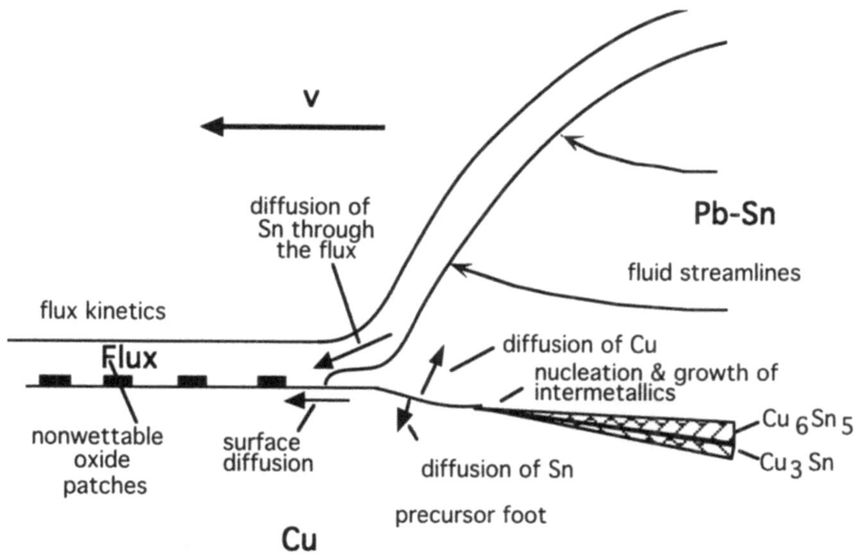

Figure 4-1 Schematic representation of possible processes occurring during the spreading of a solder droplet on a Cu substrate. The vertical scale is exaggerated.

Finally, the schematic indicates the importance of the reaction of flux with oxide on the copper substrate. Oxides are here considered as discrete, non-wettable (or poorly wettable) patches with sizes that are decreasing with time due to the flux action. The effect of these obstacles on wetting behavior is very important.

A simple observation of solder spreading is informative and will be used as a reference for subsequent analysis. Figure 4-2 shows a sequence of photographs taken from a video of the melting, shape change, and spreading of a 50 mg (4mm diameter x 0.5 mm thick) 60%Pb-40%Sn (wt.%) solder pellet resting on 1 mm thick copper substrate covered with a non-activated rosin flux. Heat was applied to the underside of the copper substrate by floating it on a large solder bath held at 235°C. Flux bubbling occurred for ~3 s followed by melting of the pellet and quick formation of a spherical cap in the next 0.5 s with a contact diameter smaller than the pellet. Over the next 3 s the contact circle diameter doubled with an average speed of approximately 0.1 cm/s. Over the next 14 s, the solder droplet spread very slowly another 0.1 mm, after which the experiment was terminated. In a similar experiment, but using a mildly activated RMA flux, the melting and spreading stages could not be visually separated. The speed of the contact line was estimated at ~1 cm/s. Our challenge is to develop an understanding of the mechanisms that limit the rate and extent of spreading of solder. A different but related question regards the spreading of solder on a substrate that has already been precoated with solder.

In this chapter we will analyze and describe some aspects of the dynamics of solder spreading, the equilibrium and the motion of contact lines over obstacles (e.g. oxide patches), the metallurgical reactions that occur when molten solder is in contact with copper, the progress being made on the Cu–Pb–Sn ternary phase diagram that governs the quantitative aspects of the metallurgical reaction, and finally, we will describe wetting balance tests on bulk samples of Cu_3Sn and Cu_6Sn_5 in comparison to results on copper.

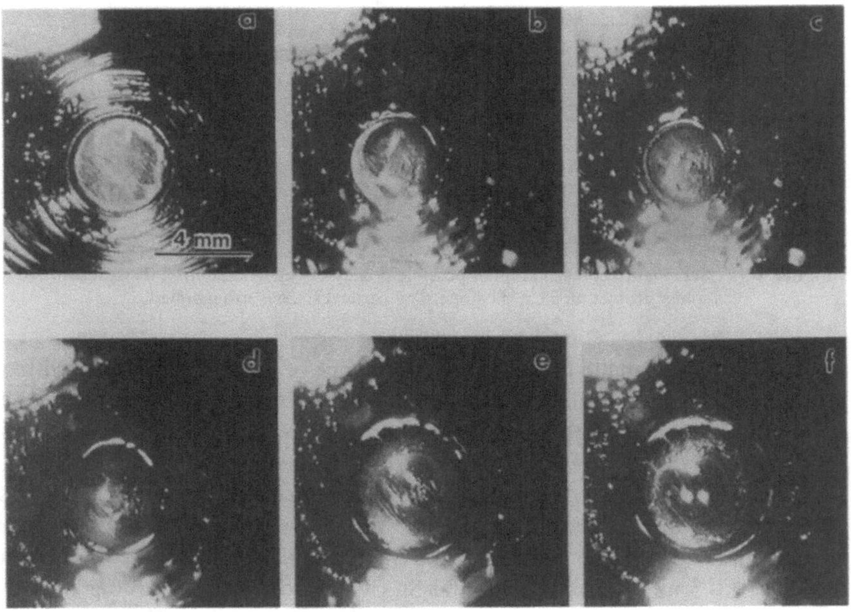

Figure 4-2 A sequence of photographs taken from a video tape of a top view of a spreading solder droplet using an R flux at times of (a-f) 0, 3, 4, 5, 6, 15 s. The 50 mg solder pellet is 4 mm in diameter. At 3 s, the pellet is about 70% molten. At 4 s, the droplet is completely molten with a diameter smaller than the initial pellet diameter. Spreading slows considerably after 6 s.

4.2 ANALYSIS OF SOLDER SPREADING KINETICS

4.2.1 Heat Flow

Heat flow clearly plays a role in practical soldering processes. The rapid rise and fall of temperature allows for variations on every phenomenon discussed in the introduction to this chapter. A particular difficulty with the modeling of heat flow is the lack of knowledge of heat transfer coefficients for given practical heating configurations. For further reference, Klein-Wassink [1989] devotes a chapter in his book, *Soldering in Electronics*, to the analysis of heat flow problems.

The experiments described in this chapter's introduction with R and RMA fluxes represent two extremes for the importance of heat flow even though the rates of heating for the experiments were identical. For the RMA flux, the wetting is so rapid that the geometry and the rate of pellet melting influence the geometry and the rate of the spreading process. Thus the spatial and temporal details of the temperature rise are important. With the R flux, melting is clearly completed before the solder spreads to any significant extent. Measurements of the temperature rise of the top surface of the 1 mm thick copper substrate indicate that the melting point of the solder (183°C) was reached in ~3 s, and that the temperature approached the bath temperature (235°C) exponentially with a time constant of ~2.2 s. Thus even for the R flux, the spreading process occurs non-isothermally and prevents an interpretation of this simple experiment based on an isothermal analysis. Generally, liquid metal viscosities and liquid diffusion coefficients are not strong functions of temperature, so their changes are not likely to be important for solder spreading. On the other hand, the variation of liquid surface energy with temperature, even though small, can have a direct role on the fluid dynamics of droplet spreading through the Marangoni Effect when temperature gradients exist (see below). It is also likely that flux activity and viscosity depend strongly on temperature. Although the temperature histories during these simple experiments are not ideal, they do represent the complexity that exists during practical soldering processes.

4.2.2 Fluid Flow

The spreading of a liquid droplet on a defect-free substrate without metallurgical reaction is a formidable fluid mechanics problem. A recent paper by Ehrhard and Davis [1991] is quite useful to understand various limiting cases of this problem. Their paper considers the influence of:

- gravity
- flow induced by a free liquid surface with non-constant curvature
- contact line kinetics
- flow induced by temperature gradients (Marangoni flow).

The effect of gravity generally can be ignored in spreading problems if the vertical height of the droplet is small compared to $(\gamma/\rho g)^{1/2}$ where γ is the liquid surface tension (~500 ergs/cm^2) [Schwaneke et al. 1978], ρ is the density of the solder (~8.3 g/cm^3), and g is the acceleration of gravity. For solder, this critical height is ~2.5 mm, which is ten times the height of the solder pellet under consideration here.

Fundamental to the approach of Ehrhard and Davis is the use of a constitutive equation:

$$v = K(\theta - \theta_e)^n \tag{4-1}$$

for the contact line kinetics, where v is the velocity of the contact line, K is a rate constant, and q and qe are the instantaneous and the equilibrium macroscopic contact angles, respectively. This equation is assumed to embody the atomistic and/or short range van der Waals forces that act at the triple line and must be obtained from either measurements or a separately derived microscopic model of the moving contact line.[1] Contact angles used in Eq. (4-1) should be considered on a scale that is appropriate for a continuum model of spreading that can be used for engineering problems in spreading dynamics. For example, the angle and position of the contact line should be considered on a scale where the volume of a small equilibrium drop can be reasonably described by the geometry of a spherical cap with that contact angle and corresponding contact circle diameter. As such, the contact angles in Eq. (4-1) are different from those that would come into consideration of the shape of the precursor foot [de Gennes 1985].

Figure 4-3 shows two limiting cases of droplet spreading for an arbitrary fluid-substrate combination. The first case is limited by the rate of fluid flow (viscosity) in the droplet and the second case is limited by the contact line kinetics. In the former case (top figure), the equilibrium contact angle is established in the vicinity of the triple junction on a time scale that is short compared to the time required for fluid flow to establish a shape with constant (spherical) curvature. The latter case

---

[1] An analogous approach is employed in solidification and crystal growth problems. An equation relating the velocity of the liquid-solid interface to the difference between its actual temperature and the equilibrium melting point is employed. Such an equation can be obtained by measurements or from models of the atomistic structure and motion of the interface. This constitutive equation is subsequently combined with macroscopic equations of heat and fluid flow to describe the system response to imposed external conditions.

(bottom figure) establishes constant curvature first, followed by an adjustment of the contact angle toward the equilibrium angle according to the contact line constitutive law, Eq. (4-1). A characteristic time, t_{vis}, for the droplet shape to adjust to constant curvature is given in terms of the initial radius of the contact line circle, b_0, (2 mm), the viscosity of the solder, μ, (0.018 poise) [Thresh and Crawley 1970], and the liquid surface tension, γ, as:

$$t_{vis} = \mu b_0/\gamma \,. \tag{4-2}$$

This characteristic time is 10^{-5} s for the size of the solder pellet considered here. Because this time is very short compared to the observations in Figure 4-2, solder spreading is not limited by fluid flow in the bulk droplet but is limited by the action at the contact line. In fact, the two limiting cases can be characterized by the dimensionless number, $K\mu/\gamma$, with large and small values corresponding to Figure 4-3 (top and bottom, respectively). Our challenge is the formulation of a generalized form of Eq. (4-1) appropriate for reactive wetting. For reactive spreading we anticipate a more complex constitutive equation for contact line kinetics:

$$v = f(\theta, \theta_e(t), T, c, s(t), \ldots) \tag{4-3}$$

that might include a time-dependent quasistatic equilibrium contact angle, $\theta_e(t)$, which changes with time due to modification of the surfaces because of the fluxing action or diffusion, the temperature T, the compositions c of the solder and/or substrate, and perhaps a time-dependent area fraction of non-wettable spots, s(t).

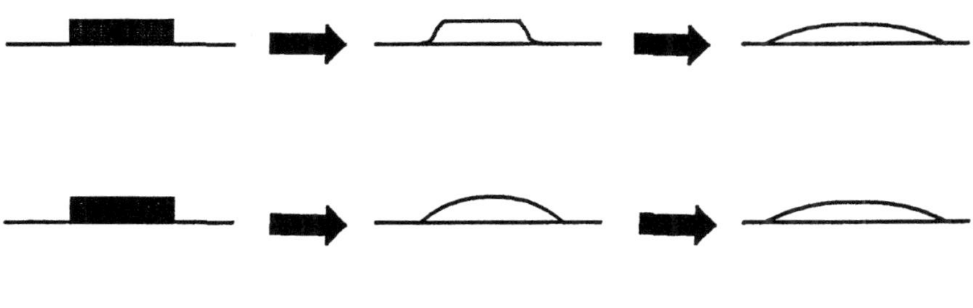

Figure 4-3 Two limiting cases for the spreading of a fluid droplet on a substrate. (Top) The droplet establishes an equilibrium contact angle more rapidly than it develops constant curvature and thus the spreading dynamics are controlled by the approach to constant curvature. (Bottom) The reverse is true and the dynamics are controlled by the contact line kinetics. Solder spreading seems to follow the bottom drawing, especially when a non-activated flux is used.

Ehrhard and Davis [1991] have also shown that Marangoni flow can either retard or promote the rate of droplet spreading. Marangoni flows result from either temperature and/or composition gradients when the liquid surface tension depends on temperature and/or composition. This dependence produces a shear stress, which acts on the free surface of the fluid in the direction of higher surface tension. For pure liquids, surface tension usually decreases with increasing temperature. Thus the Marangoni surface force acts in a direction towards the colder regions. This force then acts in a manner that either impedes or augments the normal circulation flow patterns that accomplish spreading. In a frame moving with the contact line, this normal flow is similar to the motion of a tread on a tractor vehicle. One particularly interesting result the Ehrhard and Davis paper shows is that, for a non-isothermal droplet with an equilibrium contact angle of zero, only finite spreading occurs even after infinite time. This occurs if the top of the droplet is colder than substrate and if $d\gamma/dT$ is negative.

The possibility of Marangoni flow during soldering due to the dependence of surface tension on composition has not been treated. This effect should be large because a 1 at.% increase in lead content decreases the liquid surface tension as much as a 20K increase in temperature [Schwaneke et al. 1978]. Due to the reaction of the tin in the solder with the copper, composition variations will exist in the fluid near the substrate that may be transported to the free surface. Gradients in the substrate surface energy can also lead to lateral motion of a droplet in the direction of lower substrate surface energy [Brochard 1989; Chaudry and Whitesides 1992].

4.2.3 Spreading of Spherical Droplets

For the spreading of a droplet that can be assumed to remain a spherical cap, a simple relationship exists between the exponent of the contact line kinetic law, n, and the time exponent of the contact line radius, b(t), if $\theta_e = 0$. For practical purposes the volume of the spherical liquid cap can be assumed constant. The volume of a spherical cap, V, is given by Klein-Wassink [1989]:

$$V = (1/6)\pi b^3\{(\tan(\theta/2))^3 + 3\tan(\theta/2)\} \tag{4-4}$$

and for small values of θ

$$\theta = 4V/(\pi b^3) \ . \tag{4-5}$$

After substitution into Eq. (4-1) and realizing that $v = db/dt$, one obtains

$$db/dt = (4KV/\pi)^n\{b^{-3}-b_e^{-3}\}^n \tag{4-6}$$

where b_e is the final contact radius corresponding to a contact angle of θ_e. For the case when $\theta_e = 0$ ($b_e = \infty$, infinite spreading), Eq. (4-6) is easily integrated to

$$b(t) = \{(1+3n)(4V/\pi)^n Kt\}^{1/(3n+1)} \tag{4-7}$$

where the constant of integration has been neglected as is appropriate for an evaluation of long-time behavior. For cases where $\theta_e \neq 0$, or for contact angles that are not small, the equations can be integrated numerically. Figure 4-4 shows the solutions to Eq. (4-6) for small contact angles but for $\theta_e \neq 0$ and n = 1,2,3,4. One can show that an expression valid for long times ($[b_e-b(t)]/b_e \ll 1$) for n >1 is given by

$$\frac{b_e - b}{b_e} = \left(\frac{(3\theta_e)^n(n-1)K}{b_e}t\right)^{\frac{1}{1-n}} \tag{4-8a}$$

and for n = 1 by

$$\frac{b_e - b}{b_e} = \exp\left(-\frac{3\theta_e K}{b_e}t\right) \tag{4-8b}$$

As shown in Figure 4-4, the power law behavior (n >1) converges very slowly to the equilibrium final contact angle ($b/b_e = 1$). Recently, spreading data obtained during brazing have been compared to various power laws [Ambrose et al. 1992]. It was found that the exponents were required to depend on time in order to fit data, making this approach rather unsatisfying.

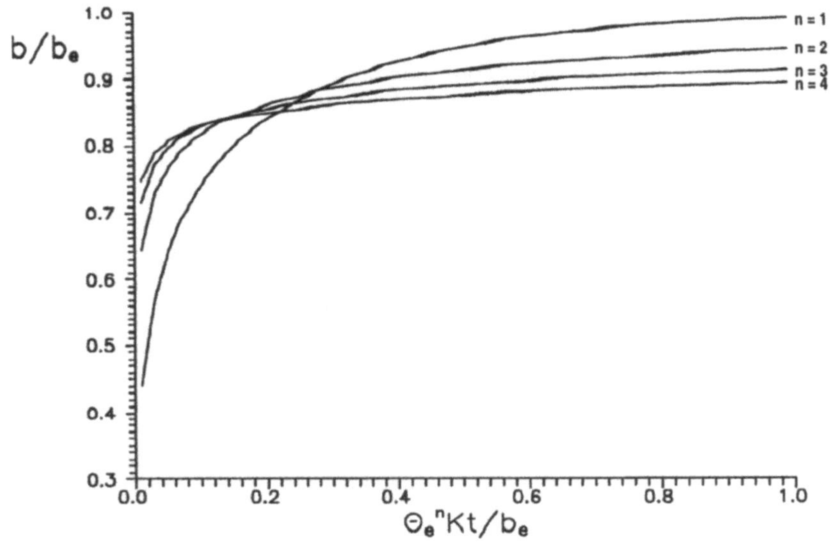

Figure 4-4 Normalized contact circle radius, b/b_e of a spreading droplet vs. dimensionless time, $\theta_e^n Kt/b_e$, Obtained by numerical integration of eq. (4-6) for various values of the contact kinetic law exponent, n.

For non-reacting systems with $\theta_e = 0$, Eq. (4-1) has been verified experimentally and is usually referred to as Tanner's Law where n is usually found to be ~3. For the case where n = 3, the time exponent in Eq. (4-7) is 1/10. The theory of de Gennes for droplet spreading [1985], which considers only the case of $\theta_e = 0$, can also be restated in the framework of a particular model for Eq. (4-1). He analyzed the viscous dissipation in the precursor foot for perfect wetting to formulate the kinetics of the moving contact line. In the context of Eqs. (4-1) and (4-7), he obtains n = 3 and $K = (1/10)(\pi/4)^3(\gamma/\mu)$.

A very simple but interesting case for spherical droplet spreading can also be obtained if θ_e is a function of time; for example, by the action of the flux. A quasistatic dynamics for solder spreading can be obtained by assuming that the actual contact angle is equal to this time-dependent θ_e and by using the volume expression, Eq. (4-5), to solve for b. One then obtains

$$b(t) = \{4V/\pi\theta_e(t)\}^{1/3}. \quad (4\text{-}9)$$

4.3 CONTACT LINE MOTION OVER OBSTACLES

Motion of the contact line during wetting can be impeded by any regions with poorer wetting characteristics than the base substrate. Such obstacles to motion include oxide patches, residual organics, scratches parallel to the moving contact line, and changes in the lead geometry (for example, solder moving around an edge of a wire with square cross-section or around an edge where a through-hole meets the surface of a board). In the following section, the effects of inhomogeneities in a flat substrate on the microscopic and macroscopic contact angles are discussed.

4.3.1 Static Contact Angles

The conditions for equilibrium of a droplet on a flat, rigid, and non-reacting surface are well known and are obtained from the calculus of variations (Figure 4-5, top). If gravity can be neglected, as described above, and the only energies that need be considered are liquid (liquid-vapor or liquid-flux), solid (solid-vapor or solid-flux), and liquid-solid surface energies, two types of conditions are obtained from the total energy minimization. The first condition requires that the free (liquid) surface have constant mean curvature. The second condition requires that the angle between the free surface and the flat rigid surface, θ_e, be given by

$$\gamma_S - \gamma_{SL} = \gamma \cos \theta_e \tag{4-10}$$

where γ_S, γ_{SL}, and γ are the solid, solid-liquid, and liquid surface tensions, respectively. It is important to remember that Eq. (4-10) arises from the energy minimization only if variations (in the sense of the calculus of variations) are possible for the contact line position. If the contact line is pinned and immobile, Eq. (4-10) is not required for energy minimization and any contact angle is possible as long as the surface has constant mean curvature.

Even if a contact line is mobile, special considerations are necessary if the contact line intersects the flat surface at a discontinuity in the solid surface energy. The different solid surfaces are designated α and β (Figure 4-5, bottom). These could represent copper (α) with thin oxide patches (β). Again the calculus of variations can be used to address the conditions of equilibrium of the contact line at the discontinuity. Variations in the position of the contact line away from the discontinuity will increase the energy if the actual contact angle, θ, lies in a range between the equilibrium contact angles that the liquid would make with homogeneous α or β surfaces, θ_e^{α} and θ_e^{β}, respectively. For droplets with contact angles in this range, the contact line will remain at the discontinuity.

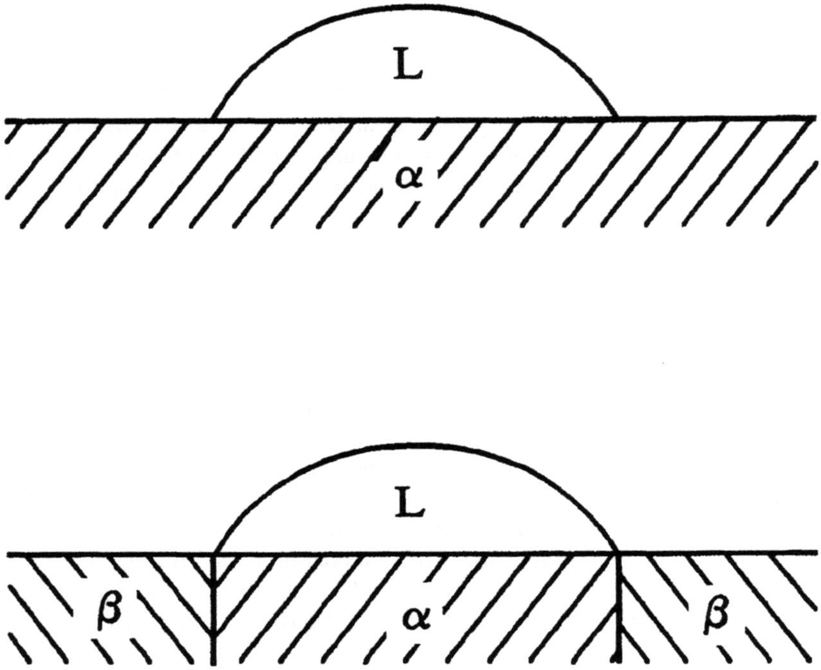

Figure 4-5 The equilibrium contact angle of a droplet whose contact line lies on a discontinuity of the substrate surface. It can have a range of angles between the equilibrium angles for the two surfaces.

A simple example [Cahn et al. 1992] of how discontinuities could affect solder spreading is given in Figure 4-6. Consider the quasistatic addition of liquid volume to a droplet lying on a surface with periodic patches of α and ß surface, as shown with $\theta_e^{\alpha} < \theta_e^{\beta}$. Assuming that the droplet spreads about a center of symmetry on the surface and that the contact line begins on the α surface, the contact radius of the droplet will increase smoothly as the volume increases until the contact line reaches the first discontinuity in the surface. The contact line will then remain at this discontinuity until sufficient liquid volume is added that the actual contact angle reaches θ_e^{β}. The contact line will then begin to spread smoothly on the β surface until it reaches the next discontinuity. At this position any additional increase in volume would require the actual contact angle to be outside the permissible range for equilibrium. Thus the contact line will jump out onto the α surface. Its position will stabilize at a contact radius appropriate for its volume with a contact angle equal to θ_e^{α}. Such a stick–slip behavior is indicated in the bottom portion of Figure 4-6, where the contact radius, b, is schematically plotted against droplet volume. The possible states, (V, b), must lie on or between the curves representing the relationship between volume and contact radius, Eq. (4-4), for the two equilibrium contact angles, θ_e^{α} and θ_e^{β}. A different path of (V, b) states would occur for a shrinking droplet. The difference between the two paths is the essence of the contact angle hysteresis. For the more general case of droplet spreading, the sticking and slipping over discontinuities can lead to equilibrium shapes and dynamics for the motion of a contact line that will depend on the spatial distribution and contiguity of the less wettable patches and the size scale of the patches relative to the droplet size. Other examples of lines and surfaces moving around obstacles abound in materials science; for example, dislocations and grain boundaries can be pinned by immobile particles.

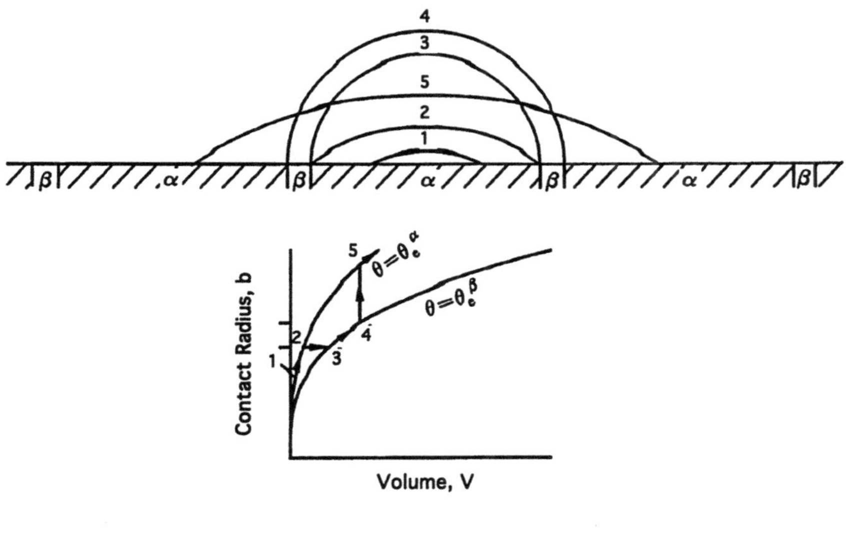

Figure 4-6 An example of discontinuous spreading of a droplet due to discontinuities, β, in the substrate surface, α, as the droplet volume increases. At the top are successive positions of the droplet (1-5) and at the bottom is a plot of the possible states, (V, b), of the droplet. The actual path lies on or between the volume vs. contact radius curves for the two equilibrium angles for homogeneous surfaces of α and β.

Figure 4-7 shows scanning electron microscope (SEM) photographs of the contact line of solder droplets after spreading and cooling. The copper substrates contained 1 or 10 volume % of 304 stainless steel spheres, ~10 to 20 μm in diameter. This material, prepared by hot isostatic pressing (HIP) of mixtures of copper and steel powders, was synthesized to simulate the presence of poorly wettable patches (e.g. oxides) that may exist on pure copper surfaces. One can observe that the contact line is distorted in the vicinity of the particles and appears to be pinned by them. Clearly the solder has spread over many other particles which lie under the solder. In the limit of very fine and non-wettable patches, the details of the contact line shape and dynamics can be incorporated into an effective contact angle and perhaps into an effective contact line kinetic law.

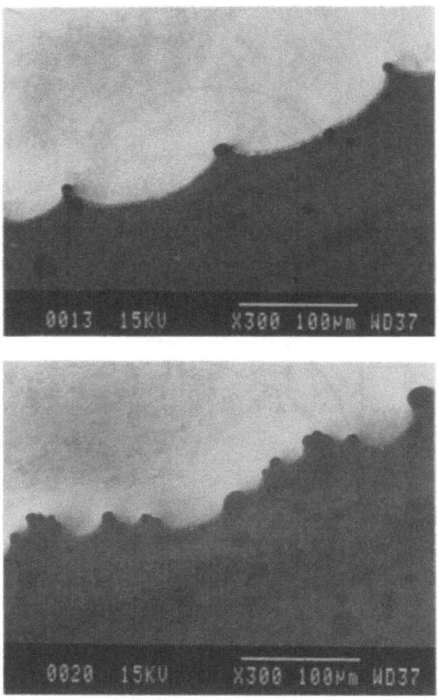

Figure 4-7 SEM views of two solidified droplets near the contact line showing the distortion of shape near less wettable patches. The patches were artificially produced by preparing bulk samples of Cu containing 1% (top) and 10% (bottom) by volume of 304 stainless steel particles ~10-20μm in diam, R flux.

4.3.2 Effective Contact Angle

For the above case one can define an effective contact angle, υ_e, by assuming that the (known) volume of the droplet can be approximated by a spherical cap with this effective contact angle and an average contact radius satisfying Eq. (4-4). This effective angle must be equal to some average of the actual contact angle integrated around the periphery of the droplet.

A simple estimate of the effective contact angle can be obtained using a linear rule-of-mixtures for a substrate of poorly wettable spots β of area fraction, s, in a matrix of α. If we construct area fraction weighted average solid-liquid and solid surface energies respectively:

$$\gamma_{LS} = (1-s)\,\gamma_{L\alpha} + s\gamma_{L\beta}$$

$$\gamma_S = (1-s)\gamma_\alpha + s\gamma_\beta \tag{4-11}$$

and substitute them into Eq. (4-10), we obtain after some rearrangement

$$\cos\upsilon_e = (1-s)\cos\theta_e^{\alpha} + s\cos\theta_e^{\beta}. \tag{4-12}$$

Figure 4-8 shows data for effective contact angle (calculated from the measured area of spread and initial droplet volume) vs. area fraction of non-wettable patches for the copper/stainless steel samples. Pure copper samples, made by hot isostatic pressing consolidation of copper powder, were used for s = 0. All surfaces were prepared by metallographic polishing; R and RMA fluxes were used to perform the spreading experiments. Also shown in Figure 4-8 are curves calculated from Eq. (4-12) using the measured values of θ_e^{α} (α = Cu) for the two fluxes and a value of $\theta_e^{\beta} = 180°$ (β = stainless steel). The trend is clearly established. The use of 180° for θ_e^{β} was substantiated by independent attempts to measure the contact angle of solder on homogeneous samples of stainless steel where the solder droplets fell off after cooling and flux removal. Clearly the linear rule-of-mixtures model used to obtain Eq. (4-12) may be too simple for other cases. Nevertheless, the idea of an effective contact angle may be useful to describe spreading on inhomogeneous surfaces.

Eq. (4-12) has two possible applications to soldering. Consider that the poorly wettable patches are oxides and that the action of a flux is to impose a time dependence that reduces the area fraction of non-wettable spots, s(t). Then the effective contact angle υ_e would decrease with time through Eq. (4-12). If a spherical cap can be maintained, the time dependence of the contact radius, b, would be given by Eq. (4-9) with $\upsilon_e(t)$ substituted for $\theta_e(t)$.

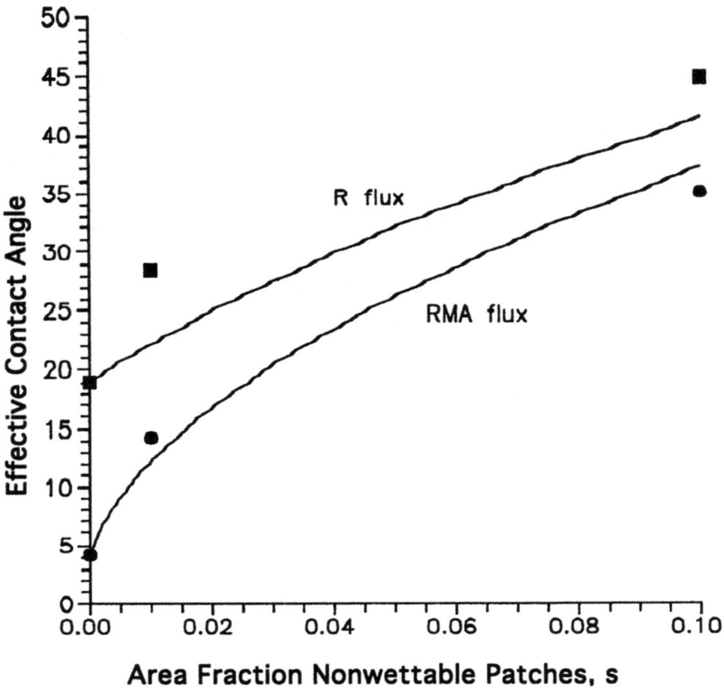

Figure 4-8 Apparent contact angles observed for solder on artificially prepared Cu containing 1 and 10% by volume of 304 stainless steel particles ~10-20μm in diam using R and RMA fluxes. The curves are plots of Eq. (4-12) using the measured values of θ_e^{Cu} for the two fluxes and a value of $\theta_e^{SS} = 180°$.

A second example comes in the area of dewetting. Suppose that a surface containing non-wettable patches has been covered, not by capillary forces, but by artificial immersion in solder or by plating. Reflowing the solder can produce instabilities around the periphery of the coated region and cause localized withdrawal of the solder (Figure 4-9) to shapes more consistent with the effective contact angle. This angle would be based on the area fraction of non-wettable patches on the surface. This idea has been previously proposed by Klein-Wassink [1989], although in a different form.

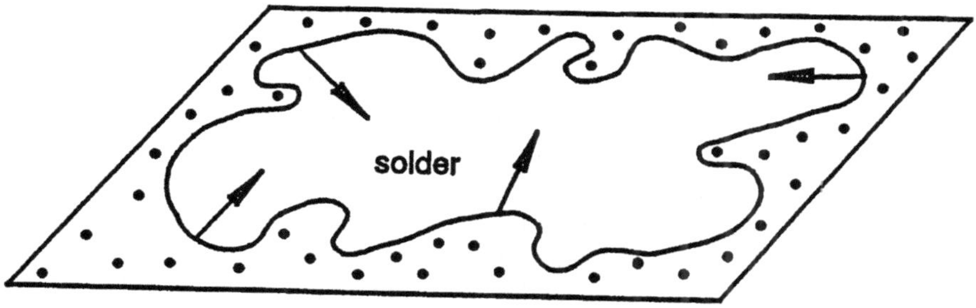

Figure 4-9 Schematic drawing of a surface containing poorly wettable patches that has been covered by solder by forced coverage, e.g. plating. An instability occurs at the exterior edge of the solder during dewetting as the solder attempts to establish the effective contact angle appropriate for that area fraction of poorly wettable patches.

Joanny and de Gennes [1984] have analyzed the distortion of the droplet surface and the contact line in the vicinity of the contact line near obstacles. An identical result to Eq. (4-12) was obtained (their equations A15a and A16a) for static equilibrium of a contact line located near a dilute array of fine rectangular or square patches (called mesa defects). Their paper goes on to analyze the hysteresis in contact line position depending on the direction of motion of the contact line. Values of effective advancing and receding contact angles are calculated for this situation. The deviation of the advancing and receding contact angles from the effective equilibrium angle described above was also found to depend on the area fraction of the less wettable patches. Clearly, for a more rigorous discussion of wetting and dewetting, the contact angle hysteresis must be included. At the present time there appears to be no dynamic theory for the motion of the contact line over obstacles.

4.4 METALLURGY OF THE MOVING CONTACT LINE

The metallurgy of a moving contact line and its influence on the contact kinetic line constitutive equation is not understood at the present time. As shown in Figure 4-1, several processes may occur which determine contact line kinetics in the absence of defects, including the volume diffusion of Cu into the liquid solder and Sn into the Cu, the nucleation and growth of intermetallic phases, Cu_6Sn_5 and Cu_3Sn, and the surface diffusion of Sn in front of the macroscopic droplet. Generally, the changes in free energy associated with alloying are large compared to surface energies. Thus it is anticipated that the rate of motion of the contact line should in some way depend on this alloying. There have been suggestions [Aksay et al. 1974; Yost and Romig 1988] that these free energy changes are coupled to changes in the surface energies and control the dynamics of spreading. At the present time, a complete theoretical basis for such coupling is not clear.

In the absence of a moving contact line, the interdiffusion and reaction between copper and pure Sn in diffusion couples is well understood when the process is controlled by volume diffusion [Kirkaldy and Young 1987; Mei et al. 1992]. If a semi-infinite slab of copper is brought in contact with a semi-infinite slab of pure Sn (liquid or solid) at a known temperature, the composition profiles in the phases and the (planar) layer thicknesses of the intermetallics can be computed using the reasonable assumption that the various interfaces have sufficient time to relax to local equilibrium. In this case the compositions on the different sides of each interface are given by the boundaries of the appropriate two-phase field of the Cu-Sn phase diagram. Figure 4-10 shows this phase diagram and the Sn content present at the interfaces between Cu(Sn)/Cu_3Sn, Cu_3Sn/Cu_6Sn_5, Cu_6Sn_5/liquid Sn(Cu) under the local equilibrium assumption. The solution to the diffusion equation yields layer thicknesses and positions of the interfaces that move as the square root of time.

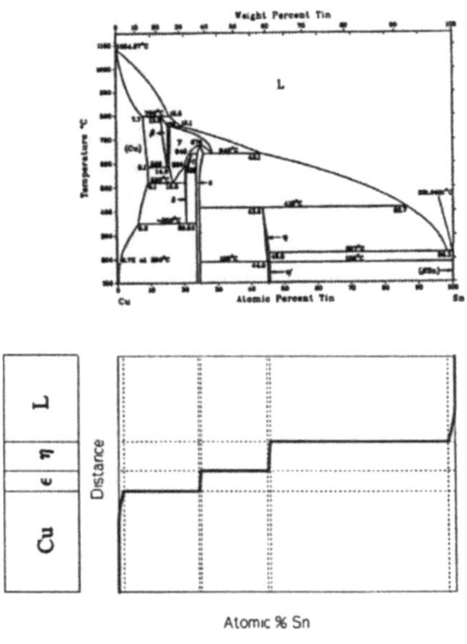

Figure 4-10 The Cu-Sn phase diagram (top) and the composition vs. distance plot (schematic at bottom) for interdiffusion in a semi-infinite geometry between Cu and liquid Sn. The compositions on both sides of each interface are obtained from the phase diagram at the temperature of the interdiffusion as shown.

For the case of Pb-Sn solder on copper, solution of the interdiffusion problem requires compositions for the interface compositions from a Cu-Pb-Sn ternary phase diagram. This diffusion calculation is in principle identical to that sketched out for binary diffusion, with the exception that the interface compositions are not fixed but can vary over different tie-lines from the ternary diagram in the two-phase field for the phases at the interface. This poses no difficulty if the solutions are obtained through numerical techniques. Figure 4-11 shows schematically the diffusion path that would be obtained during interdiffusion of a 60%Pb-40%Sn solder (here shown in atomic %) with pure copper if the interfaces remain planar. By conservation of mass, the path must cross the dotted line joining the two end compositions. The path departs the copper corner into the single phase region for the Cu(Sn) solid solution and jumps across the two-phase field between Cu(Sn) and Cu_3Sn along a tie-line. This jump corresponds to the interface between Cu(Sn) and Cu_3Sn. Next the path crosses the narrow Cu_3Sn single phase region and suffers another jump across a two-phase field, now between Cu_3Sn and Cu_6Sn_5, again following a tie-line. This corresponds to the Cu_3Sn/Cu_6Sn_5 interface. The path continues into the narrow Cu_6Sn_5 single phase region, jumps across the Cu_6Sn_5/liquid two-phase field following a tie-line, enters the single phase liquid region, and approaches the bulk solder composition at positions in the sample far from the reaction zone. Because of conservation of mass, the liquid solder is depleted in tin close to the interface with Cu_6Sn_5.

At least two modifications are required to model the interdiffusion near a moving contact line. The first clearly involves the different diffusion geometry. The diffusion problem under a spreading droplet has not been studied. At each particular location along the substrate under the spreading droplet, one might approximate the depth of diffusion and the distance that each interface moves as being proportional to the square root of the time elapsed since the spreading droplet covered that particular location.

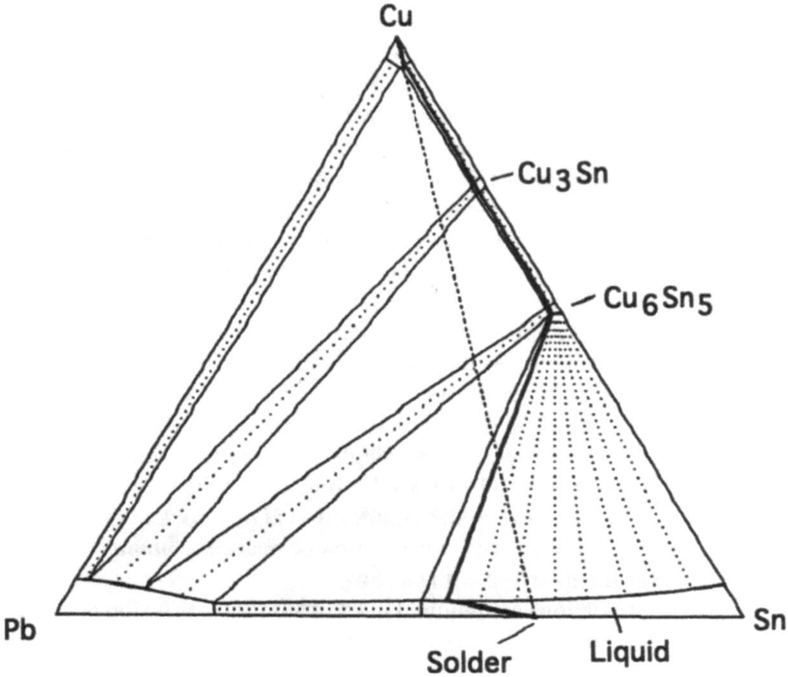

Figure 4-11 Schematic diffusion path between Cu and 60wt%Pb-40wt%Sn solder superimposed on the ternary diagram in at.%. The solubility ranges are exaggerated for clarity.

A second modification could involve a hypothesis that the intermetallic compounds are absent in the region just behind the moving contact line, as shown in Figure 4-1. Experimentally, intermetallic compounds between copper and tin or copper and Pb-Sn, which are observed under solder droplets, exist as individual grains (orientations) with no sign of columnar growth in the direction of the moving contact line. Thus, multiple nucleation of intermetallic grains occurs rather than a continuous lateral spreading of single grains of the two intermetallics. Based on this observation, we propose that the phases that contact each other just behind a moving contact line are FCC copper-rich solid solution and liquid. For such an interface to reach local equilibrium, only interdiffusion is necessary. In contrast, the formation of the intermetallics at the copper/liquid interface requires the development of supersaturation (presumably in the liquid) to overcome the nucleation barrier. This process should occur on a longer time scale further behind the moving contact line. If these notions are true, the relevant volume diffusion problem to be solved immediately behind the moving contact line requires the use of a metastable phase diagram between copper and liquid at the soldering temperature to give the interface compositions for boundary conditions. This metastable phase diagram involves a constrained equilibrium between copper and liquid where the intermetallic phases are absent. Such diagrams are successfully used throughout metallurgy to compute diffusion kinetics when some phases do not have time to nucleate [Perepezko and Boettinger 1983]. Such a phase diagram is a true equilibrium diagram in the sense that the chemical potentials are continuous across each two-phase field. Figure 4-12 shows a metastable Cu-Sn phase diagram involving only liquid, with the copper-rich and tin-rich phases constructed through thermodynamic modeling (detailed in next section). One advantage of this approach is that γ_{LS} is well defined because the two phases, liquid and copper solid solution, are at least locally in thermodynamic equilibrium. Similar ternary metastable diagrams have been calculated for liquid Pb-Sn on copper in the absence of intermetallic compounds. Experiments to test the hypothesis that intermetallics are absent immediately behind a moving contact line are under way.

Under this assumption of metastable local equilibrium, the motion of the interface between initially pure copper and liquid tin for a semi-infinite (planar) geometry can be estimated. From standard solutions to the diffusion equations for a moving boundary, the position of the interface is given by z(t) with $z > 0$ defined into the liquid as

$$z(t) = 2\lambda(D_L t)^{1/2} . \tag{4-13}$$

The general expression for λ depends on several parameters: the compositions $C_{L\infty}$ and $C_{S\infty}$, which are the liquid and solid compositions far from the interface (100%Sn and 0%Sn in our case); C_{LE} and C_{SE}, which are the liquid and solid compositions at the interface (95%Sn and 35%Sn) taken from the metastable phase diagram at the temperature of the sample (Figure 4-12); and D_L and D_S, which are the liquid and solid interdiffusion coefficients (10^{-5} and 10^{-18} cm^2/s) [Butrymowicz 1977]. In the limit applicable here where $D_S \sim 0$, λ is given by

$$\lambda\sqrt{\pi}\,(C_{LE} - C_{S\infty}) = \frac{(C_{LE} - C_{L\infty})}{1 - \mathrm{erf}(\lambda)} e^{-\lambda^2} . \tag{4-14}$$

For small values of λ, Eq. (4-14) can be approximated by

$$\lambda = \frac{-(C_{L\infty} - C_{LE})}{\sqrt{\pi}\,(C_{LE} - C_{S\infty})} \quad . \tag{4-15}$$

Note that when the diffusion in the solid is negligible, λ does not depend on C_{SE}. From either Eq. (4-14 and 4-15), $\lambda < 0$; thus the interface between the copper and the liquid moves toward the copper, i.e. the copper dissolves. In 1 s, about 1.8 μm of copper is predicted to dissolve.

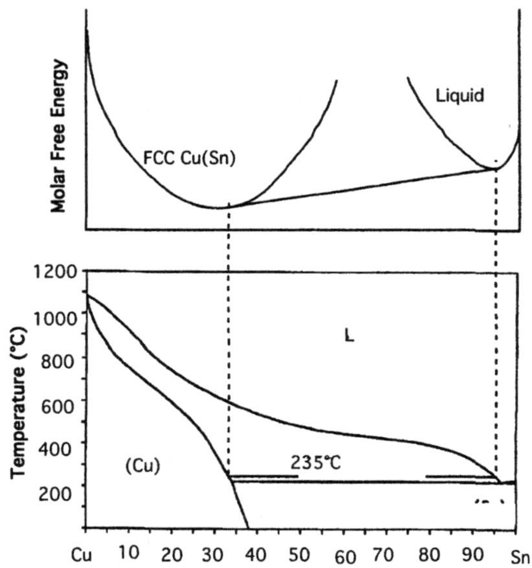

Figure 4-12 Metastable Cu-Sn phase diagram (bottom) missing all of the intermetallic compounds including Cu_3Sn and Cu_6Sn_5 calculated from the thermodynamic modeling. Such a diagram would be used to set the compositions at a reacting interface between copper and liquid tin if nucleation of the intermetallics were delayed near a quickly moving contact line. At the top is shown a schematic free–energy composition diagram indicating the common tangent that represents the metastable equilibrium.

While the above discussion has focussed on volume diffusion occurring behind the advancing contact line, changes in the unwetted copper surface adjacent to the droplet have often been observed and are usually referred to as halos [Klein-Wassink 1989]. These halos are typically observed after spreading has stopped, and usually after solidification is complete. While it is clear that retraction of liquid due to solidification shrinkage is not the cause of halo appearance, there are reports of both Sn-rich and Pb-rich regions adjacent to solder droplets under different conditions. A mechanism, where $SnCl_2$ is formed in the flux by reaction with the solder and tin is deposited on the copper, has been proposed to explain Sn-rich halos [Latin 1938]. A mechanism, where lead migrates by surface diffusion along the copper surface, has been proposed to explain Pb-rich halos [Smith and Lea 1986]. These later observations were made for spreading in vacuum without a flux. However, Pb-rich halos have also been observed for solder spreading on gold in air [Singler et al. 1992] and on copper when an RMA flux was present [Boettinger and Handwerker, 1992]. In the last case, metallographic sectioning has revealed the Pb-rich halo to lie above the substrate surface, to be ~1–3 μm thick, and to contain both Pb-rich and Sn-rich phases. Such a thick halo could not have been caused by surface diffusion nor can it be related to precursors formed in non-reacting systems, which are typically 10 nm thick. A full explanation of these thick halos is not available and its relationship to contact line kinetic law, if any, is not understood.

4.5 THERMODYNAMIC CALCULATION OF THE TERNARY Pb–Sn–Cu SYSTEM

The ternary phase diagram provides the basic metallurgical road map for understanding the reaction of Cu with molten or solid Pb-Sn solders and of the solidification of Cu-contaminated solder. Analysis of these ternary diffusion problems requires that tie-line information be available as the boundary conditions for the interfaces between the various layers. Tie-line data are rarely measured but can be calculated from an appropriate thermodynamic model. The philosophy behind these models is to develop analytical expressions for the free energy functions for all of the phases by matching the calculated diagram with the measured phase diagrams and calorimetric (heat of solution) data for the component binary systems and the ternary system where available. The phase diagram is obtained by finding the lowest common tangent planes to the functions to obtain the phase boundaries (equal chemical potentials). This approach allows a fitting of experimental data from many sources in a single self-consistent manner, and automatically handles the difficult geometry problems of three-dimensional diagrams. Diagrams then can be calculated for compositions and temperatures where data are not available or for metastable equilibrium where one or more phases are absent.

4.5.1 General Thermodynamic Description

For the analytical description of a system, the Gibbs energy of each phase must be described as a function of temperature and composition.

The binary stoichiometric compounds are assumed to have no range of homogeneity and are described by

$$G = {}^0x_1{}^0G_1 + {}^0x_2{}^0G_2 + \Delta G^f \qquad (4\text{-}16)$$

where ${}^0x_1$ and ${}^0x_2$ are mole fraction of elements 1 and 2 and are given by the stoichiometry of the compound. ${}^0G_1$ and ${}^0G_2$ are the respective reference states of elements 1 and 2, and ΔG^f is the Gibbs energy of formation. The temperature dependence of each of the Gibbs energies in the above equation (and also in equations below) is described by an expression of the form:

$$G = A - BT + CT(1 - ln(T)) \qquad (4\text{-}17)$$

with different constants, A, B, and C.

Binary solution phases such as the liquid and the terminal (Cu), (Pb), and (Sn) solid solutions are described as random mixtures of the elements:

$$G = x_1{}^0G_1 + x_2{}^0G_2 + RT(x_1\ln(x_1) + x_2\ln(x_2)) + x_1x_2\sum_{i=0}^{n}(x_1 - x_2)^i G_i \qquad (4\text{-}18)$$

where x_1 and x_2 are the mole fraction of elements 1 and 2, and ${}^0G_1$ and ${}^0G_2$ are the reference states of elements 1 and 2, respectively. The G_i are coefficients of the terms of the excess Gibbs energy; each can have the temperature dependence of Eq. (4-17).

A thermodynamic description is first developed for the three binary systems; then, the ternary system is described, using free energies for the solution phases obtained by extrapolating the binary solution phases energies using the formula of Muggianu et al. [1975]:

$$G = x_1{}^0G_1 + x_2{}^0G_2 + x_3{}^0G_3 + RT\{x_1\ln(x_1) + x_2\ln(x_2) + x_3\ln(x_3)\} + x_1x_2\sum_{i=0}^{n_{12}} (x_1 - x_2)^i G_i^{12} + x_2x_3\sum_{i=0}^{n_{23}} (x_2 - x_3)^i G_i^{23} + x_1x_3\sum_{i=0}^{n_{13}} (x_1 - x_3)^i G_i^{13} \tag{4-19}$$

where the parameters have the same values as in Eq. (4-18).

4.5.2 Binary Pb-Sn

The thermodynamic description of the Pb-Sn system was derived by Fecht et al. [1989]. They used the quasi-subregular solution model ($G_n = 0$ for $n \geq 2$ in Eq. (4-18)) to describe the liquid phase and the (Pb) solid solution, and the quasi-regular solution model ($G_n = 0$ for $n \geq 1$ in Eq. (4-18)) to describe the (Sn) solid solution. The analytical description was performed using thermodynamic and phase diagram data, including data for the metastable extension of the liquid (L) + (Pb) equilibrium.

4.5.3 Binary Cu-Pb

Several calculations of the Cu-Pb system are published in the literature [Chakrabarti and Laughlin 1984; Hayes et al. 1986; Niemelä et al. 1986; Teppo et al. 1988]. In all of these calculations the measured composition of the copper-rich liquid at the temperature of the monotectic is only approximately reproduced, despite the fact that a relatively high number of polynomial terms were used (up to $n = 4$ in Eq. (4-18)). Koa and Chang [1992] developed a description with fewer terms ($n = 2$ in Eq. (4-18)) based mainly on the available thermodynamic data that reproduced the measured diagram well, especially at temperatures below the monotectic temperature. This binary was used for the calculation of the ternary system.

4.5.4 Binary Cu-Sn

For the calculation of the Cu-Sn system, only phases which are stable below a temperature of 300 °C were considered; thus, the four high-temperature compounds, β, γ, δ, and ζ, which all decompose eutectoidally above this temperature, were ignored. The ranges of homogeneity of the remaining intermetallic compounds: ε (Cu_3Sn), and η and η' (both Cu_6Sn_5), are relatively narrow (~ 1 at.%). Therefore, these phases were assumed to be stoichiometric. The structure of η' is a superstructure of the η-structure, so the difference between the thermodynamic quantities of formation of these two phases can be expected to be fairly small; therefore, these phases were treated as one phase in the present calculation. No solubility of Cu in Sn was assumed to occur. The coefficients of the analytical description were optimized using a selected set of available experimental data [Saunders and Miodownik 1990; Kattner 1992]. The thermodynamic parameters for the binary systems are given in Table 4-1, and the three calculated binary diagrams are shown in Figure 4-13.

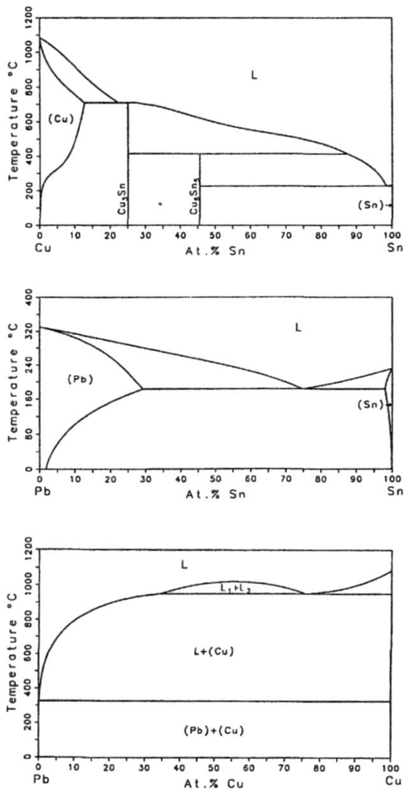

Figure 4-13 Calculated binary Cu-Sn, Pb-Sn, and Cu-Pb phase diagrams. The high temperature Cu-Sn intermetallic phases have been omitted for simplicity as they have no bearing on the phase equilibria below 350°C.

Table 4-1 Analytical Description of the Pb-Sn-Cu System
(all quantities are given in J/mol of atoms)

Lattice Stabilities of the Pure Elements	Source
${}^{0}G^{L}{}_{Pb} - {}^{0}G^{fcc}{}_{Pb} = 4773.9 - 7.9485\ T$	(Dinsdale 1991)
${}^{0}G^{L}{}_{Sn} - {}^{0}G^{bct}{}_{Sn} = 7029.1 - 13.9169\ T$	(Dinsdale 1991)
${}^{0}G^{L}{}_{Cu} - {}^{0}G^{fcc}{}_{Cu} = 13263.3 - 9.7684\ T$	(Dinsdale 1991)
${}^{0}G^{bct}{}_{Pb} - {}^{0}G^{fcc}{}_{Pb} = 489 + 3.52\ T$	(Ngai and Chang 1981)
${}^{0}G^{fcc}{}_{Sn} - {}^{0}G^{bct}{}_{S} = 5510 - 8.46\ T$	(Ngai and Chang 1981)

Stoichiometric Phases

Cu_3Sn (referred to fcc-Cu and bct-Sn)	$\Delta G^{f} = -8479.6 + 0.31836\ T$
Cu_6Sn_5 (referred to fcc-Cu and bct-Sn)	$\Delta G^{f} = -7139.4 + 0.31592\ T$

Solution Phases

Liquid:	Pb-Sn	G_0	$= 5368 + 0.97159\ T$
		G_1	$= 97.8 + 0.05608\ T$
	Cu-Pb	G_0	$= 26823.8 - 4.12985\ T$
		G_1	$= 2623 - 0.82952\ T$
		G_2	$= 4733.1 - 3.48981\ T$
	Cu-Sn	G_0	$= -15427.1 + 30.55297\ T + 4.890268\ T\ (1\text{-}\ln T)$
		G_1	$= -21335.5 + 4.86292\ T$
		G_2	$= -10300.6$
(Cu)	Cu-Sn	G_0	$= -34235.5 + 70.50539\ T$
		G_1	$= -41167 - 74.35043\ T$
		G_2	$= 64187.40$
(Pb)	Pb-Sn	G_0	$= 4758.8 + 2.4719\ T$
		G_1	$= 2293.4 - 4.9197\ T$
(Sn)	Pb-Sn	G_0	$= 19693.70 - 15.89485\ T$

Sources: Pb-Sn (Fecht et al. 1989), Cu-Pb (Kao and Chang 1992), Cu-Sn (Kattner 1992)

4.5.5 Ternary System Cu-Pb-Sn

In the present work no ternary interactions are considered. The ternary system has been predicted from the binary systems assuming no ternary solubility in the stoichiometric compounds and using extension of the solution phases into the ternary system, using Eq. (4-19). The temperatures and compositions of the phases for the invariant reactions obtained through the current approach are given in Table 4-2.

In the evaluation of the liquidus surface by Chang [1979], it was proposed that three of the high temperature compounds, β, γ, and δ from the Cu-Sn binary have liquidus fields that persist to near the Pb corner of the phase diagram. If this were the correct diagram, neglecting the high temperature Sn-rich intermetallics in the Cu-Sn binary system would cause significant error in the Pb corner of the ternary phase diagram. However, it can be assumed that it is more likely that the Cu-rich three-phase equilibria of the Cu-Sn binary will react with each other in the Cu-rich corner of the ternary, eventually resulting in a three-phase equilibrium L + (Cu) + Cu_3Sn, which then would react with the $L_1 + L_2$ + (Cu) equilibrium of the Cu-Pb binary. Experimental results (Marcotte and Schroeder 1984) indicate the occurrence of the L + $Cu_3Sn \leftrightarrow$ (Pb) + Cu_6Sn_5 invariant equilibrium with liquid composition at ~36.6 at.%Sn, ~63.1 at.%Pb and ~0.3 at.%Cu at ~270 °C, corresponding well to the calculated invariant equilibrium. In this case, neglecting the high-temperature intermetallic compounds would have no impact on the accuracy of the predicted phase equilibria on the (Pb, Sn)-rich side of the system. The ternary eutectic, L $\leftrightarrow$ Pb + Sn + Cu_6Sn_5 calculated in the present research also agrees well with the estimated value from Marcotte and Schroeder [1984] of 73.6 at.%Sn, 26.0 at.% Pb and 0.4 at.% Cu at 182°C.

Figure 4-14 shows the calculated liquidus surface of the ternary system. Figure 4-15 shows an isothermal section at 235°C. Figure 4-16 shows an isothermal section at 150°C.

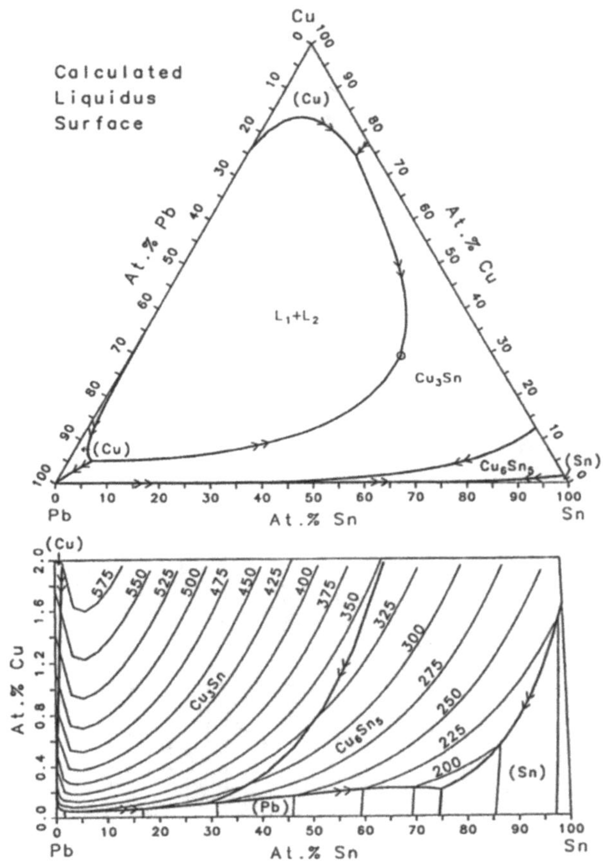

Figure 4-14 Calculated ternary Cu-Pb-Sn liquidus surface with isothermal contours. (Top) Entire composition triangle and (bottom) expanded Cu-poor region. The regions are labelled as to the identity of the primary or first phase that would freeze on cooling of an alloy in that region.

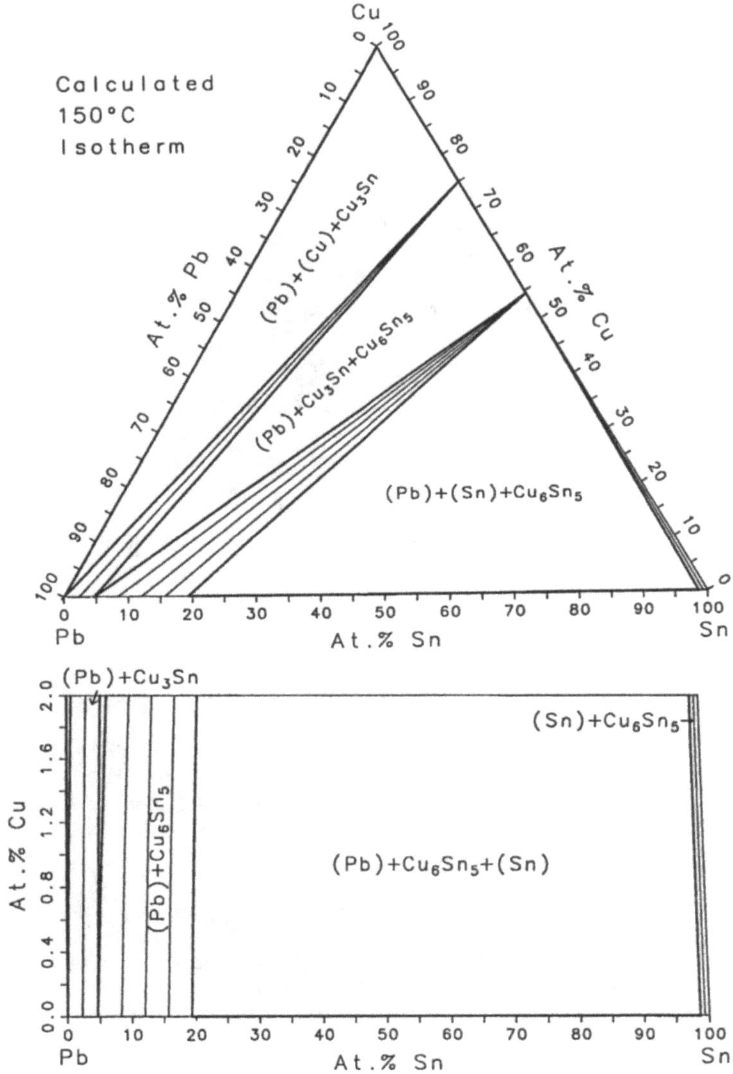

Figure 4-15 Calculated ternary Cu-Pb-Sn isothermal section at 235°C. (Top) Entire composition triangle and (bottom) Cu-poor region. This diagram is important to understand the metallurgical reactions that occur when liquid solder is in contact with Cu.

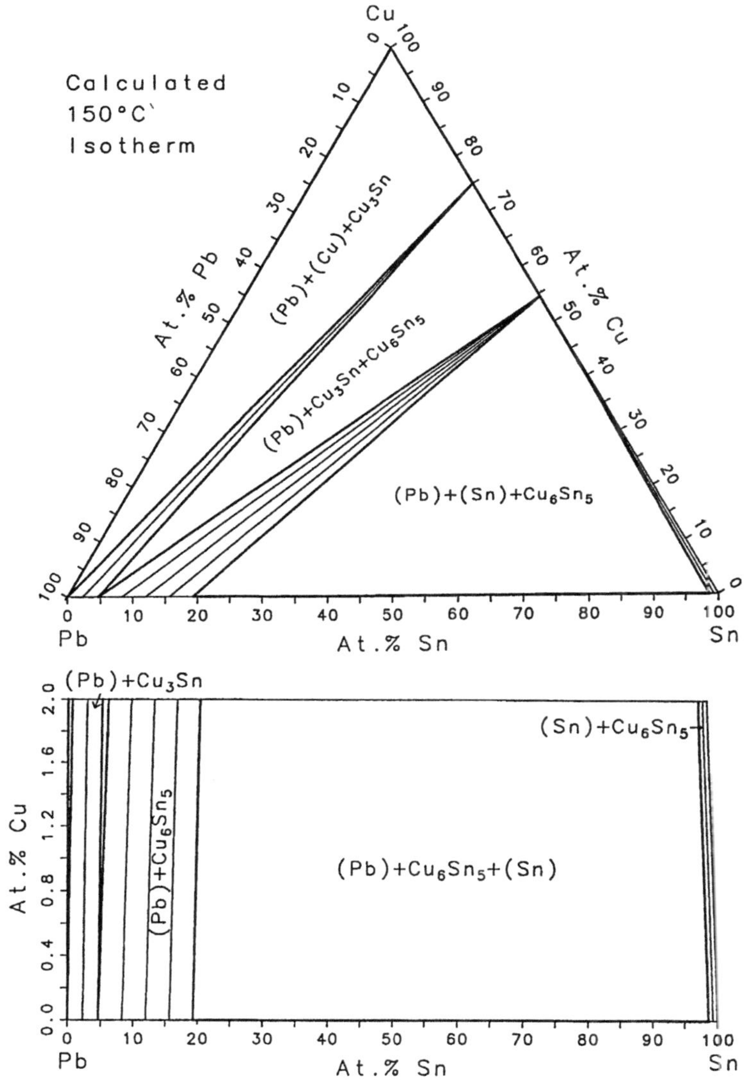

Figure 4-16 Calculated ternary Cu-Pb-Sn isothermal section at 150°C. (Top) Entire composition triangle and (bottom) Cu-poor region. This diagram is important to understand the metallurgical reactions that occur when solid solder reacts with Cu during high temperature service or extended storage

Table 4-2 Predicted Ternary Invariant Equilibria

Equilibrium	T	Phase 1		Phase 2		Phase 3		Phase 4	
	°C	at.% Pb	at.% Sn	at.% Pb	at.% Sn	at.% Pb	at.% Sn	at.% Pb	at.% Sn
$L_1 \leftrightarrow L_2$ + (Cu) + Cu_3Sn	691.8	3.95	21.55	90.26	4.54	0	12.55	0	25
L $\leftrightarrow$ (Cu) + (Pb) + Cu_3Sn	326.4	99.79	0.06	0	4.94	99.96	0.04	0	25
L + $Cu_3Sn \leftrightarrow$ (Pb) + Cu_6Sn_5	278.6	70.96	28.94	0	25	84.21	15.79	0	45.45
L $\leftrightarrow$ (Pb) + (Sn) + Cu_6Sn_5	182.2	25.03	74.77	71.09	28.91	1.98	98.02	0	45.45
$L_1 \leftrightarrow L_2$ + Cu_3Sn	694.5	3.89	21.17	90.52	4.28	0	25		
$L_1 + L_2 \leftrightarrow L$	571	22.65	54.71						

4.6 WETTING BALANCE STUDIES ON Cu_6SN_5 AND Cu_3Sn

The intermetallic compounds ε-Cu_3Sn and η-Cu_6Sn_5 are commonly found in solder joints as thin reaction layers separating copper base metal and solder. Copper that is pretinned by either hot dipping in molten Pb-Sn or by electroplating with Pb-Sn also contains intermetallic compounds at the interface. The presence of these intermetallic compounds indicates intimate metallic bonding of the solder to the copper. The intermetallic compounds continue to grow over most of the temperature range that pretinned components and solder joints experience, i.e. from room temperature to the soldering temperature, typically 250°C. During the reworking of joints or the soldering of pretinned leads, the Pb-Sn liquid must continue to wet the intermetallic compound in order to form reliable joints. Thus the wetting behavior of Pb-Sn on the intermetallic compounds themselves is of great importance.

Previous research on this topic is sparse. Kay and Mackay [1976] report that "when clean, both compounds wet if an activated rosin flux is used." No data are presented, however. Klein-Wassink [1989] states "wetting balance measurements of the wettability have revealed ε and η to be wettable by solder using a strongly activated flux." He further claims a particularly strong increase in wetting times when the intermetallic surface is covered with as little as 1.5 nm of oxide wherein the wettability is reduced "to a level unacceptable in practice." Using samples that appear to contain second phases, Fidos and Schreiner [1970] report contact angles of 65° and 180° for pure tin on ε and η, respectively, for wetting tests performed at 250°C in vacuum.

The sensitivity of the wettability of the intermetallic compounds to oxide [Klein-Wassink 1989] may explain reports in the trade literature on the lack of solderability of intermetallics compared to copper. During storage of pretinned copper leads for prolonged times at room temperature or for short times at high temperatures, the intermetallic may grow through the protective Pb-Sn layer to the surface where it is directly exposed to an oxidizing environment. This penetration is made easier by the fact that the intermetallic does not usually grow with a smooth interface but develops a scalloped interface related to the orientation of the individual grains of the intermetallic. Intermetallic may first be exposed at places where the solder coating is thin. Severely non-concentric coatings on round leads [Klein-Wassink 1989], corners of leads with rectangular cross section, or edges of through-holes are places where intermetallic will penetrate the solder coating easily. In all of these cases, oxidation of the exposed intermetallic compound occurs, leading at best to a surface with discontinuous patches of oxide surrounded by residual solder or, at worst, a uniformly oxidized intermetallic with uniformly poor wettability. During subsequent soldering of these pretinned leads, wetting difficulties ensue during final assembly where a mildly activated flux must be used to prevent subsequent corrosion problems. Other ideas about the loss of solderability [Geist and Kottke 1988; Lucey et al. 1991] involve a mechanism whereby the intermetallic is internally oxidized without having broken through to the surface. Once the pretinned layer is melted off during soldering, wetting difficulties would arise under this mechanism.

The wetting behavior of these intermetallics also may be relevant to the phenomenon of dewetting. In this process, a copper surface previously covered by molten solder becomes partially uncovered. Dewet areas usually show the presence of intermetallic compounds where the solder has withdrawn. The microstructural roughness of the exposed intermetallic surface may also be a factor in determining the effective contact angle. However, a detailed explanation of the dewetting mechanism does not presently exist.

Wetting balance tests were performed on copper, Cu_3Sn, and Cu_6Sn_5 using a solderability test system with computer control and data acquisition functions. Area-of-spread measurements also were performed as described by Boettinger et al. [1992]. Dense, bulk samples of Cu_3Sn and Cu_6Sn_5 were sliced from rods prepared by HIPing rapidly solidified powder [Schaeffer et al. 1992]. All wetting experiments were performed with 60wt%Sn-40wt%Pb solder bath at 235°C. The immersion speed was 20mm/s, the immersion depth was 3mm and the dwell time was 10 s. The copper samples were 1 mm diam wires. The intermetallic samples were rectangular cross-section rods (1mm x 1mm x 10mm) cut with a slow speed diamond saw. The as-cut surfaces of the intermetallic rods were electroetched for

40 s using 5% sulfuric acid at a voltage which just produced H_2 evolution. The copper was merely etched in the sulfuric acid solution. Samples were rinsed in distilled H_2O and alcohol. Separate tests were performed within 10 mins. of etching and after 3 days storage in air at room temperature. Two flux compositions were used for the tests: a non-activated rosin flux (SM/NA, Multicore Solders, Batch No. C708194) and an activated flux (RMA, Kester RMA-185). These fluxes are designated R and RMA, respectively.

Figure 4-17 shows the results of the tests for copper, Cu_3Sn, and Cu_6Sn_5 under the four test conditions: 1) immediately after etching, 2) after a three day storage, 3-4) with the two fluxes. With the RMA flux on freshly etched samples (Figure 4-17a), the wetting curves for all three materials are very similar. The slightly slower wetting for the intermetallics may be due to the poorer thermal diffusivity of the intermetallics compared to copper. The time to reach zero wetting force is ~ 0.5 s and the maximum wetting force is attained in ~1 s as 85% of the reference force (here set at 0.42N/m x the contact line perimeter). Thus, when clean and a RMA flux is used, the intermetallics are as wettable as copper. With an R flux, the rate of wetting of the copper is largely unchanged, but considerable slowing of the wetting of the intermetallics has occurred, as seen in Figure 4-17b.

The results after three days of storage are shown in Figures 4-17c and d. Even with the RMA flux, considerable degradation in the rate of wetting of the intermetallics is evident, while the rate of wetting of copper is largely unchanged. For the R flux, zero wetting force is never reached (contact angle > 90°) after a 6 second immersion. Thus the wetting behavior of the intermetallics is much more susceptible to degradation after storage than is copper. This confirms the sensitivity of the wetting behavior of the intermetallics to oxide.

Using samples obtained through area-of-spread tests [Boettinger et al. 1992], metallographic examination of cross-sections of the interfacial region was performed using optical microscopy with an oil immersion objective lens. Figure 4-18 shows the interface region for the copper, Cu_3Sn, and Cu_6Sn_5 samples using the RMA and R fluxes. No significant microstructural difference is observed to result from the different fluxes. The phases present in the interfacial region agree with those expected from the ternary phase diagram and with those seen by many previous investigators. For the copper samples, a scalloped Cu_6Sn_5 layer (light) is present adjacent to the solder and a very thin (~ 0.3μm) layer of Cu_3Sn (gray) appears faintly visible adjacent to the copper. Confirmation of the latter can only be resolved with transmission electron microscopy. For the Cu_3Sn samples, only the scalloped layer of Cu_6Sn_5 has formed since the Cu_3Sn is already present as the substrate. For the Cu_6Sn_5 sample, no new intermetallic layer is formed. However, the originally optically flat surface of Cu_6Sn_5 has become rough with scallops ~2μm wide and ~2μm deep. The interface between the solder and the Cu_6Sn_5 was examined near the contact line so that the position of the virgin surface could be determined. Intrusion of the molten solder into the intermetallic, and also growth of the intermetallic substrate into the solder, has occurred. The Cu_6Sn_5 phase is in equilibrium with liquid solder of this lead content. However, a small amount of Cu_6Sn_5 must dissolve to establish the small equilibrium concentration of copper in the liquid solder. In this case, the intermetallic dissolves preferentially along grain boundaries (Figure 4-18). For tests on Cu_3Sn, a thin layer of Cu_6Sn_5 forms and for copper, layers of Cu_3Sn and Cu_6Sn_5 form. In all three materials the size scale of the roughness appears to correspond to the grain size of the Cu_6Sn_5.

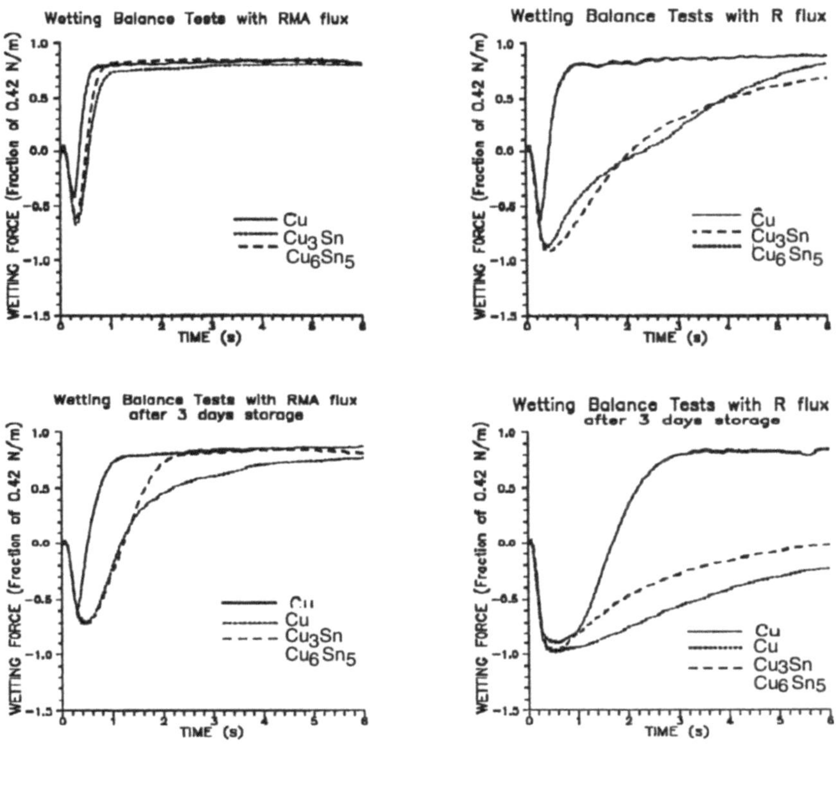

Figure 4-17 Wetting force vs. time curves obtained from wetting balance tests on Cu, Cu_3Sn Cu_6Sn_5 immediately after etching and after 3 days' storage at room temperature using RMA and R fluxes.

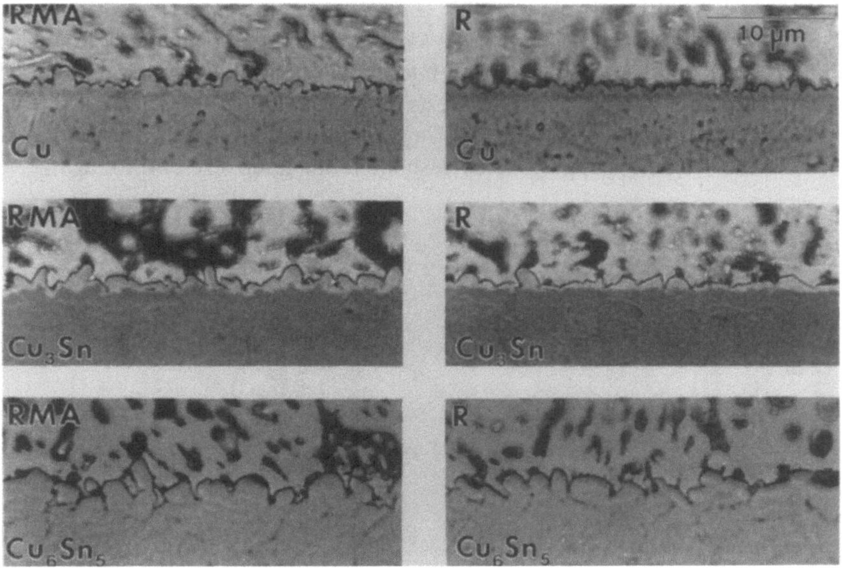

Figure 4-18 High magnification optical micrographs of the interface between the solder and Cu, Cu_3Sn, and Cu_6Sn_5 base metals with the two fluxes

4.7 CONCLUSION

Solder spreading is not simple. Its study requires a consideration of a disparate collection of fields including heat flow, fluid flow, capillarity, alloy thermodynamics, interdiffusion, and oxidation. We have attempted to outline some of the factors that need be considered in these various fields and anticipate that, through careful analysis and experimentation, considerable progress can be made toward solving technical soldering problems.

4.8 ACKNOWLEDGMENTS

The authors would like to thank S. R. Coriell for helpful discussions, M. E. Williams for conducting the spreading tests, L. H. Smith and A. Shapiro for the metallographic observations, H. L. Lukas of the Max-Planck-Institut für Metallforschung (Stuttgart, Germany) for providing the phase diagram software; and R.

J. Schaefer, F. S. Biancaniello, and R. D. Jiggetts for the hot isostatic pressing. This research was partially supported by the U. S. Army-Harry Diamond Laboratory through G.K. Lucey, Jr.

REFERENCES

Aksay, I. A., C. A. Hoge, and J. A. Pask. 1974. *J Phys. Chem.* 78:1178.

Ambrose, J. C., M. G. Nicholas, and A. M. Stoneham. 1992. *Acta Metall. Mater.* 40:2483.

Bailey, G. L. J., and H. C. Watkins. 1951-2. *J Inst. Metals.* 80:57.

Boettinger, W. J., and C. A. Handwerker. 1992. Unpublished research.NIST

Boettinger, W. J., C. A. Handwerker, and L. C. Smith. 1992. In: *The Metal Science of Joining.* Edited by M. J. Cieslak, J. H. Perepezko, S. Kang, and M.E. Glicksman. Warrendale, PA.: The Minerals, Metals, and Materials Soc.

Brochard, F. 1989. *Langmuir.* 5:432.

Butrymowicz, D. B. 1977. *Diffusion in Cu Alloys.* INCRA Monograph Series V. New York: The International Copper Research Assn., Inc.

Cahn, J. W., W. J. Boettinger, and C. A. Handwerker. 1992. Unpublished research. NIST

Chakrabarti, D. J., and D. E. Laughlin. 1984. *Bull. Alloy Phase Diagrams.* 5:503-510.

Chang, Y. A., J. P. Neumann, A. Mikula, and D. Goldberg. 1979. *Phase Diagrams and Thermodynamic Properties of Ternary Copper-Metal Systems.* INCRA Monograph Series VI. New York: The International Copper Research Assn., Inc.

Chaudry, M. K., and G. M. Whitesides. 1992. *Science.* 256:1539.

de Gennes, P. G. 1985. *Rev. Mod. Phys.* 57:827.

Dinsdale, A. T. 1991. *CALPHAD.* 15:317.

Ehrhard, P., and S. H. Davis. 1991. *J Fluid Mech.* 229:365.

Fecht, H. J., M.-X. Zhang, Y. A. Chang, and J. H. Perepezko. 1989. *Metall. Trans.* 20A:795.

Fidos, H., and H. Schreiner. 1970. *Zeit. Metall.* 61:225.

Geist, H., and M. Kottke. 1988. *IEEE Trans. on Components, Hybrids, and Manufacturing Tech.* 11:270.

Hayes, F. H., H. L. Lukas, G. Effenberg, and G. Petzow. 1986. *Z. Metallkd.* 77:749-754.

Joanny, J. F., and P. G. de Gennes. 1984. *J Chem. Phys.* 81:552.

Kao, R., and Y. A. Chang. 1992. Private communication. University of Wisconsin—Madison.

Kattner, U. R. 1992. Unpublished research. NIST.

Kay, P. J., and C. A. Mackay. 1976. *Trans. Inst. of Metal Finishing*. 54:68.

Kirkaldy, J. S., and D. J. Young. 1987. *Diffusion in the Condensed State*. London, U. K.: The Institute of Metals.

Klein-Wassink, R. J. 1989. *Soldering in Electronics*. 2nd ed. Ayr, Scotland: Electrochemical Publications, Ltd.

Latin, A. 1938. *Trans Faraday Soc.* 34(2):1383.

Lucey, G., J. Marshall, C. A. Handwerker, D. Tench, and A. Sunwoo. 1991. *NEPCON '91 West Proc.* Des Plaines, IL: Cahners Exposition Group. pp. 3-10.

Marcotte, V. C., and K. Schroeder. 1984. In: *Proceedings of the Thirteenth North American Thermal Analysis Society*. Edited by A. R. McGhie. North American Thermal Analysis Society. p. 294.

Mei, Z., A. J. Sunwoo, and J. W. Morris, Jr. 1992. *Met. Trans.* 23A:857.

Muggianu, Y.-M., M. Gambino, and J.-P. Bros. 1975. *J Chem. Phys.* 72:85.

Ngai, T. L., and Y. A. Chang. 1981. *CALPHAD*. 5:267.

Niemelä, J., G. Effenberg, K. Hack, and P. J. Spencer. 1986. *CALPHAD*. 10:77.

Perepezko, J. H., and W. J. Boettinger. 1983. In: *Alloy Phase Diagrams*. Edited by L. H. Bennett, T. B. Massalski, and B. C. Giessen. *Mat. Res. Soc. Symp. Proc.* Elsevier North Holland, NY. 19:223-240.

Saunders, N., and A. P. Miodownik. 1990. *Bull. Alloy Phase Diagrams*. 11:278-287.

Schaefer, R. J., R. Jiggets, and F. S. Biancaniello. 1992. In: *The Metal Science of Joining*. Edited by M. J. Cieslak, J. H. Perepezko, S. Kang, and M. E. Glicksman. Warrendale, PA: The Minerals, Metals, and Materials Soc. p. 175.

Schwaneke, A. E., W. L. Falke, and V. R. Miller. 1978. *J Chem. and Eng. Data*. 23:298.

Singler, T. J., J. A. Clum, and E. R. Prack. 1992. *Trans. ASME*. 114:128.

Smith, G. C., and C. Lea. 1986. *J Surf. Interf. Anal.* 9:145.

Teppo, O., J. Niemelä, and P. Taskinen. 1988. Espoo, Finland: Helsinki Univ. of Technology.

Thresh, H. R., and A. F. Crawley. 1970. *Met. Trans.* 1:1531.

Turkdogan, E. T., and S. Zador. 1961. *J Iron and Steel Inst.* 197:233.

Yost, F. G., and A. D. Romig. 1988. *Mat. Res. Soc. Symp.* 108:385.

5

Loss of Solderability and Dewetting

Krishna Rajan

5.1 INTRODUCTION

The loss of solderability and dewetting is a problem that is often observed in the manufacturing environment but not well understood. This lack of understanding as to the exact origins of this phenomenon and the conditions which lead to it make it difficult to ensure a reliable means of predicting the conditions for limiting dewetting. This chapter will aim to address the fundamental materials science issues that we feel may govern the dewetting or partial wetting phenomenon. The emphasis of this chapter is those aspects of the extensive literature on wetting phenomena in general that may have a direct bearing on the causes of dewetting of solders. Specifically, the issues of wetting instabilities and their relationship to the dynamics of wetting will be stressed. It is proposed as well to juxtapose the continuum descriptions of wetting phenomena with microstructural parameters that may affect the partial wetting or dewetting process. Particular attention will be paid to the possible role of intermetallic formation at the solder substrate interface on the stability of the wetting phenomena (Figure 5-1).

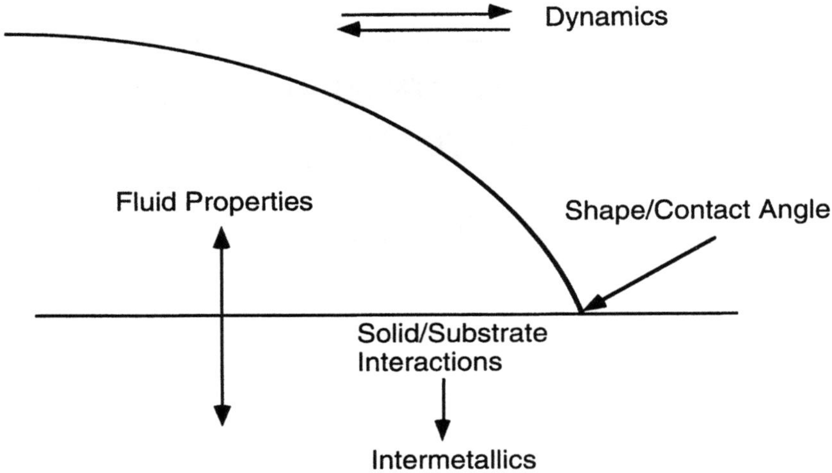

Figure 5-1 Schematic diagram outlining some of the parameters affecting the wetting behavior of solders

5.2 CHARACTERIZATION OF DEWETTING

The issue of interest in this chapter is to develop a unified description of wetting and dewetting. There are three cases to consider in the spreading of a liquid droplet on a solid substrate: complete wetting, partial wetting, and dewetting. These classes of wetting behavior can be described in terms of the temporal evolution of the contact angles, as shown in Figure 5-2. If the liquid droplet spreads completely, the contact angle eventually reaches zero. If the droplet only partially spreads, then the contact angle reaches some finite non-zero value. When the droplet first spreads and then recedes, that can be described as a case where the contact angle first diminishes and then increases again. Hence we will define dewetting as the case where an initially decreasing contact angle (wetting) later changes to a situation of an increasing contact angle.

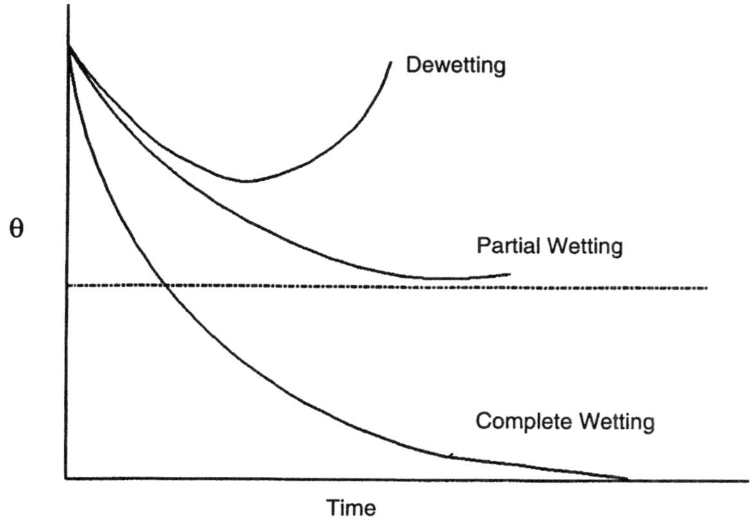

Figure 5-2 Dynamics of wetting behavior in terms of contact angle

Keeping this definition in mind, an examination of the literature shows different examples of where this non-monotonic character in contact angle has been observed. Nogi and Ogino [1988] found such a behavior for liquid Fe on a SiC substrate, depending on the means by which the SiC was fabricated. The implication of this work is that the surface property of the SiC was altered in some manner as the fabrication procedure was altered (Figure 5-3). The work of Kawakatsu and Osawa [1973] is another example where a wetting/dewetting phenomenon occurs as the chemistry of the spreading liquid changes (Figure 5-4). A more dramatic example of non-monotonic changes in contact angle is shown in some studies of the wetting behavior of liquid tin on solid copper [Yokota et al. 1980; Hasouna et al. 1988]. Here the control parameter was temperature, and it was found that a well defined bifurcation existed in the contact angle as the temperature reached a critical value. It is interesting to note that this temperature corresponds to the solidus temperature for the incongruently melting phase Cu_3Sn (Figure 5-5).

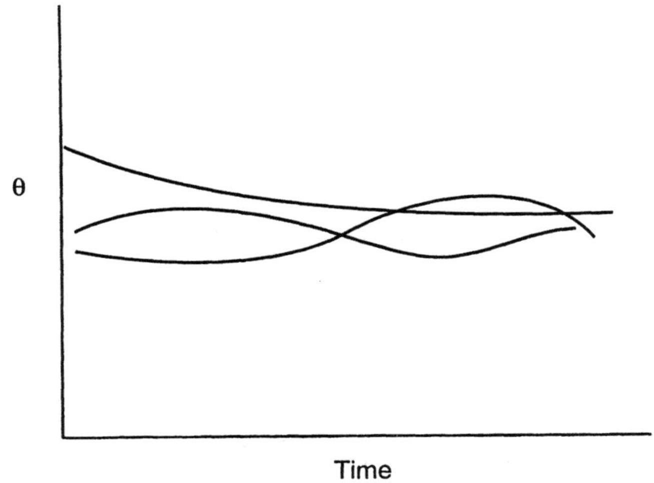

Figure 5-3 Non-monotonic behavior of contact angle for liquid Fe on SiC for different types of SiC fabrication procedures [Nogi and Ogino 1988]

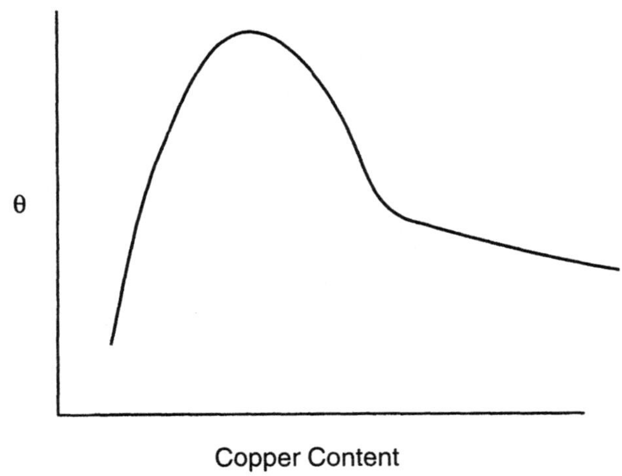

Figure 5-4 Contact angle hysteresis induced by changes in the alloy content of Cu-Sn liquid on a Cu substrate [Kawatsu and Osawa 1973]

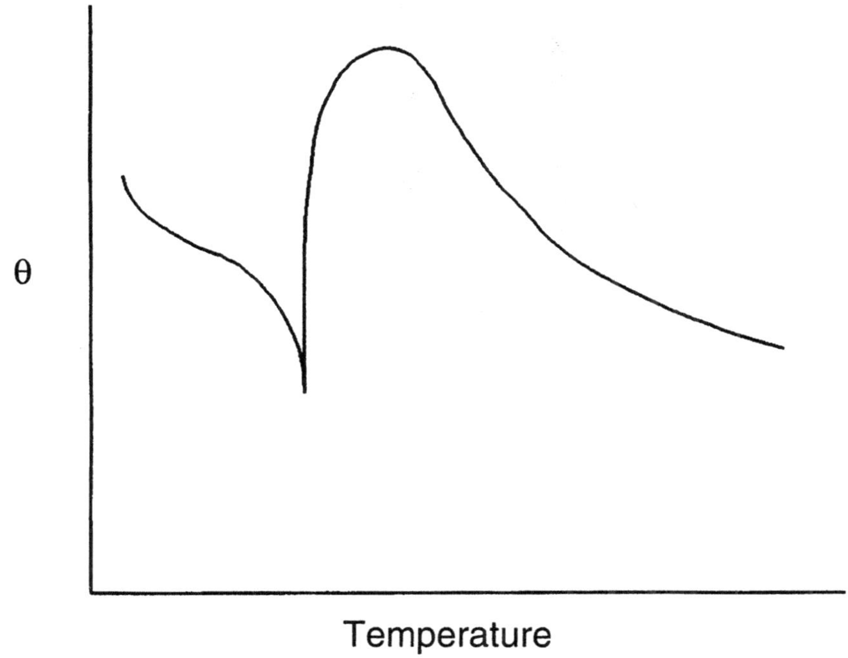

Figure 5-5 Critical point behavior in wetting dynamics for liquid Sn on solid Cu [Yokota et al. 1980 and Hasouna et al. 1988]

5.3 WETTING STABILITY DIAGRAMS

At present there exists no quantitative formulation describing the dynamics of wetting in solder alloys. However, it is worthwhile to examine the nature of wetting dynamics in other systems which may provide a useful insight into this problem. De Gennes [1990] has suggested that based on empirical observations of wetting behavior of non-volatile liquids, the dynamics of the contact angle (θ) may be described by a relationship of the form:

$$\text{velocity} \ \alpha \ \theta 3 \ . \tag{5-1}$$

The contact angle in turn is defined at any instant by the classical Young-Dupre equation:

$$\gamma_S = \gamma_{LS} + \gamma_L \cos\theta \quad . \tag{5-2}$$

Changes in the contact angle in turn are associated with changes in the interfacial energy associated with interfacial reactions, changes in temperature, or other mechanical forces. Hence the changes in contact angle may be viewed as a reflection of the dynamics of this equilibrium condition. A potential function for this equilibrium condition can be defined, given as:

$$\nabla_T \phi = 0 \tag{5-3}$$

where the subscript T indicates what the gradient is with respect to temperature. This surface is made up of all the equilibrium conditions defined by the Young-Dupre equation. Based on the suggestion of de Gennes, we may relate our potential function to his velocity function as:

$$\phi(\theta, T) = \theta^3 + T\theta \quad . \tag{5-4}$$

The phase portrait is two-dimensional, and equilibrium surface is the curve given by the first order condition, resulting in:

$$3\theta^3 + T = 0 \quad . \tag{5-5}$$

The critical condition for dewetting is given as the singularity set defined by the second order condition for which the equation $6\theta = O$ is also satisfied. Thus, a wetting stability diagram may be plotted (say with respect to temperature), as shown in Figure 5-6.

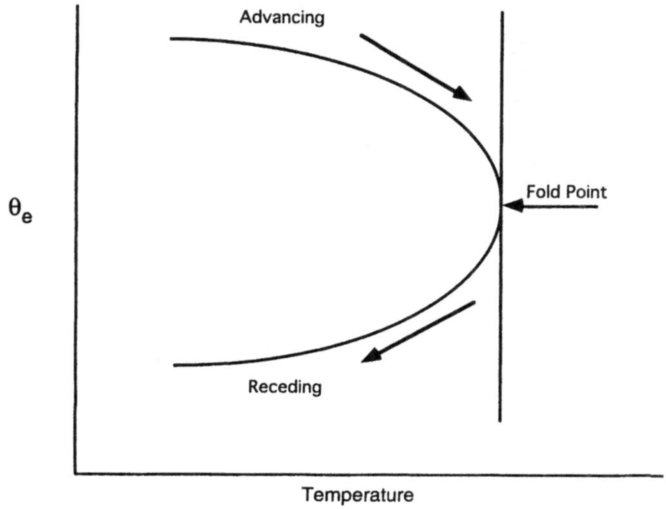

Figure 5-6 A wetting stability diagram based on the critical point behavior illustrated in Figure 5-5

This diagram mathematically describes a fold bifurcation which delineates the regions of advancing and receding contact angles. The equilibrium surface defined by this curve is, at any given instant, locally satisfying the conditions for an equilibrium contact angle. For the purpose of illustration, a simple functional form to characterize the dynamics of the contact angle was chosen. The relationship suggested by de Gennes in itself does not imply any instabilities in the contact angle. However, his formulation relating spreading velocity to contact angle does make a fundamental connection between the dynamics of spreading and the shape or curvature of the droplet as defined by the contact angle.

The forces associated with spreading, such as capillarity and viscosity effects, can be countered by the presence of surface heterogeneities. It is well known, for example, that surface roughness can induce hysteresis in contact angle dynamics [Bickermann 1950; Bartell and Shepard 1953]. Localized changes in contact angle reflect the varying energetics and forces at the liquid solder/solid substrate/air triple junction. If the advancing forces are retarded enough, the surface tension forces of the liquid, as defined by the curvature of the liquid droplet, can become sufficiently dominant to control the dynamics of the wetting behavior. In such a situation, the tendency of the liquid droplet to minimize its surface area will effectively lead to dewetting as the droplet recedes from its spreading motion.

Modelling curvature-driven motion is a mathematically complex problem resulting in an equation of motion described by a class of partial differential equations known as Hamilton-Jacobi equations [Osher and Sethian 1988; Gage and Hamilton 1986]. Similar problems appear in other areas of materials science such as grain growth. Hamilton-Jacobi equations have been numerically solved, imposing the constraints of triple junctions (in this case grain boundary triple points). The appropriate algorithms have been used to accurately simulate the *shrinkage* of grains, matching well with experimental observations [Fradkov et al. 1992; Rajan et al. 1992]. Similar approaches can also be used to model the shrinkage of dewetting droplets with comparable forms of constraints at triple point junctions.

5.4 DYNAMICS OF WETTING INSTABILITIES: PHYSICAL MECHANISMS

The preceding discussion has focussed on relating the dynamics of a spreading liquid solder to its shape as defined by the contact angle with the substrate. The advantage of this approach is that it does not require any *a priori* assumption about its causes, yet it intrinsically captures mathematically the physics of the instabilities as well. Next, the specifics of these physical mechanisms will be addressed. Given that soldering systems involve reactive wetting where the reaction kinetics are very rapid compared to the spreading velocities, the instabilities associated with wetting are associated with changes at the solder/substrate interface.

From a fluid mechanics perspective, the dynamics of the spreading liquid are controlled by viscosity of the solder. There exists a large body of literature on the viscosity of liquid metals. One of the most widely used formulations for the viscosity (η) of liquid metals at the melting point is that proposed by Andrade [1934]:

$$\eta(T_m) = C_A \frac{\sqrt{A T_m}}{V^{2/3}} \tag{5-6}$$

where C_A is a constant, A is the atomic weight, and V is the molar volume at the melting point T_m. Andrade derived this expression by considering that momentum would be transferred from one layer of fluid to another with a different velocity by contact between atoms in neighboring layers as a result of vibrational displacement from their mean positions. Bottezzati and Greer [1989] have critically reviewed the available experimental data for a wide variety of intermetallic compounds and found good agreement with the Andrade correlation. While this formulation was derived for the behavior of liquid metals at the melting point, from a soldering perspective the question of interest is how this viscosity will change as a function of temperature and composition. Andrade has proposed an equation for the temperature-dependent viscosity of the form:

$$\eta(T) = \frac{C_1}{v^{1/3}} exp\left(\frac{C_2}{vT}\right) \tag{5-7}$$

where C_1 and C_2 are constants and v is the specific volume. This equation is of the basic Arrhenius form:

$$\eta(T) = \eta_0 exp\left(\frac{E}{RT}\right) \quad \cdot \tag{5-8}$$

The pre-exponential viscosity η_0 empirically reflects the intrinsic physical properties of the fluid. It is suggested here that the highly reactive nature of the solder/substrate interface leads to rapid changes in the local chemistry of the liquid solder near the interface. These changes in turn reflect changes in the viscosity of the solder near the substrate. The compositional dependence of η_0 for Cu-Sn and Ag-Sn alloys is listed in Table 5-1. As can be seen, variations up to a factor of 2 in η_0 are observed depending on which intermetallic composition of Cu-Sn forms. In the case of Ag-Sn, variations occur over two orders of magnitude.

This discussion has thus far focussed on how the shear viscosity of intermetallic forming liquid alloys can change significantly depending on which line composition forms. Hence, depending on which intermetallic compositions form at the interface and the sequence in which they form, the liquid metal viscosity can be altered significantly. Other parameters such as oxide formation and surface roughness can affect the motion of liquid solders on a substrate. While not discounting these effects, it is also important to consider the more microscopic parameters of correlating viscosity changes to intermetallic compositions. The changes in intermetallic composition cannot be easily predicted and in fact can be unstable. Instabilities in the compositional evolution of solder can lead to instabilities in fluid viscosity, and, in turn, wetting behavior. In the next section, some direct observations of intermetallic formation in the Cu-Sn system will be detailed to illustrate microstructural instabilities which reflect the complexities of the reactive nature of solder/substrate interface.

Table 5-1 Viscosity Data for Liquid Alloys at Compound-Forming Compositions

Compound	E kJ/mole^{-1}	η_o mPa s	Reference
β-AgSn	21.8	0.404	Gebhardt et al.
γ-AgSn	0.541	14.5	"
γ-CuSn	0.386	22.8	Gebhardt et al.
δ-CuSn	0.655	17.6	"
ε-CuSn	0.414	22.0	"
η-CuSn	0.756	10.7	"

5.5 INTERMETALLIC FORMATION

Most studies in the published literature dealing with intermetallic formation at solder joints involve optical microscopy and scanning electron microscopy. However, such approaches fail to capture microstructural changes associated with initial stages of the reactive wetting process that can significantly affect the fluid solder properties. Hence, this author and others have been extensively studying the early stages of intermetallic formation in solder alloys using transmission electron microscopy (TEM) [Dakshinamurthy 1992]. Here a portion of the work is described to illustrate the point that the process of diffusion associated with a reactive interface cannot be explained by simple continuum descriptions.

Using the binary Cu-Sn system, the microstructure near the interface after liquid Sn wets a Cu substrate is studied by TEM. It has been found that upon immediate quenching the copper diffuses into the tin grains to form fine, oriented precipitates of Cu_6Sn_5, ranging in size from about 10–30 nm [Felton et al. 1991]. With longer reaction times a complex morphological transition occurs. Rather than a continuous coarsening, these precipitates exhibit shape transitions from the cuboidal shape to rod and plate-like morphologies (Figure 5-7a-b). These shape transitions are often a result of changes in the relative contributions of the interfacial and elastic energies of the total energy of the system. Such size-induced shape transitions in the absence of an external stress is often viewed as a bifurcation problem. Such microstructural instabilities reflect the non-linearities in the diffusion process at the solder/substrate interface. Again it should be pointed out that while macroscopic parameters can influence wetting instabilities, the presence of microstructural non-linearities must not be ignored.

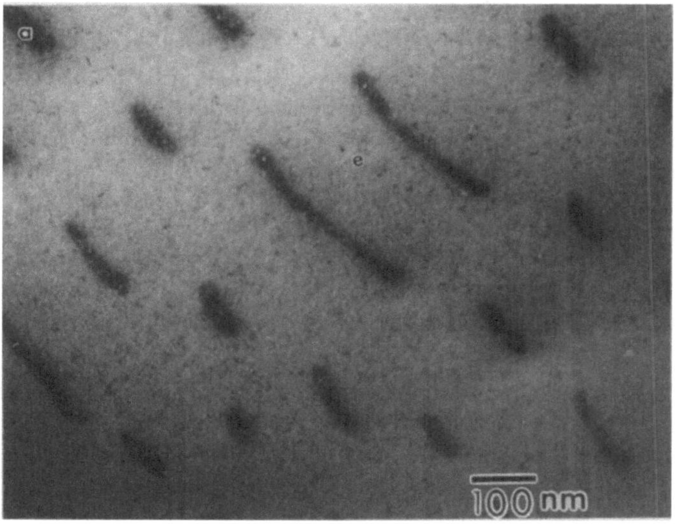

Figure 5-7a TEM micrograph of cuboidal intermetallic precipitates after initial wetting at Cu/Sn interface

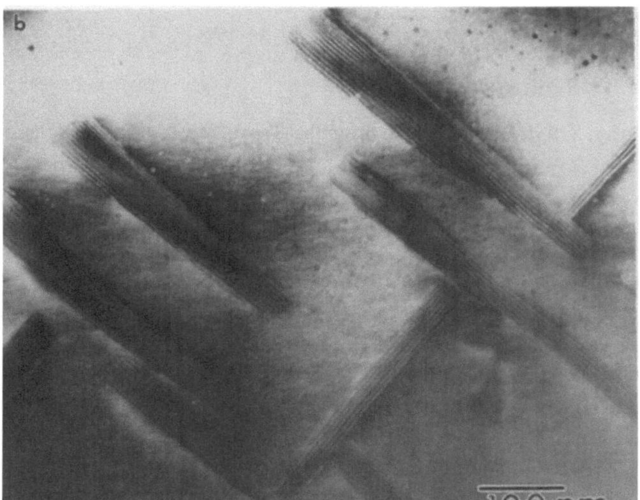

Figure 5-7b Shape transition associated with intermetallic precipitate after further aging

5.6 CONCLUSIONS

The issue of dewetting in solders is interpreted in terms of both continuum and microstructural descriptions. In the case of the former we have applied for the first time the mathematical techniques of differential geometry to the problem of wetting and dewetting. In terms of physical mechanisms of dewetting we have suggested the sources of microstructural instabilities which can affect the dewetting process.

5.7 ACKNOWLEDGMENTS

This work has been supported by Agreement No. MHVU-2109 under sponsorship through the IBM Shared University Research Program. The valuable support of Dr. P. J. Singh of IBM is gratefully acknowledged, along with the work of Mr. S. Dakshinamurthy of Rensselaer Polytechnic Institute.

REFERENCES

Andrade, E.N. da C. 1934. *Phil. Mag.* 17:497.

Bartell, F.E., and J.W. Shepard. 1953. *J Phys. Chem.* 57:211.

Bottezzati, L., and A.L. Greer. *1989. Acta Metall.* 37:1791.

Bickerman, J. 1950. *J Colloid Sci.* 5:349.

Dakshinamurthy, S. n.d. PhD Thesis in progress. Troy, New York: Rensselaer Polytechnic Institute.

deGennes, P.G. 1990. In: *Liquids at Interfaces*. Edited by J. Charvolin, J.F. Joamy, and J. Zinn-Justin. Elsevier Science Pub. BV. p. 273.

Felton, L., K. Rajan, and P. J. Singh. 1991. *Scripta Metall.* 25:2329.

Fradkov, V.E., M. E. Glicksman, M. Palmer, J. Nordberg, and K. Rajan. n.d. *Physica D.*

Gage, M., and R. S. Hamilton. 1986. *J. Diff. Geom.* 23:69.

Gebhardt, E., M. Becker, and Schafer. 1952.*Z. Metallk.* 43:292.

Gebhardt, E., M. Becker, and Z. Tragner. 1953. *Metallk.* 44:379.

Hasouna, A.T., K. Nogi, and K. Ogino. 1988. *Trans Jap. Inst. Metals.* 29:748.

Kawakatsu, I., and T.Osawa. 1973. *Trans Jap. Inst. Metals.* 14:114.

Nogi, K., and K. Ogino. 1988. *Trans Jap. Inst. Metals.* 29:742.

Osher, S., and J. A. Sethian. 1988. *J Computational Phys.* 79:12.

Rajan, K., M. Glicksman, V. Fradkov, M. Palmer, and J. Nordberg. n.d. In: *Coarsening and Grain Growth,* Edited by S. P. Marsh and C. S. Pande. TMS.

Yokota, M., M. Nose, Y. Takano, and H. Mitani. 1980. *J Jap. Inst. Metals.* 44:770.

6

Oxidation of Solder Coatings

D. Morgan Tench

6.1 INTRODUCTION

Solder coatings are widely used to protect copper printed wiring board (PWB) circuitry and electronic component leads from oxidation that can lead to loss of solderability. The coating is usually eutectic Sn-Pb that is either applied directly by hot dipping or electroplated and then densified by reflowing (melting). Numerous studies and ample practical experience have shown that, when properly applied and sufficiently thick, such coatings remain solderable even after several years of normal storage. In practice, however, the coating composition/morphology and storage conditions may not be optimum, allowing solderability degradation to occur. In addition, molten solder tends to flow away from curved areas in the substrate, resulting in localized thinning during solder dipping or reflow. Exposure of the underlying Cu-Sn intermetallic layer in the thinned region can lead to loss of solderability. Thinned solder is a particular problem at PWB through-hole rims and is commonly referred to as "weak knees."

Oxidation of the coating or substrate is the primary cause of solderability degradation but under some conditions ancillary factors may also be important. Many common metallic contaminants in solder are known to degrade solderability, primarily via formation of intermetallic compounds and tenacious oxides. Maximum allowable concentration limits and control methods for these have been established

[Ackroyd et al. 1975; MacKay 1983; Steen and Becker 1986]. The two most troublesome are copper, which is difficult to avoid; and gold, which forms a particularly brittle intermetallic compound with tin. Although such extraneous metals are usually introduced during the soldering process, codeposition of copper with electroplated Sn-Pb has also been shown to degrade solderability [Klein-Wassink 1989]. It is also widely recognized that excessive amounts of organic materials included in the coating or substrate during electroplating can seriously degrade solderability and must be avoided by proper control of bath additive and contaminant levels [see Lea 1986]. In addition, sulfate and chromate residues from etching solutions can precipitate insoluble lead compounds on the substrate surface in solder plating baths [Klein-Wassink 1989; Spiliotis 1978]. Such contaminant effects are usually avoided with available procedures and will not be considered further.

6.2 TIN-LEAD-COPPER SYSTEM METALLURGY

In addition to composition, the morphology of substrates can exert a pronounced effect on the nature of the surface oxide formed. Crystallinity and texture (preferred orientation) are important since reactivity of surface atoms depends on the degree of crystal field stabilization, and the steric viability of a given oxide species may depend on substrate interatomic spacing. Such epitaxial effects are more pronounced in the earlier stages of oxide formation. Substrate grain size is likely to exert a predominant effect as the oxide layer grows since the primary oxide species formed often depends on the available supply of reactants, and the rate of atomic diffusion/migration is greatly enhanced along grain boundaries. It is therefore of interest to consider the metallurgical aspects of the system.

6.2.1 Tin-Lead Morphology

Reflow (fusing) of plated solder is generally considered to yield a denser coating providing superior protection, and is widely employed in the industry. This practice is supported by considerable common experience and by one relatively thorough study [Davis et al. 1982 and 1983] in which the properties of reflowed and unreflowed Sn-Pb coatings were compared. On the other hand, a recently developed electroplating process provides extremely fine-grained Sn-Pb deposits that have been termed "age-resistant" because they offer outstanding protection against solderability degradation [Edington and Lawrence 1988]. Unfortunately, this process operates at extremely high current densities and cannot provide the throwing power needed for PWB plating. Also, the age-resistant properties of such coatings are destroyed by reflowing so that they cannot be used for leads to components that are processed at high temperatures. Nonetheless, the exceptional oxidation resistance attained via grain refinement demonstrates the important role that substrate morphology plays in solder oxidation.

After solidification of Sn-Pb solder, the microstructure coarsens markedly at room temperature as β-Sn precipitates from the lead-rich α-phase, which has an equilibrium composition of 97%Pb-3%Sn [Lampe 1976]. Most of the excess tin precipitates in less than 1 hour, whereas the remainder comes out over a period of up to 60 days. Thermal cycling has been shown to hasten the coarsening process and to lead to cracks that initiate in the tin-rich phase [Frear et al. 1988]. Tin apparently has an appreciable mobility in Sn-Pb solder even at room temperature, and the diffusion coefficient increases with the tin content [Mei et al. 1987].

6.2.2 Intermetallic Underlayer

A duplex Cu-Sn intermetallic layer, comprised of Cu_6Sn_5 (η-phase) in contact with the solder and Cu_3Sn (ε-phase) in contact with the copper substrate, is practically always formed. After a more rapid initial reaction, the thickness increases as the square root of time for a given temperature [Klein-Wassink 1989; Unsworth and MacKay 1973; Kay and MacKay 1976; LeFerre and Barczykowski 1985]. It is clear that Cu_6Sn_5 forms first [Kay and MacKay 1976; Simic and Marinkovic 1980] and that the copper-rich Cu_3Sn phase results from tin depletion at the reaction site. Thus, as tin is depleted, the copper-rich phase, Cu_3Sn, increases in thickness at the expense of the Cu_6Sn_5 phase. Undoubtedly, tin is the mobile species in Sn-Pb solder, but both tin [Kay and MacKay 1979] and copper [Starke and Wever 1964] have been reported to be mobile in the intermetallic phases. Although brittle, interfacial Cu-Sn intermetallics, at least when not excessively thick, apparently have no direct effect on the solder joint integrity since failure during thermal cycling generally occurs in the bulk solder [Frear et al. 1988].

Intermetallic growth rate increases exponentially with temperature [Klein-Wassink 1989; Unsworth and MacKay 1973; Kay and MacKay 1976; LeFevre and Barczykowski 1985] and depends on the microstructure of the base metal, the solder lead content, and whether the solder coating is bright, matte, or fused [Klein-Wassink 1989]. At temperatures below 170°C, the growth rate is faster on hard copper than on soft copper substrates [Kay and MacKay 1976]. As might be expected, the presence of steam has no effect on the intermetallic growth rate [Steen and Bengston 1987]. The room temperature growth rate has been reported as 0.3 [Klein-Wassink 1989] and 0.5 μm/year [Kay 1981], with 0.8 μm/year given as the maximum value observed for a large number of specimens in one study [Kay and MacKay 1976]. A typical burn-in of 24 hours at 125°C consumes 1–2 μm of tin [Bowlby 1987].

The Cu-Sn intermetallics are apparently solderable to some extent when unoxidized, but this issue has not been completely resolved. It is clear that clean Cu_6Sn_5 is adequately wettable since, except in some extreme cases, it is always present beneath the solder coating. On the other hand, Cu_3Sn might be poorly wettable because of its low surface energy [Billot and Clement 1981], but the extent to which this is true remains obscure because of the difficulties involved in preparing and maintaining an oxide-free surface. In addition, much of the work in this area has involved Sn-Pb coatings for which lead enrichment in the coating is a complicating factor. The effects of the two oxidized intermetallics on solderability are also difficult to separate unambiguously because of the possibilities of non-stoichiometry and Cu_3Sn porosity. It is clear, however, that the intermetallic structures normally obtained oxidize within a few days in clean air at room temperature [Dunn 1980] and become unsolderable [Davis et al. 1983; Kay 1981; Bowlby 1987; Billot and Clement 1981; Dunn 1980; Davis 1971; MacKay 1979] except with strongly activated fluxes [Klein-Wassink 1989; Kay and MacKay 1976; Davis 1971].

6.3 ROLE OF OXIDES IN SOLDERABILITY LOSS

6.3.1 Wettability and Heat Transfer Effects

The effects of oxides during soldering of a fusible solder coating are illustrated schematically in Figure 6-1. Oxides are typically poor thermal/electrical conductors (compared to metals) and act as effective heat transfer barriers, slowing the rate of heating and delaying fusion of the coating. The situation is exacerbated because oxides are not wettable by solder so that penetration of the molten solder into small pinholes and cracks, which would improve overall heat transfer and produce local melting that might undercut the oxide film, does not readily occur. The thermal contact resistance of electroplated/reflowed eutectic Sn-Pb solder on copper has been shown to increase as the logarithm of the oxide thickness [Di Giacomo 1986].

Because they are not intrinsically wettable, oxides on the substrate, either present before application of the coating or formed subsequently, can cause dewetting during the soldering process. When a fusible coating is not involved, any surface oxide must be fully dissolved or displaced during the soldering operation to permit intimate contact between the advancing solder front and the substrate (or an unfusible finish). In this case, poor heat transfer through the oxide slows heating of the substrate and inhibits the wetting process.

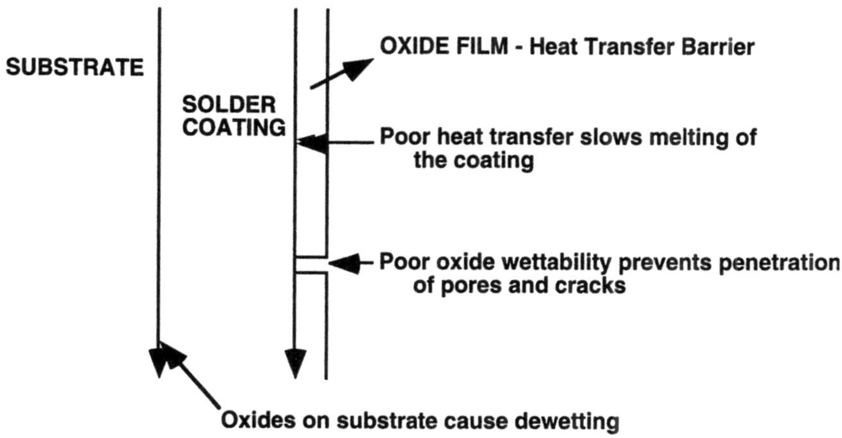

Figure 6-1 Schematic illustration of the effects of surface oxides on the solderability of fusible coatings

6.3.2 Copper Substrate Effects

It is generally recognized that oxidation under ambient conditions rapidly degrades the solderability of unprotected copper. A general discussion of copper oxides and the methods used to prevent oxidation is beyond the scope of the present work, which deals only with copper substrates protected with tin or Sn-Pb coatings. In the latter case, dewetting can result from application of the coating over a pre-existing oxide layer or oxidation of the substrate through porous coatings (prior to reflow). The ease with which copper oxidizes is illustrated by the finding of McCarthy, MacKay, and Thwaites [1980] that excessive exposure to pure rinse water results in a surface oxide that reduces solderability, but these authors also found that such oxidation does not occur in slightly acidic solutions (in which copper oxides are soluble). Since tin and Sn-Pb plating baths are generally highly acidic, and copper oxides are also readily reduced electrochemically under such conditions, electrodeposition of the coating over an oxide layer is unlikely unless the substrate is very highly oxidized. More care is required, however, when the coating is applied by a dry process, e.g. hot roller tinning, for which oxide entrapment is likely. Oxidation of the substrate through a porous coating is typically prevented by fusing the coating shortly after deposition.

6.3.3 Tin and Lead Oxides

Although various thin oxide films have been detected at solder surfaces [Farrell 1976; Geist and Kottke 1986], it is apparently the lead oxides, which form less often but tend to be thicker, that are most detrimental to solderability [YiYu et al. 1987]. This is evident since suitably thick tin and tin-rich coatings usually perform well [see Bowlby 1987], whereas lead-rich deposits are particularly prone to loss of solderability during storage [Costello 1978; Boyer 1979]. Likewise, an electroless tin overcoat has been shown to retard solderability degradation of Sn-Pb coatings [Bernier 1974]. In addition, 100 nm-thick tin oxide films formed on tin-rich solder coatings during thermal cycling at high humidity have been shown to have no effect on solderability [Penzek 1977]. On the other hand, reflowed tin and 85% Sn-Pb have been found to have relatively poor resistance to steam aging compared to solders containing less tin [Wild 1987]. Note that electroplated solder is usually targeted at the eutectic composition but the actual Sn/Pb ratio often varies widely because of current density variations, especially across circuit board surfaces, and lack of proper control of bath additives.

6.3.4 Relative Effects for Tin-Lead-Copper System

The principal cause of solderability degradation with time for near-eutectic or tin-rich solder is insufficient coating thickness to prevent oxidation of the underlying intermetallic, or to sustain intermetallic growth without excessive tin depletion [MacKay 1979; Thwaites 1959]. This is supported by ample evidence that properly applied eutectic or tin-rich coatings of at least 8 μm thickness consistently retain solderability after accelerated aging and long-term storage, whereas thinner coatings degrade under the same conditions.[1] Tin-depletion can result in either total conversion of the intermetallic to easily oxidized (and possibly unwettable) Cu_3Sn, or formation of a lead-rich coating that is unfusible or forms an unwettable lead oxide [MacKay 1979]. A secondary effect that may exacerbate the situation is increased concentration of contaminants in the thinning solder coating as the intermetallic layer advances [Thwaites 1972]. For example, oxygen from additional agents present in acid copper electrodeposits is rejected by the growing intermetallic, and forms oxide particles that can cause solder adhesion failure [Kumar and Moscaritolo 1981].

[1] See Davies et al. 1983; Kay 1981; Bowlby 1987; MacKay 1979; Wild 1987; Thwaites 1959, 1965, 1972; Wilson 1977, 1978; Nagel 1985.

From this discussion, it would appear that long-term solderability is ensured by sufficiently thick coatings of uncontaminated, reflowed eutectic or tin-rich solder applied to clean copper substrates. However, the methods that have generally been used to assess solderability lack sensitivity because they are based on an attempt to simulate the actual soldering process, whose variables are difficult to quantify. Even for the wetting balance, which provides a sensitive measure of the wetting force as a function of time, the results depend on flux-type, pre-heat procedure, equipment design, part thermal inertia, and data interpretation procedure [Lea 1991] so that proper calibration with respect to actual soldering processes is tenuous. The widely used area-of-spread and dip-and-look tests are subjective and operator-dependent as well. As discussed below, improved solderability test methods based on assessment of inherent wettability via oxide detection are able to measure solderability degradation associated with oxidation of the solder coating itself. The effects of such oxidation, which are often relatively subtle, must be minimized if the current industry goal of defect rates in the low parts per million range is to be attained.

In practice, reflowed solder coatings almost always exhibit thinned areas since molten solder tends to flow away from regions of high substrate curvature (so as to minimize the surface energy of the system). Such thinned solder is a particular problem at the rims of circuit board through-holes and is referred to as a "weak knee." It is not unusual for the coating thickness at weak knees to be less than 1 μm, the absolute minimum required to avoid solderability degradation associated with intermetallic exposure [Kay 1981]. Therefore, oxidation of both the solder coating and underlying intermetallic layers must be considered in practical systems.

6.3.5 Overall Degradation Factors

Figure 6-2 illustrates solderability degradation of near-eutectic Sn-Pb solder on copper in terms of the variation in the relative thicknesses of oxide and intermetallic layers as the coating thickness decreases. A normal/protective coating (Figure 6-2a) is sufficiently thick that both the intermetallic outer surface and the adjacent lead-rich region resulting from tin depletion by intermetallic growth are protected from oxidation by the remaining coating of near-eutectic solder. The thin layer of tin oxide is always formed on Sn-Pb surfaces by preferential oxidation of tin [Steen and Bengston 1987] and exerts a relatively small effect on solderability for short-term aging under normal conditions. Substantial solderability degradation (Figure 6-2b) occurs when the coating is too thin to adequately protect the intermetallic layer from oxidation and the easily oxidized and possibly nonfusible lead-rich region extends practically throughout the coating. In this case, solderability is strongly dependent on the storage environment and is rapidly degraded by steam aging. For very thin coatings (Figure 6-2c), there may be a thin layer of mixed Sn-Pb oxides at the outer surface and some Cu_6Sn_5, but solderability is very poor since the bulk of the coating is comprised of Cu_3Sn with a highly lead-rich overlayer, both of which readily form poorly solderable oxide layers.

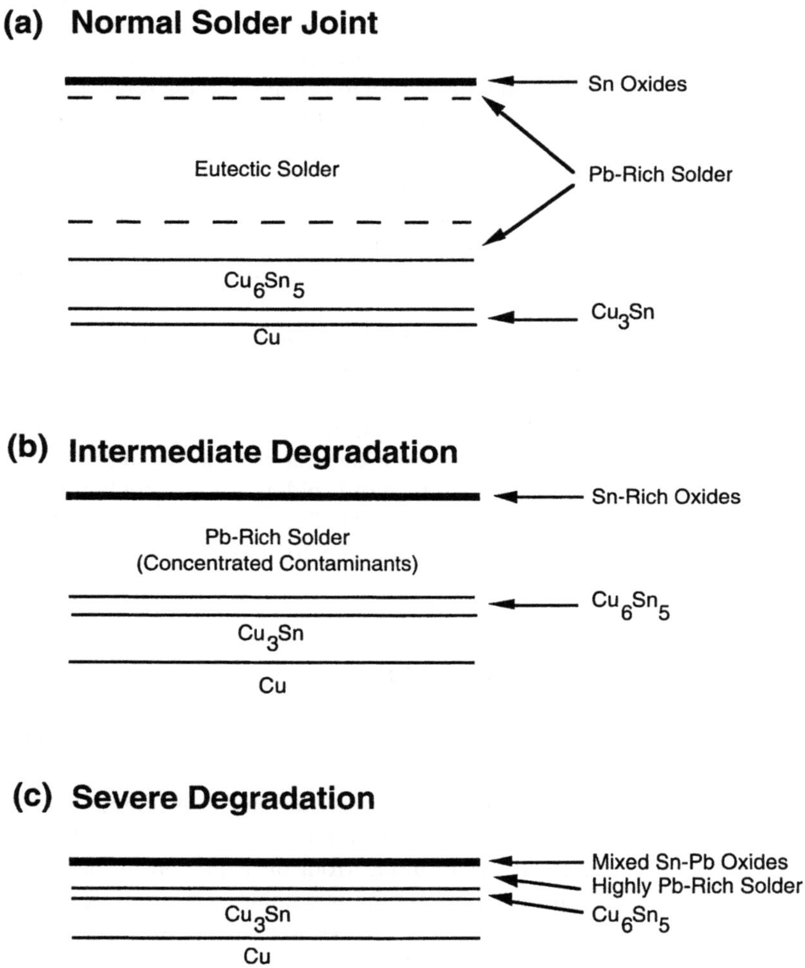

Figure 6-2 Schematic illustration of the effect of coating thickness on the relative thicknesses of intermetallic and oxide phases for eutectic Sn-Pb/Cu system

6.4 TIN-LEAD OXIDATION

6.4.1 *Thermodynamic Considerations*

Table 6-1 compares the free energies of formation per mole of oxygen consumed for the more stable tin, lead, and copper oxides [Pourbaix 1974]. It is clear that formation of tin oxides is thermodynamically favored in the Sn-Pb-Cu system so that preferential oxidation of tin would be expected. Table 6-1 also includes data for formation of the tin hydroxides [House and Kelsall 1984], which are thermodynamically unstable with respect to the oxides, particularly SnO_2, and would probably not form under ambient conditions. Thus, in considering oxidation of solder coatings, attention is focused on the unhydrated tin oxides.

Preferential oxidation of tin at Sn-Pb surfaces has been demonstrated. Early studies showed that lead additions have an insignificant effect on the tin oxidation rate [Boggs et al. 1963] and only tin oxidizes on molten Sn-Pb solder [Kurz and Kleiner 1971]. Extending this work, Bird [1973] applied x-ray photoelectron spectroscopy (XPS) and sputter profiling, and inferred that oxidized tin is enriched at the surface of Sn-Pb by at least a factor of 20 over the bulk concentration. These results have been qualitatively supported by other XPS results [Farrell 1976] and Auger electron spectroscopy (AES) studies [Frankenthal and Siconolfi 1980; Konetzki et al. 1989] but the magnitude of the tin enrichment is in question since lead is to some extent preferentially sputtered at Sn-Pb surfaces [Frankenthal and Siconolfi 1981]. More recent studies [Konetzki et al. 1990] involving AES microprobe analysis show that the tin enrichment is most pronounced around grain boundaries, along which tin apparently diffuses to the surface. Preferential oxidation of tin has been shown to occur even for Sn-Pb alloys containing only 2.9% tin [Konetzki and Chang 1988], indicating that lead oxidation normally occurs only when tin is severely depleted at the surface.

Table 6-1 Standard Free Energies for Formation of Sn, Pb, and Cu Oxides and Sn Hydroxides

Reaction	ΔG° (kJ/mole)
$4\,Cu + O_2 = 2\,Cu_2O$	–292.9
$2\,Cu + O_2 = 2\,CuO$	–254.6
$2\,Pb + O_2 = 2\,PbO$ (orthorhombic)	–378.9
$2\,Sn + O_2 = 2\,SnO$	–515.0
$Sn + O_2 = SnO_2$	–515.8
$SnO + H_2O = Sn(OH)_2$	2.5
$SnO_2 + 2\,H_2O = Sn(OH)_4$	37.3

6.4.2 Oxides Formed

Since the free energies of formation for the various Sn(II) and Sn(IV) oxide/hydroxide species are very close in value (see Table 6-1), the surface oxide formed is governed primarily by kinetic parameters and is likely to be a mixture. Consequently, some disagreement among investigators concerning the nature of tin surface oxides is to be expected. When the limitations of the analytical methods used are taken into account, however, a consistent picture emerges.

Application of low energy electron loss spectroscopy (LEELS) has shown conclusively that SnO_2 is involved in even the very early stages of tin oxidation in the gas phase and is enriched in the outer surface as the oxide film grows [Bevolo et al. 1983]. As shown in Figure 6-3, the bulk plasmon transition for SnO_2 (peak A) is well defined and occurs in a region (around 18 eV) devoid of transitions associated with either SnO or unoxidized tin. Peak B reflects the bulk plasmon transition of SnO (at 1.6 eV) and is not observed for unsputtered SnO_2, showing that 1 keV Ar^+ sputtering reduces the surface of SnO_2 to SnO. These data are generally in agreement with the earlier results of Powell [1979]. By using a relatively low incident energy (75 eV), however, Bevolo et al. [1983] were able to attain a penetration depth of one monolayer of oxide and take advantage of bulk and surface plasmon effects to distinguish between partial and full coverage. They showed that the oxide formed on both polycrystalline and single crystal tin from room temperature to nearly the melting point is continuous after one monolayer of coverage and is sharply enriched in SnO_2 at the outer surface. For tin exposed to the laboratory environment, they found that the outer surface was completely covered with an SnO_2 layer.

Auger electron spectroscopy and x-ray photoelectron spectroscopy have often been used to study tin oxidation, but it is not possible to unambiguously distinguish between SnO and SnO_2 with these techniques [Bevolo et al. 1983; Powell 1979; Lin et al. 1977]. As shown in Figure 6-4, the peaks observed in AES derivative spectra for SnO and SnO_2 are identical in energy, the tin oxides yielding a doublet at 425/432 eV in each case (compared to 430/437 eV for metallic tin). There is a difference in the tin peak heights (asymmetry in the electron energy distribution), but utilization of this to distinguish the two would be tenuous and require precise calibration using pure SnO and SnO_2 standards. Preparation of the former is difficult since SnO readily forms an outer layer of SnO_2 (apparently by disproportionation) and this reaction is enhanced by sputtering [Lin et al. 1977]. Thus, literature inferences [see Farrell 1976; Fidos and Piekarski 1973] about the nature of tin surface oxides that are based on shifts in AES or XPS tin peaks should be viewed with extreme skepticism.

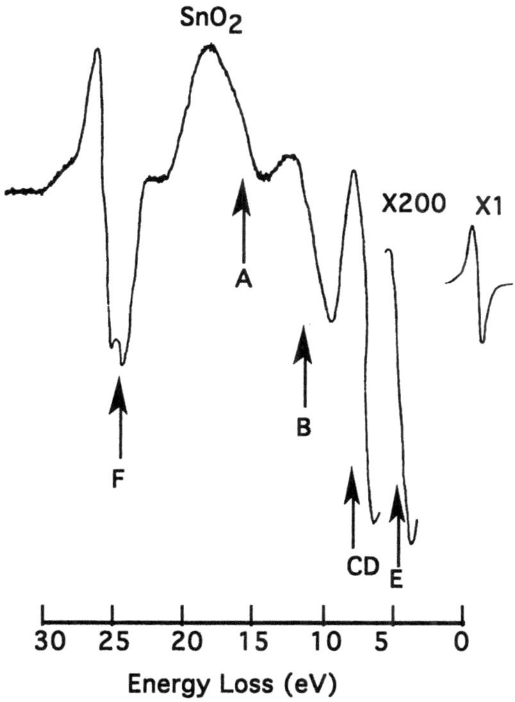

Figure 6-3 First derivative of 75 eV LEELS spectrum of ion sputtered SnO_2 powder [from Bevolo et al. 1983]

On the other hand, Lin, Armstrong, and Kuwana [1977] showed that the ratio of the AES or XPS signal intensities for oxidized tin and oxygen (peak at 510 eV in Figure 6-4) can be used to infer the oxidation state of the tin. The correct value of 2.0 for the O/Sn ratio of an SnO_2 standard specimen was obtained by correcting the AES intensities for the electron impact ionization cross sections. Argon ion sputtering (even at below 1 keV) was found to produce disproportionation of SnO to SnO_2 (O/Sn ratio increased and metallic tin was detected) so that a reliable calibration for SnO could not be obtained. Nonetheless, these results show that properly corrected O/Sn intensity ratios from AES or XPS analysis can distinguish between SnO_2 and a partially reduced oxide. These authors found the oxide on an untreated specimen of pure tin foil to be SnO_2.

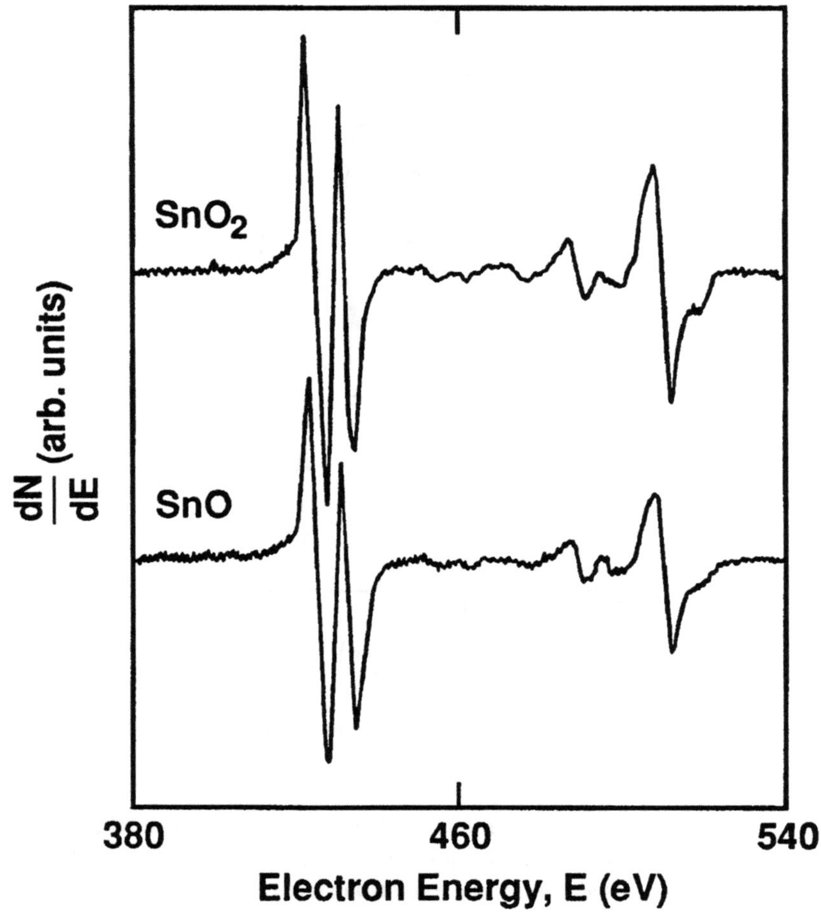

Figure 6-4 Comparison of AES derivative spectra for SnO and SnO_2 powders between 380-540 eV [from Powell 1979]

Electron diffraction has also been used to study oxides formed on tin. The multiplicity of peaks observed makes positive identification of the oxide somewhat subjective. For example, Boggs et al. [1961] concluded that SnO is the only oxide formed on tin under a variety of conditions; but, in the author's opinion, the data are more consistent with a mixed SnO/SnO_2 oxide, especially when the diffraction lines for orthorhombic SnO_2 are included for comparison.

6.4.3 Oxidation Rate

The amount of oxide formed on tin and Sn-Pb has generally been found to increase with the square root [Konetzki et al. 1989; Nagasaka et al. 1975; Miller and Bowles 1990] or the logarithm of time [Konetzki et al. 1989; Fidos and Piekarski 1973; Britton and Bright 1957; Luner 1960]. The activation energy would be expected to depend on the substrate composition and morphology, and the oxidizing atmosphere and temperature range if a mechanism change is involved, but a systematic study of these factors has not been performed. Konetzki et al. [1989] found that oxidation of mechanically cleaned eutectic and single-phase (2.5 at.% Sn) Sn-Pb materials in air is parabolic from 22 to 175°C, and logarithmic at higher temperatures. In spite of the low tin content of the single-phase material, only tin oxide was formed in both cases, and the activation energies for oxidation measured in the parabolic range were comparable (67.1 and 69.5 kJ/mole, respectively).

From a solderability loss standpoint, oxidation under ambient conditions is of primary importance. In original work, Britton and Bright [1957] applied coulommetric electrochemical reduction to show that oxides formed on tin and tinplate at room temperature for periods of up to 5 years are generally less than 5 nm (50 Å) thick. They observed that about 1.5 nm of oxide (assumed to be SnO) forms within a few minutes and grows rapidly to about 2 nm within a week, and then much more slowly to a maximum thickness of 7 nm after as much as 20 years. These values are approximate since SnO_2 forms in appreciable amounts as the oxide ages and requires 4 electrons instead of 2 for reduction, but may not be fully reduced electrochemically [see Britton and Bright 1967 and discussion below]. Similar studies performed on production printed wiring boards [Tench et al. (a)] indicate that the amount of oxide formed on reflowed Sn-Pb solder during storage is comparable. Britton and Bright [1967] also found that the oxides on tin grow 50% thicker when the relative humidity is very high (80%). Oxide thickness measurements by other techniques are generally in agreement with these results. For example, Konetzki et al. [1989] found that the oxide thickness on eutectic solder measured by AES depth profiling increased from about 1.5 to 2.5 nm in 7 days in air, and to about 3 nm after 40 days.

6.4.4 Electrochemical Oxidation

Electrochemical studies of tin oxidation are particularly germane to solderability degradation since the latter is enhanced in the presence of moisture, implying that an electrochemical mechanism is involved. Electrochemistry can also be used to quantitatively form and detect surface oxides as a means of establishing correlations between surface oxide structure and solderability.

Based on turbidimetric analysis of the solution for dissolved tin, Shah and Davies [1963] proposed that oxidation of tin foil in 0.5 M sodium borate (pH 9.3) up to 0.2 V vs SCE (saturated calomel electrode) involves simultaneous tin dissolution and formation of SnO according to:

$$Sn + 5\,OH^- \longrightarrow SnO + HSnO_2^- + 2\,H_2O + 4\,e^- \quad (6\text{-}1)$$

This hypothesis was supported by the observation that the ratio of the anodic to the cathodic charge is 2.0 in this voltage region, which also indicates that the oxide is quantitatively reduced. These authors attributed a negative voltage dip observed in cathodic chronopotentiograms to reduction of an SnO_2 overlayer, which apparently forms at potentials between 0.2 and 1.0 V, inhibits tin dissolution, and causes an increase in the anodic/cathodic charge ratio. An important finding from this work is that tin dissolution does not occur unless the native oxide is removed prior to anodization of the specimen. This indicates that the naturally formed oxide inhibits direct oxidation of the underlying tin metal.

There is general agreement among workers who have investigated tin passivation in borate buffer solution[2] and various other electrolytes[3] that SnO_2, or the hydrated species tin$(OH)_4$, is formed at least at more positive potentials. Negative peaks observed in oxide reduction chronopotentiograms also have generally been attributed to the presence of an SnO_2 overlayer.[4] The formation of SnO_2 in electrochemical systems is supported by electron diffraction [Hampson and Spenser 1968; Britton and Sherlock 1974] and Mössbauer spectroscopy [Vértes et al. Varsányi 1985] studies. The blocking nature of SnO_2 is evident from the results of Nagasaka et al. [1975] who found that oxidation of SnO at high temperatures (350–600°C) completely ceases after formation of a thin SnO_2 surface layer.

---

[2] See Shah and Davies 1963; El Wakkad et al. 1954; Pugh et al. 1967; Vertes et al. 1978; Ammar et al. 1988; Varsanyi et al. 1985.

[3] See Ammar et al. 1983, 1985, 1988; Ammar et al. 1988 (ref. 2), 1990; Davies and Shah 1963; Shams El Din and Abd El Wahab 1964; Hampson and Spenser 1968; Stirrup and Hampson 1976; Gabe 1977; Ansell et al. 1977; Do Duc and Tissot 1979; Burleigh and Gerischer 1988; Giannetti et al. 1990; Drogowska et al. 1991.

[4] See Shah and Davies 1963; Davies and Shah 1963; Shams El Din and Abd El Wahab 1964; Do Duc and Tissot 1979; Britton and Sherlock 1974.

6.5 SOLDERABILITY ASSESSMENT VIA OXIDES

6.5.1 *Sequential Electrochemical Reduction Analysis*

A sequential electrochemical reduction analysis (SERA) method for non-destructively assessing solderability by evaluation of surface oxides has recently been described [Tench and Anderson 1990]. In the SERA method, a small constant cathodic current is applied to the part in a deaerated electrolyte via an inert counter electrode, and the cathode voltage vs. a reference electrode is followed as a function of time. The SERA method can detect oxides, and in some cases sulfides, on a variety of metals and can be applied to almost any part geometry, including printed wiring board through-holes and surface pads [Tench et al. (b)].

Oxide Reduction Potentials

Table 6-2 gives the equilibrium potentials for reduction of the more stable oxides and hydroxides of Sn, Pb, Cu, Ag, Fe, and Ni [Pourbaix 1974] to the respective metals at pH 8.4 (that of the borate buffer electrolyte typically used for SERA analysis). For Sn, Pb, and Cu, the difference in equilibrium potential for complete reduction of the oxide compared to the hydroxide (hydrated form) is very small (< 40 mV) except for $Sn(OH)_4$, whose reduction potential of –0.746 V is 0.1 V positive of that for SnO_2 (–0.844 V). Note that $Sn(OH)_4$ can also be reduced to either $Sn(OH)_2$ or SnO at more positive potentials (about –0.65 V), but is thermodynamically unstable with respect to SnO_2 and would not be formed under ambient conditions [House and Kelsall 1984].

Table 6-2 Equilibrium Voltages vs SCE for Reduction of Various Hydroxides and Oxides at pH 8.4

Hydroxides	Oxides
AgOH 0.767	Ag_2O 0.435
$Cu(OH)_2$ –0.129	CuO –0.168
$Pb(OH)_2$ –0.461	PbO –0.488
$Ni(OH)_2$ –0.622	NiO –0.628
	FeO –0.785
	Fe_3O_4 –0.823
$Fe(OH)_3$ –0.679	Fe_2O_3 –0.789
$Sn(OH)_2$ –0.829	SnO –0.842
$Sn(OH)_4$ –0.746	SnO_2 –0.844

Actual reduction potentials depend on the applied current density and are generally more negative than the equilibrium values because of kinetic factors. The potential ranges for reduction of copper and lead oxide species in pH 8.4 borate buffer are usually –0.3 to –0.6 V and –0.5 to –0.6 V, respectively. Although the equilibrium potentials for the Sn(II) and Sn(IV) species are practically equivalent, reduction of the latter is kinetically very inhibited and occurs at much more negative potentials. Surface tin oxides are generally mixtures of the Sn(II) and Sn(IV) species in varying amounts (depending on the conditions of formation) and reduce over the broad range from –0.85 to –1.3 V vs SCE. Therefore, it is convenient to classify tin oxide films as "lower" or "higher" oxides [Tench et al. (b)]. The lower tin oxide is reduced relatively easily, between –0.85 and –1.0 V vs SCE, and it is predominately SnO which exerts a relatively small effect on solderability. The higher tin oxide reduces between –1.0 and –1.4 V and contains appreciable amounts of SnO_2, which is a good thermal/electrical insulator [Samsonov 1973] and is highly detrimental to solderability [Tench et al. (b)].

SERA Method

Figure 6-5 schematically illustrates the features normally observed for SERA curves of electrode potential vs. charge density (current density x time), and delineates the voltage regions for the various oxides pertinent to Sn-Pb coatings on copper substrates. For the ideal case involving well-defined surface compounds that are all exposed to the electrolyte (solid curve in Figure 6-5), the electrode potential initially decreases to a plateau corresponding to reduction of the most easily reduced oxide. After the latter is completely reduced, the voltage again decreases to the value required for reduction of the next oxide. This process is repeated until all oxides are reduced and a steady voltage corresponding to hydrogen evolution from water electrolysis is attained. For each plateau, the voltage identifies the type of oxide and the associated charge density yields a measure of the amount present. The point of complete reduction for a given oxide may be taken as the intersection between linear extrapolations of the steepest part of the curve (in the transition region) and the plateau for the next reduction process (Figure 6-5), or as the inflection point (determined by differentiating the curve). For uniform, non-porous oxides of known chemical composition, the oxide thickness can be calculated; 1 mC/cm^2 is equivalent to about 1.1 nm of SnO and 0.5 nm of SnO_2. Since oxide mixtures are usually involved, it is less ambiguous to utilize charge density as a relative measure of the oxide amount, keeping in mind that 1 mC/cm^2 corresponds to roughly 1 nm of SnO-rich oxide.

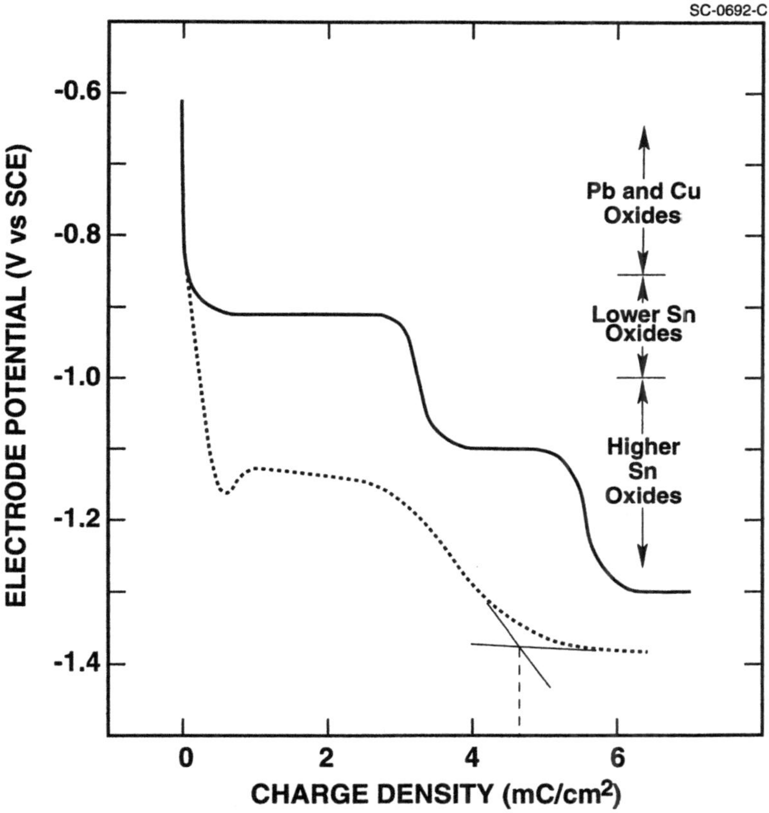

Figure 6-5 Schematic illustration of ideal and typical SERA curve features [Tench et al. (b)].

In practice, SERA curves are often distorted since surface oxides may be comprised of ill-defined compounds, intimate mixtures, and/or layers, and all species may not be exposed to the electrolyte initially. Older solder coatings with degraded solderability often yield curves like the lower curve in Figure 6-5 (dotted). The oxide in this case contains large amounts of the Sn(IV) species, as indicated by the sloping plateau with a relatively negative voltage, and the slow voltage decay between 3 and 5 mC/cm^2 (tailing). The negative peak, which is not unusual, indicates the presence of a blocking overlayer (crust) of Sn(IV) oxide that must be reduced before reduction of the underlying oxide can proceed (at less negative voltages).

6.5.2 Solderability Loss Mechanism

The type and amount of oxides detected for tin coatings has been found to correlate with solderability determined by the wetting balance method. Data obtained for tin and Sn-Pb coatings on copper substrates show that SnO_2 is much more detrimental to solderability than SnO. During natural aging, SnO is converted to SnO_2, apparently by disproportionation.

Anodized Tin

Figure 6-6 shows SERA curves obtained for hot-dipped tin specimens anodized at constant current (20 $\mu A/cm^2$) to various voltage limits. Up to 0.2 V, a single plateau at about –0.85 V is observed, reflecting the presence of a Sn(II) species, SnO or $Sn(OH)_2$, which forms under mildly oxidizing conditions [see MacKay 1983]. As the anodization voltage limit is increased positive of 0.2 V, a negative voltage dip, associated with formation of an overlayer of SnO_2 or the hydrated species $Sn(OH)_4$, becomes increasingly distinct, and the voltage plateau shifts toward more negative values. The increase in the final hydrogen evolution potential for anodization voltage limits positive of 0.4 V has been shown to reflect residual oxide [Tench et al. (b)].

Figure 6-7 shows the effect of anodization voltage limit on the amount of oxide determined by SERA analysis (Figure 6-6) and the wetting times measured for hot-dipped tin specimens. The wetting time increases relatively little (0.4 – 0.7 s) up to the voltage at which formation of SnO_2 commences (0.2 V), and then increases sharply to a peak of more than two seconds at 0.7 V. Although this change in slope would be expected to reflect a significant change in the surface oxide, the amount of oxide detected increases linearly over the entire voltage range. This is surprising since SnO_2 formation beginning at 0.2 V should involve double the charge compared to that for formation of an equal amount of SnO and cause the slope of the oxide amount-voltage limit curve to increase. The most reasonable explanations of why this does not occur are: (1) the amount of SnO_2 is only a small fraction of the total oxide present but has a large effect on the wetting time because it forms a dense layer that is highly resistive, or (2) steric factors associated with the growing oxide layer cause SnO to disproportionate chemically to Sn and SnO_2 so that the average oxidation state of tin in the oxide film remains constant. Another alternative explanation that cannot be ruled out from existing data [Tench et al. (b)] is that formation of SnO_2 involves simultaneous Sn dissolution analogous to SnO formation via Eq. (6-1).

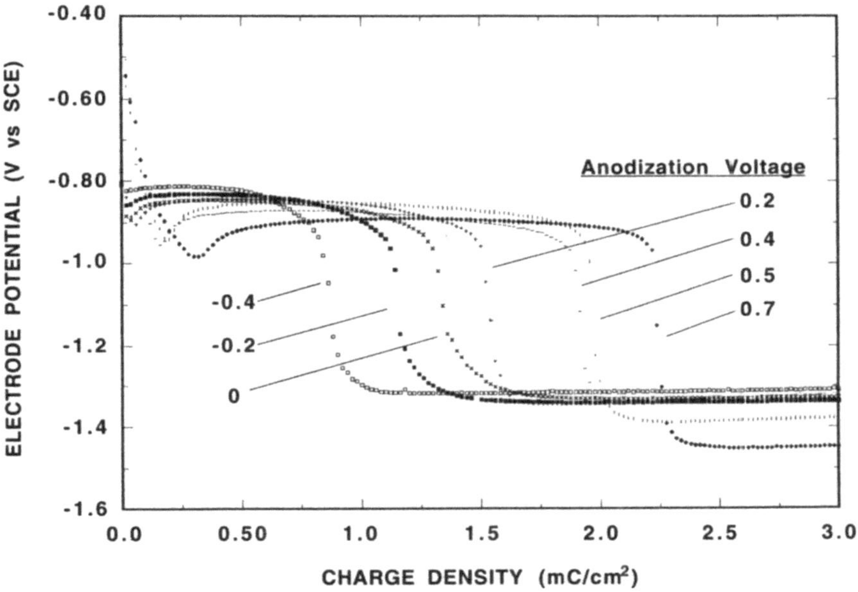

Figure 6-6 SERA curves for pre-reduced tin specimens (hot-dipped copper wires) anodized at 20 $\mu A/cm^2$ to various voltage limits from –0.4 to 0.7 V vs SCE in pH 8.4 borate buffer solution [from Tench et al. (b)]

From Figure 6-7, it is evident that SnO_2 is much more detrimental to solderability than SnO. When the oxide is predominantly SnO (up to 0.2 V), doubling the oxide amount from 0.85 to 1.7 mC/cm^2 increases the wetting time by only 0.3 s (from 0.4 to 0.7 s). On the other hand, a 35% increase in the total amount of oxide (from 1.7 to 2.3 mC/cm^2) in the voltage region where SnO_2 is formed (0.2 – 0.7 V) increases the wetting time by 1.5 s (from 0.7 to 2.2 s).

Since SERA analysis has consistently been found to restore solderability, the oxides detected are obviously those primarily responsible for the degradation. For example, the average wetting times (and standard deviations) after SERA analysis for Sn/Cu specimens (5 each) anodized to –0.2, 0.7, and 1.0 V were found to be 0.38 (2%), 0.41 (2%), and 0.47 s (8%), respectively [Tench et al. (a)]. It should be mentioned that oxides produced at the more positive voltages are not completely reduced, but the small residue has minimal effect on the wetting time and does not denigrate the utility of the analysis. Note that the tin surface oxide formed at less positive anodization voltages, which is more like that formed during natural aging, is apparently reduced fully during SERA analysis.

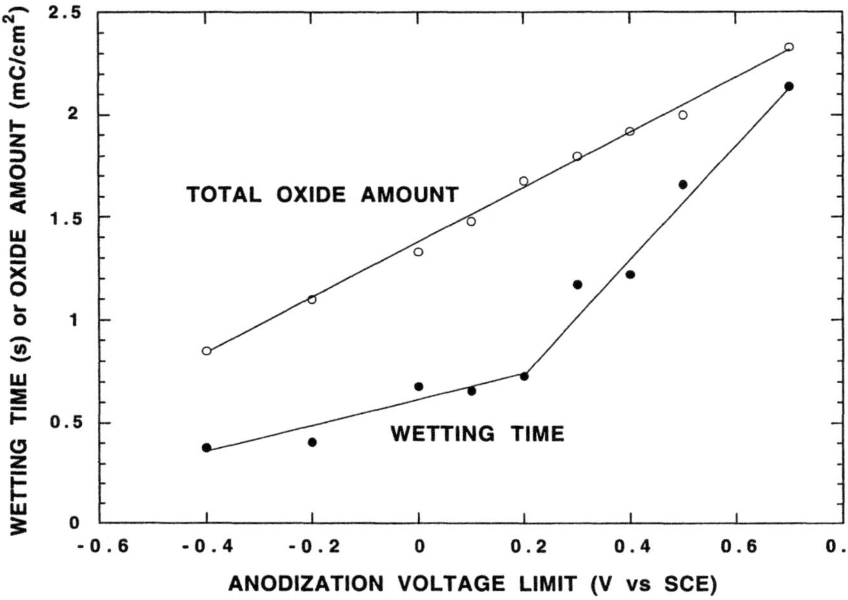

Figure 6-7 Plots of the total amount of oxide determined from SERA analysis and wetting time at 235°C vs. the positive voltage limit for hot-dipped Sn specimens anodized at 20 μA/cm^2 in pH 8.4 borate buffer solution [Tench et al. (b)]

Printed Wiring Boards

It has generally been assumed that solderability loss occurs because the surface oxide thickens with time, but recent SERA production data [Tench et al. (a)], supported by the results discussed above for anodized Sn [Tench et al. (b)], indicate that the type of oxide is more important than the amount. As shown by the representative SERA curves for PWB through-holes in Figure 6-8, the number of soldering defects generally increases as the plateau voltage for the higher Sn oxide becomes more negative, reflecting an increase in the amount of SnO_2 relative to SnO in the surface oxide. The fact that the total amount of charge remains constant (at about 3 mC/cm^2) despite this increase in oxidation state suggests that SnO_2 is formed by disproportionation of SnO so that the overall charge is conserved.

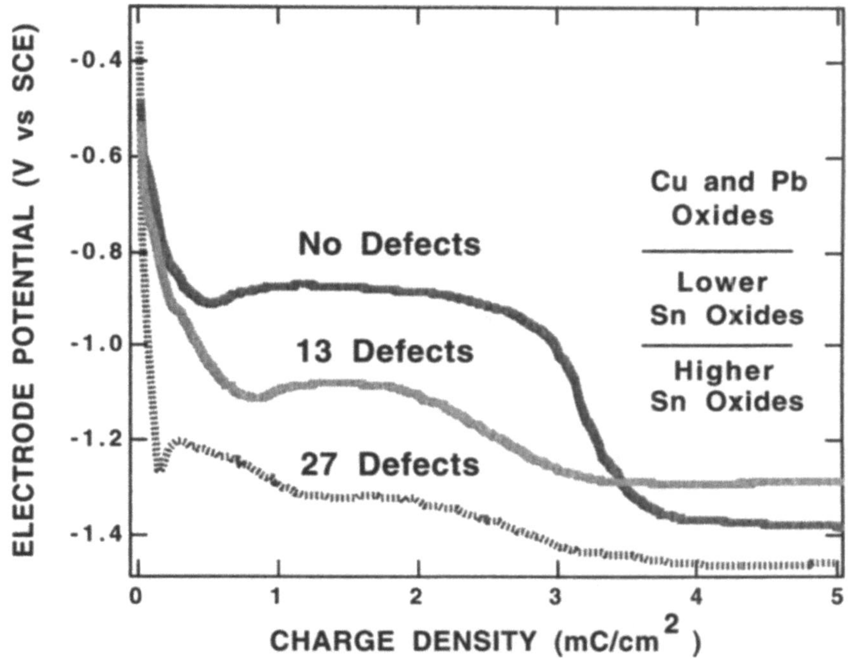

Figure 6-8 Typical SERA curves for circuit boards exhibiting 0, 13, and 27 defects after wave soldering [from Tench et al. (a)]

6.5.3 Solderability Prediction

In general, the percentage of defects on printed wiring boards is small enough that the best predictive capability attainable with any solderability test method is in terms of the probability of defect occurrence. Figure 6-9 summarizes the results of a test program [Tench et al. (a)] in which SERA results for 940 circuit boards tested just prior to wave soldering were correlated with soldering defect occurrence rates. The total number of board-related soldering defects increases sharply for values of the plateau voltage for tin oxide reduction more negative than about −1.07 V. Thus, the probability that a given PWB will exhibit a large number of soldering defects can be defined in terms of the SERA tin oxide plateau voltage. The tradeoff between the cost of the board and rework costs can be determined by reference to the integral giving the fraction of PWBs exhibiting more negative plateau values (dashed line in Figure 6-9).

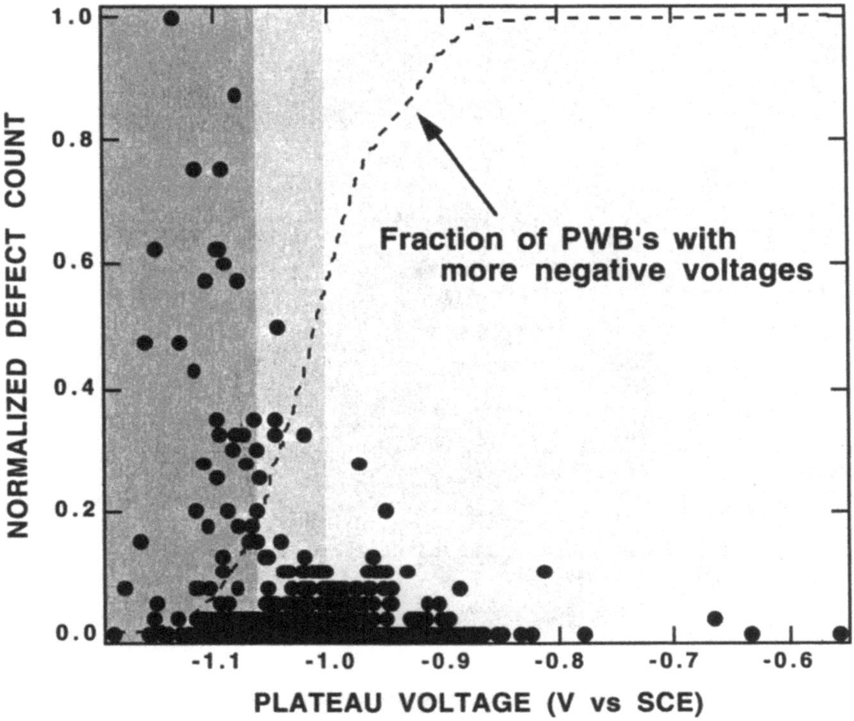

Figure 6-9 Distribution of defects per board (normalization factor = 40) and overall board population (solid line) with respect to SERA tin oxide plateau voltage [from Tench et al. (a)]

Scatter in the data given in Figure 6-9 is to be expected since component leads, which were not SERA tested in this case, can appreciably affect the overall solderability of the system (comprised of the through-hole and component lead). Even better predictive capability should be attainable by SERA testing both boards and component leads, which must be viewed as parts of an interactive system. In this case, the plateau voltage rejection limit could probably be set at a more positive value to take advantage of synergistic effects derived by assuring good solderability of both halves of the board-component system.

6.5.4 Solderability Degradation Factors

It is clear from the discussion above that the major factor leading to solderability loss, at least in the short term, is exposure and oxidation of Cu-Sn intermetallics through excessively thin or porous coatings. Other factors operate over a longer time frame and result in the conversion of oxides on tin and Sn-Pb coatings to more detrimental forms. It is generally recognized [Costello 1978] that excessively Pb-rich or Sn-rich coatings are more prone to solderability loss. However, coarse-grained coatings also lose solderability much more readily than fine-grained finishes [Geiger 1986]. The importance of deposit morphology on solderability loss is illustrated by the age-resistant solder that retains solderability under a wide variety of conditions by virtue of its fine-grained structure [Edington and Lawrence 1988]. Wetting balance studies and SERA testing in the author's laboratory have shown that appreciable oxide growth and solderability degradation of such coatings is not effected by even 16 hours of steam aging, which produces very thick oxides (several hundred Angstroms) on reflowed solder coatings. Likewise, the oxidation rate of polycrystalline tin has been found to be five times faster than that of single crystal tin [Bevolo et al. 1983].

6.5.5 Intermetallic Effects

Oxidation of Cu-Sn intermetallics has not been studied extensively because pure specimens have not been readily available, and complete oxide removal is difficult to attain. Nonetheless, one study indicated that these intermetallics passivate in fluorosilic acid electrolytes and that the passivation process is reversible [Orchard et al. 1982]. Oxides formed on Cu-Sn intermetallics apparently involve intimate mixtures of copper and tin oxide species since such compounds exhibit ill-defined SERA waves in both the copper and tin oxide potential regions [Tench and Anderson 1990]. Recent results obtained in the author's laboratory for both Cu_6Sn_5 and Cu_3Sn prepared by hot isostatic pressing reinforce the earlier results and show that Cu_3Sn, in particular, rapidly forms a relatively thick oxide layer upon exposure to air. From the SERA curves in Figure 6-10, it is evident that the oxide formed on Cu_3Sn, and to a lesser extent on Cu_6Sn_5, contains appreciable amounts of the higher tin oxide.

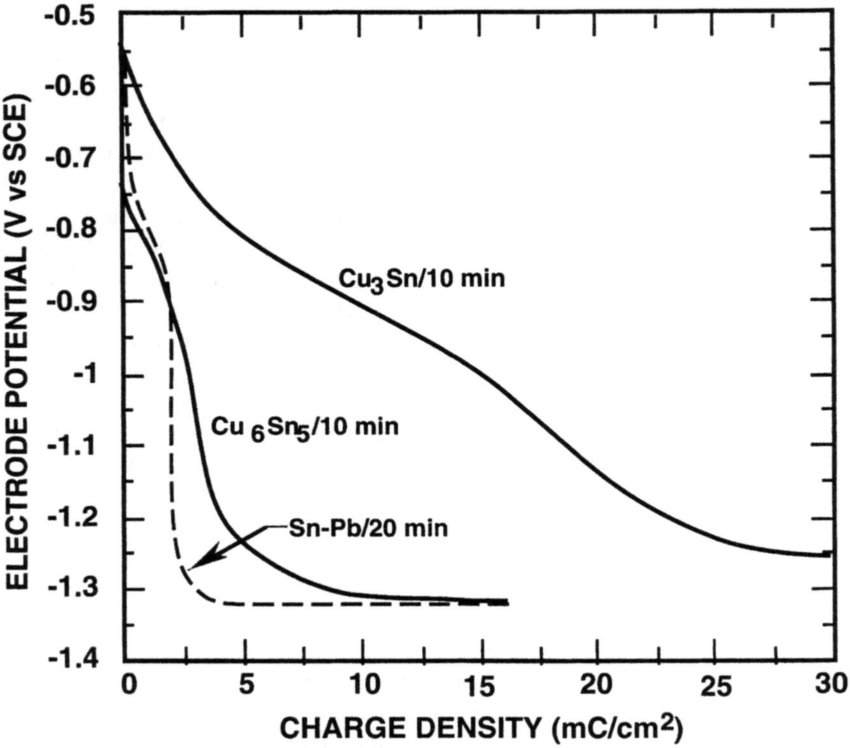

Figure 6-10 SERA curves for pre-reduced Sn-Pb and Cu-Sn intermetallic specimens exposed to oxygen in borate buffer solution (pH 8.4) for the indicated time

Since lead oxides usually do not form on Sn-Pb surfaces [Kurz and Kleiner 1971; Bird 1973; Frankenthal and Siconolfi 1980; Konetzki et al. 1989], the presence of a shoulder positive of –0.8 V in SERA curves for Sn-Pb (or tin) coatings on copper may prove to be a good indicator of exposed/oxidized Cu-Sn intermetallic. Based on limited data [Tench et al. (a)], SERA analysis can apparently detect the presence of PWB "weak knees," i.e. thinned solder coating at through-hole rims which exposes underlying Cu-Sn intermetallic layers and causes solderability loss. As illustrated in Figure 6-11, circuit boards that exhibited unfilled plated through-holes (no component lead) were all found to exhibit the low voltage shoulder (onset at about –0.6 V) expected for oxidized Cu-Sn intermetallics. Additional data are needed to substantiate this correlation.

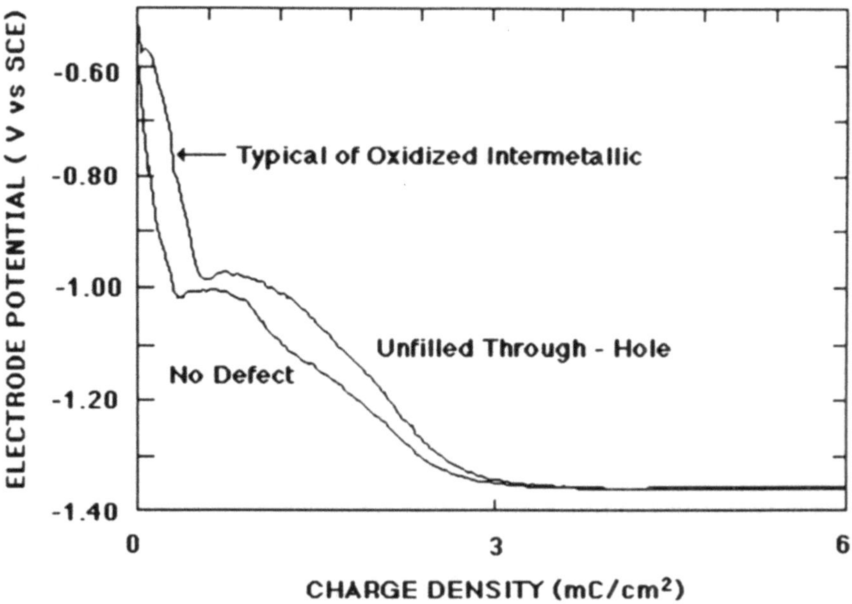

Figure 6-11 Comparison of SERA curves for a PWB exhibiting unfilled through-holes and one with no defects [from Tench et al. (a)]

6.6 ACCELERATED AGING

As the name implies, the intent of accelerated aging is to increase the rates of the processes involved in solderability degradation so that their effects can be ascertained in a short period of time. Generally, solderability tests are performed after representative parts have been exposed to elevated temperature and/or humidity for a specified time period. The goal is to predict the solderability of the remaining parts for a given storage time. The degree of success depends on the extent to which the accelerated aging procedure simulates the natural aging process in terms of the thickness and composition of the oxide produced, and on the sensitivity and predictive capability of the solderability test employed. Since there seems to be some confusion in the literature, it should be emphasized that accelerated aging is not a solderability test *per se*, but rather an associated procedure or process. The combination might be termed a "solderability retention test."

6.6.1 Accelerated Aging Procedures

Three general types of accelerated aging procedures have received significant attention: ambient heating at a specific temperature in the 125 – 155°C range for several hours (dry heat), closed atmosphere heating at 35 – 65°C and 75 – 95% relative humidity for 10 days or more (humid heat), and exposure to steam (and oxygen) at 90 – 100°C for up to 48 hours (steam aging). Most investigations have focused on tin and/or Sn-Pb coatings on copper or copper-coated substrates. Early work with steam aging involved suspending the part above boiling water in an open beaker, whereas modern commercial equipment provides an enclosure for the specimens and better temperature control. In either case, the oxygen content of the vapor, which has a pronounced effect on the rate of solderability degradation [Geist 1983], is determined by the relative rates of steam production and leakage of air into the system. This is probably one reason why steam aging results have been found to vary from laboratory to laboratory even when ostensibly standardized procedures have been used [IPC/EIA Steam Aging Task Group Report 1992]. A steam aging temperature of about 95°C is now recommended since displacement of oxygen at higher temperatures has been found to substantially lower the effectiveness (harshness) of the treatment [Geist 1983; Russel et al. 1989].

Although many workers have relied heavily on subjective dip-and-look solderability tests and have not adequately controlled the oxygen content during steam aging, there is nonetheless general agreement on several important points. Solderability degradation occurs much more rapidly in the presence of water vapor, for both artificial [Edington and Lawrence 1988; Steen and Bengston 1987; Geist 1983] and natural aging [Stoneman and MacKay 1977], and is most severe for the steam aging procedure [Edington and Lawrence 1988; Hagge and Davis 1985]. Insufficient coating thickness has consistently been found to be the most important factor leading to solderability loss. It is clear that a coating thickness of 8 μm provides optimum protection,[5] whereas a thickness of less than 3 μm is grossly inadequate to prevent oxidation of the underlying intermetallic layer.[6] Compared to eutectic Sn-Pb, pure tin and high-tin coatings are apparently somewhat more prone to solderability loss [Wild 1987; Bader and Baker 1973].

---

[5] See Twaites 1959, 1965; Wilson 1978; Stoneman and MacKay 1977; MacKay 1977.

[6] See Bernier 1974; Hagge and Davis 1985; DeVore 1964; Ackroyd 1976; Bader and Baker 1973; Mather 1984.

6.6.2 Predictive Capability

Many investigators have attempted to establish a correlation between accelerated and natural aging to provide the basis for predicting the extent of solderability degradation that will occur in storage during a given time period. In pioneering work, Thwaites [1965] determined the solder area-of-spread and solder immersion time required for wetting of tin and Sn-Pb coated specimens after long-term storage (6 months), 24 hours of steam aging (95–100°C), and 21 days damp heat (90–95%RH, 38–42°C). He concluded, "it is not possible to simulate normal storage by steam aging or by long-term damp-heat treatments with all coatings although there is some similarity in behavior in one or two instances." This statement is still more or less true since subsequent investigations have failed to establish any quantitative correlations that have general validity.

Figure 6-12 shows plots from the work of Edington and Lawrence [1988] comparing the effects of steam and natural aging on wetting time (measured by the wetting balance method) for age resistant, electroplated, and hot-dipped eutectic solder coatings (5–10 μm thick) on copper. Except for the age resistant coating that was not degraded under any of the conditions studied, the curves drawn only roughly follow the data points, which exhibit large scatter. In particular, wetting times are highly variable for steam aging times up to 30 hours, and actually tend to decrease with natural aging (especially for the electroplated coating). Likewise, Stoneman and MacKay [1977] found that the wetting time measured by the globule test for pure tin and Sn-Pb coatings on copper after accelerated aging exhibited very large standard deviations and correlated with those for natural aging only in the narrow region for which solderabilities were marginally altered from their starting values. Hence, even for studies involving quantitative solderability measurements, data scatter has made correlations tenuous.

Since accelerated aging procedures have generally been found to appreciably degrade solderability only when protective coatings are excessively thin,[7] they may provide go/no-go information for weeding out very poor materials. However, statements concerning the equivalency between a given number of hours of steam aging and so many months of natural storage [Boyer 1979; Wild 1987; Wilson 1977, 1978] should be considered very rough approximations, especially in view of the low sensitivity of the solderability tests upon which they are usually based. It should be kept in mind that an aggressive procedure like steam aging may actually reduce discriminating ability by masking solderability differences discernible before the treatment [Edington and Lawrence 1988; Reed et al. 1991].

---

[7] See Boyer 1979; Wild 1985, 1987; Thwaites 1965; Wilson 1977, 1978; IPC/EIA Steam Aging Task Group Report 1992; Mather 1984.

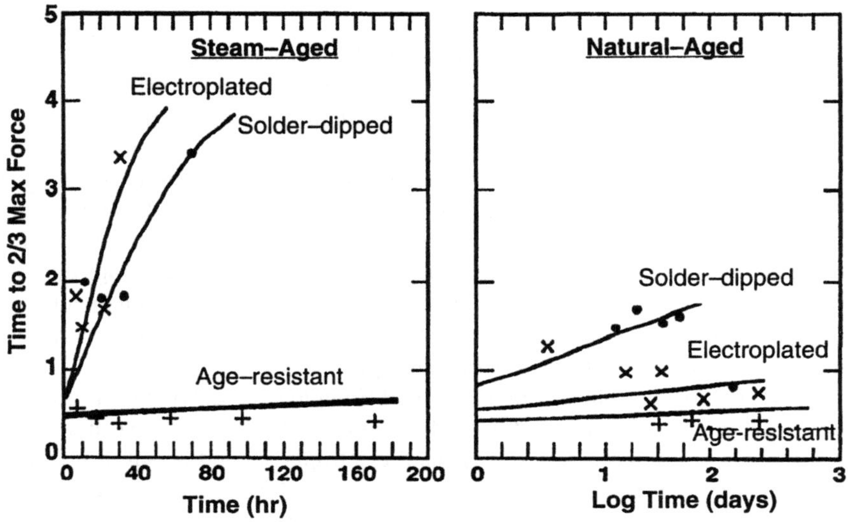

Figure 6-12 Wetting time for age resistant, electroplated, and hot-dipped eutectic solder coatings vs. time of steam and natural aging [Edington and Lawrence 1988]

6.6.3 Oxides Produced by Steam Aging

Figure 6-13 shows SERA curves for special test PWBs subjected to steam aging at 93° C for 4, 8, and 16 hours [Tench et al. (a)]. All of these boards exhibited very negative plateau voltages (about –1.2 V) and more than an order of magnitude more oxide than normally formed during natural aging. From these results, it is clear that steam aging does not simulate natural aging in terms of either the nature or quality of oxide produced.

Based on the data obtained for naturally aged PWBs, these steam-aged boards would be expected to exhibit very poor solderability. However, the number of defects per board ranged from 0 to 6, not substantially higher than the 0 – 2 defects observed for the untreated boards. Consistent with this, an earlier study [Hagge and Davis 1985] found that circuit boards steam aged for 4 hours exhibited fewer soldering defects than typical production boards. A reasonable explanation for the good performance observed is that the very thick oxides produced by steam aging can crack during wave soldering and be removed as scale, allowing the underlying solder coating to melt. Note that it is a general principle that the toughness of thin films decreases as the thickness increases. Alternatively, the oxides produced by steam aging may be more hydrated than those formed naturally, so water vapor released at soldering temperatures causes the oxide film to crack.

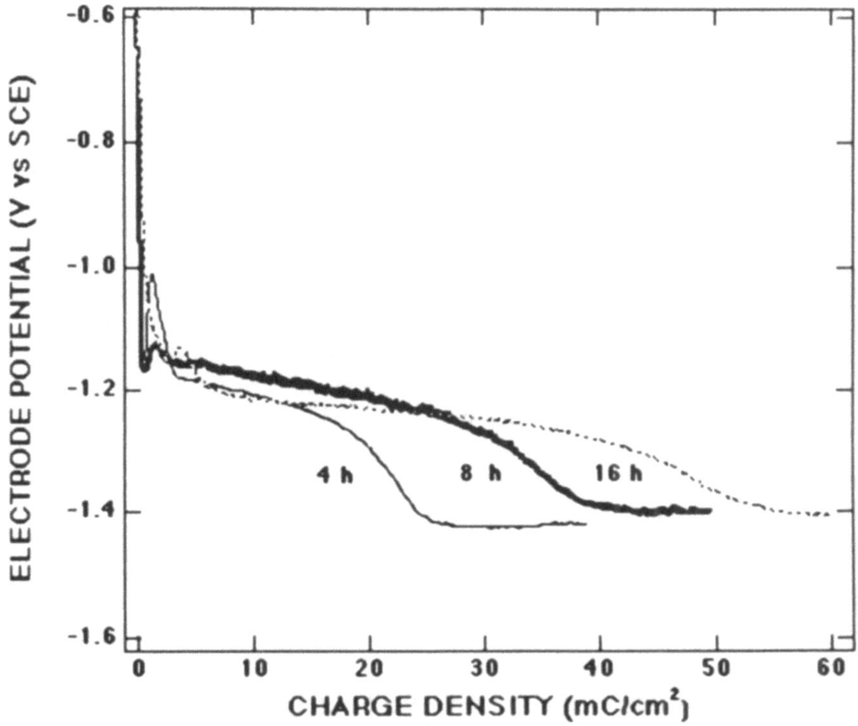

Figure 6-13 Typical SERA curves for PWB through-holes after steam aging (93°C) for 4, 8, and 16 hours [from Tench et al. (a)]

6.7 FUTURE DIRECTIONS

Significant progress has been made in understanding how oxides on tin and Sn-Pb coatings and Cu-Sn intermetallics affect solderability. The pressing need now is to delineate the factors that determine the nature (and amount) of oxide formed so that they can be controlled. Solder morphology is obviously important, but exactly how it influences oxidation and is itself influenced by processing parameters is unclear. One key factor that has gone unexplored is the rate of cooling after reflow, which would be expected to determine the coating grain size and texture (crystallographic orientation) — both known to affect solderability retention. The more subtle effects associated with variations in the coating Sn/Pb ratio have also not been quantitatively established. A related area that needs attention is improved means of controlling tin and Sn-Pb plating baths so that the coating composition can be held constant, and inclusion of detrimental amounts of bath additives and breakdown products can be avoided.

Methods for directly improving the solderability retention of coatings are also needed. In particular, a means for minimizing solder thinning during reflow so as to avoid exposed intermetallics would represent a major advance. A recent patent [Tench and Anderson 1992] describes a copper etching procedure that may prove to be a solution to this problem. Since alloying with other elements, e.g. In and P [Britton and Bright 1957], is known to inhibit oxidation of tin, alternate solder compositions might yield coatings that are more resistant to solderability loss.

Progress in these areas would be greatly accelerated by the availability of a reliable method for non-destructively predicting solderability retention. Such an alternative to accelerated aging must involve a small perturbation since oxidation of tin involves several energetically equivalent reaction paths for which a change in mechanism is likely to occur as the oxidation rate is increased.

Ultimately, if low soldering defect rates are to be attained, we must develop methods for handling the oxides that we are unable to avoid. This is rendered more difficult and more urgent by the current trend toward the use of milder fluxes designed to obviate the need for cleaning procedures involving ozone-depleting chemicals. One approach is to completely remove oxides just prior to the soldering operation. A reduced oxide soldering activation (ROSA™) process that involves electrochemical regeneration of a reducing agent used to reduce metallic oxides back to the metals has recently been reported [Tench and Anderson 1993]. A recently reported alternative [Eishon and Bobbio 1990] is to convert the oxide to an oxyfluoride which breaks up in contact with molten solder and permits soldering without a flux. The oxyfluoride film formed by this plasma-assisted dry soldering (PADS) process apparently prevents reoxidation of the part and may enable fluxless soldering to be performed in air.

REFERENCES

Ackroyd, M. L. 1976. *Proc. INTERNEPCON 1976.* Oct. 1976. p. 214.

Ackroyd, M. L., C. A. MacKay, and C. J. Thwaites. 1975. *Met. Technol.* 2(pt.2):73.

Ammar, I. A., S. Darwish, M. W. Khalil, and A. Galal. 1983. *Z. Werkstofftech.* 14:330.

—. 1985. *Z. Werkstofftech.* 16:194.

—. 1988. *Mat.-wiss. u. Werkstofftech.* 19:294.

Ammar, I. A., S. Darwish, M. W. Khalil, and S. El-Taher. 1988. *Mat.-wiss. u. Werkstofftech.* 19:271.

—. 1990. *Corros. J.* 46:197.

Ansell, R. O., T. Dickinson, A. F. Povey, and P. M. A. Sherwood. 1977. *J. Electrochem. Soc.* 124:1360.

Bader, W. G., and R. G. Baker. 1973. *Plating.* March 1973. p. 242.

Bernier, D. 1974. *Plating.* 61(9):842.

Bevolo, A. J., J. D. Verhoeven, and M. Noack. 1983. *Surf. Sci.* 134:499.

Billot, M., and S. Clement. 1981. *Tin and Its Uses.* 131:1.

Bird, R. J. 1973. *Met. Sci. J.* 7:109.

Boggs, W. E., P. S. Trozzo, and G. E. Pellisier. 1961. *J. Electrochem. Soc.* 108:13.

Boggs, W. E., R. H. Kachik, and G. E. Pellisier. 1963. *J. Electrochem. Soc.* 110:4.

Bowlby, R. 1987. *Circuits Manufacturing.* Nov. 1987. p. 76.

Boyer, A. P. 1979. *Electron Packaging Prod.* Oct 1979. p. 48.

Britton, S. C., and K. Bright. 1957. *Metallurgia.* 56:163.

Britton, S. C., and J. C. Sherlock. 1974. *Br. Corros. J.* 9:96.

Burleigh, T. D., and H. Gerischer. 1988. *J. Electrochem. Soc.* 135:2938.

Costello, B. J. 1978. *Electron. Packaging. Prod.* July 1978. p. 64.

Davies, D. E., and S. N. Shah. 1963. *Electrochim. Acta* 8:703.

Davis, P. E. 1971. *Plating.* 58(7):694.

Davis, P. E., M. E. Warwick, and P. J. Kay. 1982. *Plating Surf. Fin.* 69(9):72.

Davis, P. E., M. E. Warwick, and S. J. Muckett. 1983. *Plating Surf. Fin.* 70(8):49.

DeVore, J. A. 1964. *GE Tech Information Series.* General Electric Company. Report No. R64ELS-44. May 1964.

Di Giacomo, G. 1986. *Proc. Int. Symp. Microelectronics.* Atlanta, GA. p. 322.

Do Duc, H., and P. Tissot. 1979. *Corros. Sci.* 19:179.

Drowgowska, M., H. Ménard, and L. Brossard. 1991. *J. Appl. Electrochem.* 21:84.

Dunn, B. D. 1980. *Trans. Inst. Met. Finish.* 58:26.

Edington, R. J., and L. A. Lawrence. 1988. *Printed Circuit Assembly.* 2(6):13.

Eishon, G., and S. M. Bobbio. 1990. "Fluxless Soldering Process." U. S. Patent No. 4,921,157. Issued 1 May 1990.

El Wakkad, S. E. S., A. M. Shams El Din, and J. A. El Sayed. 1954. *J. Chem. Soc. (London).* p. 3103.

Farrell, T. 1976. *Met. Sci.* March 1976. p. 87.

Fidos, H., and K. Piekarski. 1973. *Inst. Metals.* 101:95.

Frankenthal, R. P., and D. J. Siconolfi. 1980. *J. Vac. Sci. Technol.* 17:1315.

—. 1981. *Surf. Sci.* 104:205.

Frear, D., D. Grivas, and J. W. Morris, Jr. 1988. *J. Electron. Mater.* 17(2):171.

Gabe, D. R. 1977. *Surf. Technol.* 5:463.

Geiger, A. L. 1986. *Proc. 10th Annual Soldering/Manufacturing Seminar.* Naval Weapons Center, China Lake. NWC TP 6707. p. 111.

Geist, H. E. 1983. *Electrochem. Soc. Extended Absts.* 83(2):314.

Geist, H., and M. Kottke. 1986. *Proc. 36th IEEE Electronic Conf.* p. 636.

Giannetti, B. F., P. T. A. Sumodjo, and T. Rabockai. 1990. *J. Appl. Electrochem.* 20:672.

Hagge, J. K., and G. J. Davis. 1985. *Circuit World.* 11(3):8.

Hampson, N. A., and N. E. Spencer. 1968. *Br. Corros. J.* 3:1.

House, C. I., and G. H. Kelsall. 1984. *Electrochem. Acta.* 29:1439.

IPC/EIA Steam Aging Task Group Report. April 1992.

Kapusta, S. D., and N. Hackerman. 1980. *Electrochim. Acta.* 25:1625.

Kay, P. J. 1981. *The Joining of Metals: Practice and Performance.* vol. 2. London, U.K.: Inst. of Metallurgists. p. 131.

Kay, P. J., and C. A. MacKay. 1976. *Trans. Inst. Met. Finish.* 54:68.

—. 1979. *Trans. Inst. Met. Finish.* 57:169.

Klein-Wassink, R. J. 1989. *Soldering in Electronics.* Ayr, Scotland: Electrochemical Publications, Ltd.

Konetzki, R.A., Y. A. Chang, and V. C. Marcotte. 1989. *J. Mater. Res.* 4:1421.

Konetzki, R. P., and Y. A. Chang. 1988. *J. Mater. Res.* 3:466.

Konetzki, R. P., M. X. Zhang, D. A. Sluzewski, and Y. A. Chang. 1990. *J. Electron. Packaging.* 112:175

Kumar, K., and A. Moscaritolo. 1981. *J. Electrochem. Soc.* 128(2):379.

Kurz, R., and E. Kleiner. 1971. *Z. f. Werkstofftechnik/J. Mater. Technol.* 2:418.

Lampe, B. T. 1976. *Welding Res. Suppl.* 330-s.

Le Fevre, B. G., and R. A. Barczykowski. 1985. *Wire J. Int.* 18(1):66.

Lea, C. 1986. *DVS.* 104:180.

—. 1991. *Soldering Surf. Mount Technol.* 7:10; 8:4.

Lin, A. W. C., N. R. Armstrong, T. Kuwana. 1977. *Anal. Chem.* 49:1228.

Luner, C. 1960. *Trans. Metallur. Soc. AIME.* 218:572.

MacKay, C. A. 1979. *Welding Met. Fab.* Jan/Feb. 1979. p. 53.

1983. *Electronics.* April 1983. p. 41.

Mather, J. C. 1984. *PC Fab.* April 1984. p. 34.

McCarthy, T. J., C. A. MacKay, and C. J. Thwaites. 1980. *Circuit World.* 6(4):6.

Mei, S., H. B. Huntington, C. K. Hu, and M. J. McBride. 1987. *Scripta Metal.* 21:153.

Miller, R. G., and C. Q. Bowles. 1990. *Oxid. Metals.* 33:95.

Nagasaka, M., H. Fuse, and T. Yamashina. 1975. *Thin Solid Films.* 29:L29.

Nagel, V. 1985. *ZIS-Mitteilungen.* 27:440.

Orchard, S. W., R. L. Paul, and J. E. C. Timm. 1982. *S. Afr. J. Chem.* 35(4):161.

Penzek, E. S. 1977. *NEPCON Proc.* March & May 1977.

Pourbaix, M. 1974. *Atlas of Electrochemical Equilibria in Agueous Solutions.* 2nd ed. NACE (Cebelcor). Houston, TX. (Brussels).

Powell, R. A. 1979. *Appl. Surf. Sci.* 2:397.

Pugh, M., L. M. Warner, and D. R. Gabe. 1967. *Corros. Sci.* 7:807.

Reed, J. R., N. E. Thompson, and R. K. Pond. 1991. *IPC Tech. Review.* June 1991. p. 16.

Russel, W. R., B. M. Waller, P. S. Barry, and W. M. Wolverton. 1989. *Soldering Surf. Mount. Technol.* Oct. 1989. p. 25.

Shah, S. N., and D. E. Davies. 1963. *Electromchim. Acta.* 8:663.

Shams El Din, A. M., and F. M. Abd El Wahab. 1964. *Electrochim. Acta.* 9:883.

Simic, V., and Z. Marinkovic. 1980. *J. Less Common Met.* 72:133.

Spiliotis, N. J. 1978. *Insulation/Circuits.* June 1978. p. 33.

Starke, E., and H. Wever. 1964. *Z. Metallkde.* 55(3):107.

Steen, H. A. H., and G. Becker. 1986. *Brazing and Soldering.* 11:4.

Steen, H. A. H., and A. Bengston. 1987. *Brazing and Soldering.* 13:28.

Stirrup, B. N., and N. A. Hampson. 1976. *J. Electroanal. Chem.* 67:45; 73:189.

Stoneman, A. M., and C. A. MacKay. 1977. *Proc. INTERNEPCON.* Oct. 1977. p. 49.

Tench, D. M., and D. P. Anderson. 1990. *Plating Surf. Finish.* 77(8):44; U. S. Patent pending (filed 28 May 1991).

—. 1992. "Method of Restoring Solderability." U. S. Patent No. 5,104,494. Issued 14 April 1992.

—. 1993. "Roughened Substrate for Uniform Solder Coating." U. S. Patent No. 5,178,965. Issued 12 Jan. 1993.

Tench, D. M., M. W. Kendig, D. P. Anderson, D. D. Hillman, G. K. Lucey, and T.J. Gher. (a) (in press). "Production Validation of SERA Solderability Testing." *Soldering Surf. Mount Technol.*

Tench, D. M., D. P. Anderson, and P. Kim. (b). "Solderability Assessment via Sequential Electrochemical Reduction Analysis," submitted to *J. Appl. Electrochem.*

Samsonov, G. V., ed. 1973. *The Oxide Handbook.* New York: IFI/Plenum. p. 263.

Thwaites, C. J. 1959. *Trans. Inst. Met. Finish.* 36:203.

—. 1965. *Trans. Inst. Met. Finish.* 43:143.

—. 1972. *Welding J.* Oct. 1972. p. 702.

Unsworth, D. A., and C. A. MacKay. 1973. *Trans. Inst. Met. Finish.* 51:85.

Varsányi, M. L., J. Jaén, A. Vértes, and L. Kiss. 1985 *Electrochim. Acta.* 30:529.

Vértes, A., H. Leidheier, Jr., M. L. Varsányi, G. W. Simmons, and L. Kiss. 1978. *J. Electrochem. Soc.* 125:1946.

Wild, R. N. 1985. *Brazing and Soldering.* 9:5.

—. 1987. *Brazing and Soldering*. 13:42.

Wilson, G. C. 1977. *New Electronics*. 10(6):99.

—. 1978. *Circuit World*. 4(3):39.

YiYu, Q., F. Hongyuan, C. Dinghua, F. Fuhua, and H. Lixia. 1987. *Brazing and Soldering*. 13:39.

7

Surface and Interface Energy Measurements

R. Trivedi and J.D. Hunt

7.1 ABSTRACT

Surface and interface energies are often critical in the success of a soldering operation; thus, the basic concepts of interface energies and their variations with temperature and composition are briefly discussed. The effects of curvature and anisotropy of interface energy on the equilibrium conditions between two phases are presented. Finally, the major techniques used to measure interfacial energies are described for solid-vapor, liquid-vapor and solid-liquid interfaces.

7.2 INTRODUCTION

Surface and interface energies play an important role in wetting and dewetting phenomena. In soldering, one encounters different interfaces such as solid-liquid, liquid-vapor, and solid-vapor interfaces. Thus it is important to obtain experimental values of surface and interface energies so that the relative effects of solid-liquid, liquid-vapor, and solid-vapor energies can be determined. Some of the common techniques for measuring these interfacial energies will be briefly described in this chapter. However, before these techniques are presented, it is important to examine the concept of interfacial energies, develop basic equations that are used for interface energy measurements, and examine all the variables that influence the value of the interfacial energy. We shall therefore briefly discuss the basic properties of interfaces and then describe experimental techniques for measuring interfacial energies.

7.3 BASIC CONCEPTS

7.3.1 *Definition of the Interface*

In order to develop the thermodynamics of interfaces, it is important to define the interface so as to examine its effects on the adjacent phases. Each of the adjoining phases far from the interface is homogeneous, although close to the interface their properties can be influenced by the presence of the interface. First, the two phases may be assumed to be homogeneous right up to the interface (i.e. density, entropy, energy, etc. of each of the phases is uniform right up to the plane of contact). However, as pointed out by Gibbs [1961], this cannot be true: for example, if the densities of the masses were uniform up to the surface of contact, the energy would not be continuous because of the finite range of atomic interaction. Because of the equilibrium conditions in each phase and a short range of atomic interaction, each phase can be considered to be homogeneous throughout its interior except in the vicinity of the other phase where a transition from one to the other takes place within a thin layer. The thickness of the transition layer is not arbitrary but is uniquely determined from the equilibrium constraint that the chemical potential of each species present be constant throughout the system.

In order to obtain a very simple phenomenological treatment of the problem, Gibbs replaced the transition layer by a hypothetical geometrical surface which he called a "dividing surface." Each phase is assumed to be homogeneous right up to this geometrical surface so that thermodynamics of homogeneous phases can be applied to each phase. Also, any extensive thermodynamic property ϕ can be considered as sum of ϕ_1 and ϕ_2 which the two phases would have if they remained homogeneous right up to the dividing surface, plus an additional term ϕ_S, which gives the necessary correction arising from the actual presence of a transition layer. Thus, one can write:

$$\phi = \phi_1 + \phi_2 + \phi_S . \tag{7-1}$$

Such a division, though artificial, enables us to define various thermodynamic properties for an interface since ϕ, ϕ_1, and ϕ_2 are well defined for equilibrium between the two phases. The properties of the thin transition layer, thus determined by the dividing surface, can be characterized by the area and curvature of the dividing surface.

In order to assign definite volumes V_1 and V_2 to the two phases, and a definite value of the curvature to the dividing surface, a precise location of such a dividing surface is required. The dividing surface can be constructed to lie within the transition layer such that each point on the surface has nearly the same surroundings as the neighboring points of the surface. Such a geometrical surface uniquely determines only the normal to the dividing surface so that any other surface in the transition layer which is parallel to this surface will also satisfy the definition of the dividing surface. Thus, some other convention is needed to define uniquely the geometrical surface. The actual position of the interface is not critical if the interface is flat (curvature equal to zero) since the interface properties are characterized by area only, and any normal displacement will not change the area. However, for a curved interface, the precise position of the interface is essential since parallel curved interfaces will have different areas, and thus different interface properties.

The location of the dividing surface is critical if the radius of curvature of the interface is comparable to the thickness of the transition layer. However, when the radii of curvature are much larger than the thickness of the interface, the error introduced by the arbitrary location of the dividing surface is negligible, so that any convenient convention can be chosen in locating this surface. Gibbs suggested a unique division by locating the dividing surface such that the surface density of atoms is zero in a one-component system. Such an interface location is given by the condition $N_1 + N_2 = N$, or $N_s = 0$, where N is the total number of particles in the system and N_1 and N_2 are the number of particles in each phase. As we shall see later, this convention leads to the equality of the surface energy and the Helmholtz free energy for one-component systems. For a multi-component system, the dividing surface is located so that the surface density of any one component (usually the principal component) is zero.

Once the dividing surface is defined, the volumes V_1 and V_2 of the two phases are uniquely defined, and they satisfy the relationship $V = V_1 + V_2$. All thermodynamic extensive properties can now be expressed as the sum of volume and surface contributions.

7.3.2 Surface or Interface Energy

Having established the basic criteria for the thermodynamic description of a system containing an interface, we shall now define surface energy and derive basic relationships which show the dependence of surface energy on system variables. Surface energy is then defined as the reversible work required to create a unit area of the surface at constant temperature, volume, and chemical potentials.

Consider a small system in the vicinity of the dividing surface which is described by a truncated cone whose sides are normal to the interface and which contains an area A of the interface S, as shown in Figure 7-1. Let the cone extend into the homogeneous parts of each phase. If dW is the reversible work at constant T, V, and μ_i required to increase the area by dA without changing the volumes or the states of each phase, then surface energy is defined as:

$$\gamma = \left(\frac{dW}{dA}\right)_{T,V\mu_i} = \left(\frac{d\Omega_d}{dA}\right)_{T,V,\mu_i} \tag{7-2}$$

where $d\Omega = d(F - G)$ is the free energy which characterizes reversible work at constant T, V and μ_i, where F and G are the Helmholtz and Gibbs free energies, respectively [Landau and Lifshitz 1958; Herring 1951; Mullins 1963; Trivedi 1975]. Thus, total surface work is given by the expression:

$$\Omega_s = \int_S \gamma dA \tag{7-3}$$

The equilibrium shape of the interface will therefore be given by the minimum value of the above integral under the condition that no work is done on the bulk phases. In the case of solids, surface energy also depends on orientation of the interface which must be taken into account. This will be discussed in a later section.

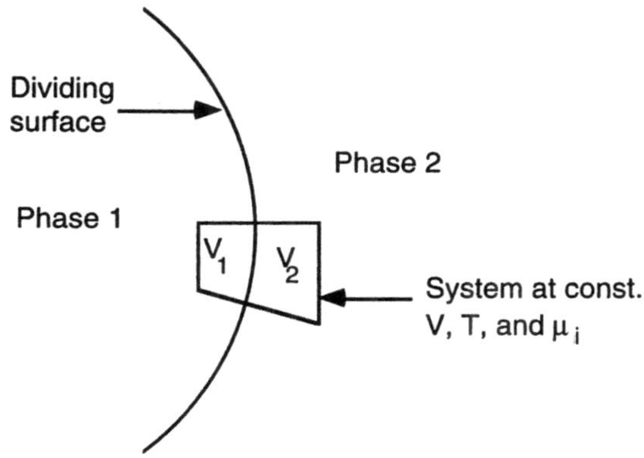

Figure 7-1 Conical sub-system which contains an interface as well as parts of two homogeneous phases. The sides are chosen to be conical since the variation in thermodynamic properties in the direction normal to the conical sides will be zero from the definition of the dividing surface

The creation of the interface under the constraint of constant chemical potentials requires flow of ith species, dN_i, to the interface from the bulk phases outside of the conical system under consideration here. The quantity dN_i/dA is defined as surface excess Γ, of the ith component, and it can have a positive or a negative value depending upon whether at equilibrium the ith species segregates or depletes at the interface. Combining Eq. (7-2) with the definition of Ω_s, i.e. $\Omega_s = F - G$, we obtain

$$d\Omega_S = \gamma dA = dF_S - \Sigma dN_i \tag{7.4a}$$

or

$$\gamma = f_S - \Sigma\mu_i\Gamma_i \tag{7.4b}$$

where f_s is the Helmholtz free energy per unit area of the interface. Since the dividing surface is chosen such that one of the $\Gamma_i = 0$, surface energy is equal to the Helmholtz free energy per unit area of the surface in one-component systems only. For more than one-component systems, the Helmholtz free energy can be visualized as the reversible work required to create a surface at constant T,V, and N_i. However, for equilibrium between the surface and the bulk phases, some of the

dN_i moles will migrate towards or away from the surface to maintain constant chemical potential, thereby reducing some of the reversible work given by the Helmholtz free energy.

The variation of surface energy with temperature and chemical potentials can be obtained from the Gibbs-Duhem equation, written as:

$$d\gamma = -\,s dT - \Sigma \Gamma_i \, d\mu_i \tag{7-5}$$

where s is the entropy per unit area of the surface. The above equation shows that a surface energy gradient will exist when temperature or composition gradients are present along the surface, and this surface energy gradient can give rise to Marangoni flow [Adamson 1967].

7.4 EQUILIBRIUM CONDITIONS FOR A CURVED INTERFACE

Consider a subsystem of fixed volume V, as shown in Figure 7-1, in which two incompressible fluids are separated by a curved interface. Let T and μ_i be equal in the two phases. At equilibrium the free energy Ω is stationary for any infinitesimal change in surface configuration so that

$$\begin{aligned} d\Omega &= 0 = d\Omega_1 + d\Omega_2 + d\Omega_s \\ &= -\,P_1 dV_1 - P_2 dV_2 + d(\gamma A)\,. \end{aligned} \tag{7-6}$$

However, $dV = 0 = dV_1 + dV_2$, so that

$$(P_1 - P_2)\, dV_1 = d(\gamma A)\,. \tag{7-7}$$

For isotropic surface energy, γ is constant, and $dA/dV_1 = K$, the mean curvature of the interface. The mean curvature is defined as $K = (1/r_1) + (1/r_2)$, where r_1 and r_2 are the principal radii of curvature of the surface. Substituting these values in the above equation gives

$$P_1 - P_2 = \Delta P = \gamma K\,. \tag{7-8}$$

This result shows that at equilibrium, the pressures in two phases separated by an interface are not the same unless the interface is flat, i.e. $K = 0$. For a spherical particle, $K = 2/r$, where r is the radius of the sphere, so that

$$\Delta P = 2\gamma / r\,. \tag{7-9}$$

Eq. (7-9) is known as the Laplace equation.

The increase in pressure in phase 1 will increase the molar free energy of phase 1 compared to the molar free energy of phase 1 at pressure P_2. For a change involving no transfer of material (i.e. μ_i = constant), and T = constant, dg = vdP:

$$g_1(T, P_1) = g_1(T, P_2) + \int_{P_2}^{P_1} v\,dP \tag{7.10a}$$

or

$$\Delta g(T, P_2) = \int_{P_1}^{P_2} v\,dP \tag{7.10b}$$

where $\Delta g = g_1(T, P_1) - g_1(t, P_2)$, is the increase in the free energy of phase 1 due to higher pressure P_1 over the pressure P_2 (i.e. the pressure for a flat interface):

$$\Delta g = v\,\Delta P = \gamma v K\,. \tag{7-11}$$

For a spherical phase 1, the increase in the free energy of phase 1 due to the presence of a curved interface will be given by

$$\Delta g = 2\,\gamma v/r\,. \tag{7-12}$$

The above equation is known as the Gibbs-Thomson equation, and it refers to the phase 1 only.

We now consider the equilibrium condition between the two phases, which is given by the equality of chemical potential of each component in the two phases:

$$\mu_1(T, P_1) = \mu_2(T, P_2) \tag{7-13}$$

For a one-component system, the chemical potential is equal to the molar free-energy so that

$$g_1(T, P_1) = g_2(T, P_2) \tag{7.14a}$$

or

$$g_1(T, P_2) + \gamma v_1 K = g_2(T, P_2)\,. \tag{7.14b}$$

Figure 7-2 illustrates the effect of curvature on the equilibrium temperature between the two phases. Note that the equilibrium temperature is lowered by ΔT at the curved interface (the value of Δg can be obtained from Figure 7-2):

$$\Delta g = \Delta T\left[\left(\frac{\partial g}{(\delta T)}\right)_2 - \left(\frac{\partial g}{\partial T}\right)_1 \right] = \Delta T\,[S_2 - S_1] \tag{7-15}$$

Substituting $\Delta g = \gamma v K$ and $\Delta S_v = \Delta S/v$ (the entropy change per unit volume) results in

$$\Delta T = (\gamma / \Delta S_v)\, K\,. \tag{7-16}$$

For a solid particle in its own melt, the lowering of the equilibrium freezing temperature is given by the above equation. In this case ΔS_v is the entropy change per unit volume for melting. For freezing of pure nickel, where T_o is 1728K, $\gamma =$ 225 mJ/m^2 and $\Delta S_v = 0.381$ cal/deg cm, we obtain $T \sim 1728 - 0.000015$ K. Thus, surface effects become significant only when $K > 10^5$ cm^{-1}.

For a two-component system, the change in equilibrium compositions can be examined from the free-energy composition diagram. Note that the free-energy curve for the phase 1 at pressure P_1 is simply shifted upwards (for molar volume independent of composition) compared to the free-energy curve for pressure P_2 which is taken as 1 atm., as shown in Figure 7-3a. The equilibrium condition is given by the common tangent, i.e. μ_{i1} (T, P_1) = μ_{i2} (T, P_2), where the subscript i refers to the two components.

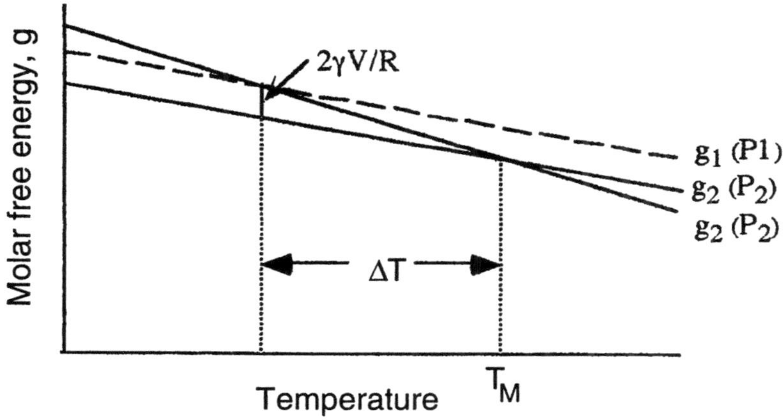

Figure 7-2 The effect of curvature on the temperature dependence of molar free energy of the solid showing the depression in the equilibrium temperature between the solid and the liquid in a one-component system

A binary system shall now be considered in which the change in composition due to the curvature of the interface can be obtained from Figure 7-3. For simplicity, consider the case in which phase 1 is a solid sphere of radius r and the phase 2 is liquid. Let x_L and x_S be the equilibrium mole fractions of solute in the solid and liquid phases, respectively, for a curved interface. Let $x_L{}^0$ and $x_s{}^0$ be the corresponding equilibrium concentrations for a flat interface, i.e. K = 0. Let the chemical potentials of solute for a curved interface and a flat interface be represented as $\mu_r{}^B$ and $\mu_\infty{}^B$, respectively. For small changes and assuming approximately similar triangles on the free-energy composition diagram:

$$\mu_r^B - \mu_\infty^B = \left[\frac{1 - x_L^0}{x_S^0 - x_L^0}\right]\left[\frac{2\gamma v}{r}\right] \tag{7-17}$$

Since $\mu = \mu_o + RT \ln a$, where a is the activity which is the product of activity coefficient and concentration, one obtains

$$\mu_r^B - \mu_\infty^B = RT\ln\left[\frac{x_L}{x_L^0}\right] = RT\ln\left[1 + \frac{\Delta x_L}{x_L^0}\right] \tag{7-18}$$

where $\Delta x_L = x_L - x_L^0$. For small values of Δx_L,

$$\mu_r^B - \mu_\infty^B = RT\mathit{ln}\left[\frac{x_L}{x_L^0}\right] \tag{7-19}$$

Equating equations (7-17) and (7-19):

$$\Delta x_L = \left[\frac{x_L^0 (1 - x_L^0)}{x_S^0 - x_L^0}\right]\left[\frac{2\gamma v}{rRT}\right] \tag{7-20}$$

The change in liquid composition due to curvature at constant temperature may be expressed as a change in temperature at constant composition using the liquidus slope, i.e. $\Delta T = m\Delta x_L$. The change in equilibrium liquidus line for a curved interface can then be calculated and is shown in Figure 7-3b.

7.5 CAPILLARITY EFFECT IN SOLIDS

Surface energy concepts have so far been discussed only for incompressible fluids. When one or both phases are solid, additional effects become important which must be included in the capillarity models. Two primary concepts for solids are: a) surface stress (or surface tension, σ) is not equal to surface energy as is the case with fluids, and b) surface energy is generally anisotropic in solids. For the sake of simplicity, the term surface will be used to represent a "surface" or an interface.

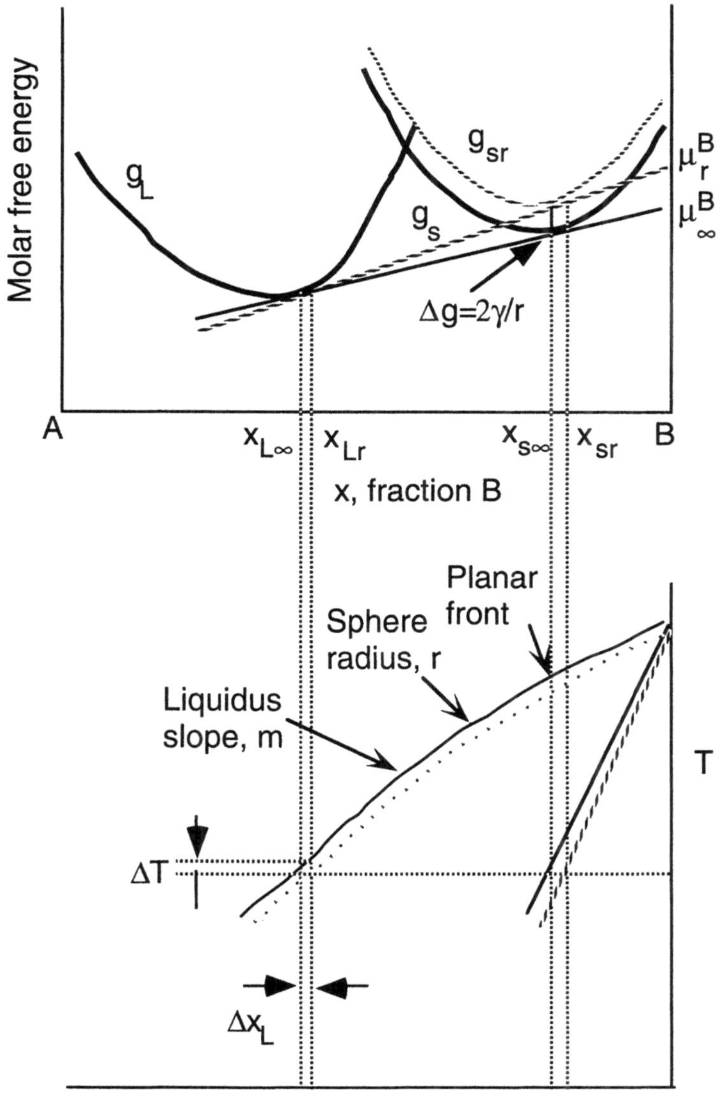

Figure 7-3 The free energy-composition diagram showing the effect of interface energy on the equilibrium between the solid and the liquid and the change in the liquidus and solidus lines for a spherical interface.

7.5.1 Surface Energy and Surface Stress

The term "surface tension," σ, will be used to denote the force per unit length in contrast to surface energy, γ, which is the reversible work required to create a unit area of the surface (at constant T, V, and μ_i). For surfaces with a solid phase, σ and γ are not equal, but they are related by the expression [Mullins 1963; Trivedi 1975; Shuttleworth 1950]:

$$\sigma ij = \delta ij\ \gamma + (\partial\gamma/\partial\varepsilon_{ij}),\ i, j = 1, 2 \tag{7-21}$$

where δij is Kronecker delta whose value is unity if $i = j$ and zero if $i \neq j$. The orientation of the surface is assumed to be constant in calculating $\partial\gamma/\partial\varepsilon_{ij}$. The above result is general and is valid for solid as well as liquid surfaces. It shows that surface stress is equal to surface energy only if γ does not change with the stretching process. This result for the fluid and solid surfaces will now be examined.

In order to visualize the surface tension force, consider a one-component fluid film stretched over a rectangular wire frame in which one side of the frame of length l is movable [Adamson 1967]. Then work done in reversibly stretching the film by extending the movable side a distance dx is

$$\text{Work} = 2\ \sigma_{xx}\ l\ dx \tag{7-22}$$

where surface stress, σ_{xx}, acts normal to the edge of the frame so as to contract the film, and the factor of two comes from the two surfaces of the film. Note that the surface configurations before and after the film was stretched will be the same, and equal to the equilibrium surface configuration with surface energy γ. Once the film is stretched, the surface configuration will tend to maintain the equilibrium configuration by moving atoms from the bulk of the film to the surface so that the increased surface area has the same surface density of atoms as the unstretched film. For fluid phases, the atomic mobility is large so that bulk atoms can readily move to the surface to maintain equilibrium configuration. Also, the thickness of the film can adjust freely to prevent any volume strains in the liquid. Since the extended surface of the film has precisely the same surface configuration as the unextended film, $\partial\gamma/\partial\varepsilon_{ij} = 0$, we have in effect done work to create an extra surface of area 2 l dx and energy γ. Thus, from the definition of surface energy, work done in extending the film can also be written as

$$\text{Work} = 2\ \gamma\ l\ dx\ . \tag{7-23}$$

Equating Eqs. (7-22) and (7-23) gives $\sigma_{xx} = \gamma$. Using a similar argument, by stretching the film in the y direction, under the constraint that the length in x direction remain constant, σ_{yy} would also be equal to γ. Surface tension, σ, is generally defined as the average of the surface stresses in two mutually perpendicular directions, and is equal to $\sigma = (\sigma_{xx} + \sigma_{yy})/2 = \gamma$. Thus, surface tension and surface energy are identical for a fluid-fluid interface. Note that the definition of surface tension in two dimensions is analogous to the definition of hydrostatic pressure in three dimensions.

The case of a solid surface which can increase its area either by the addition of atoms to the surface or by merely stretching the bonds of existing surface atoms will now be considered. Since solids can sustain stresses, the work required to stretch a surface (surface stress) is quite different from the work required to create additional surface having the same configuration as the original surface. In order to examine the distinction between the surface stress and surface energy, we now consider two examples: first, consider the equilibrium configuration at absolute zero of a two-dimensional surface plane only. The equilibrium spacing of atoms in this two-dimensional lattice will be different from that in the bulk of a crystal since the number of neighbors will be different in the two cases. If this two-dimensional plane is to become the surface plane, then some force must be applied to the edges of this plane so that there is a matching of the lattice spacing. These surface forces can be tensile or compressive depending upon whether the atomic spacing in the two-dimensional lattice is smaller or larger than that in the equivalent plane in the bulk of the crystal. The applied force necessary to produce matching is reduced slightly by the adjustments in atom spacings in the second and succeeding layers below the surface. Besides the surface plane, some small tangential forces will also need to be applied to the successive planes below the surface to keep these planes in equilibrium. The sum of all these forces per unit length of edge gives the surface tension of the solid surface.

Second, that the surface stress is not usually the same as surface energy can be clearly seen by considering an ideal case of the nearest neighbor pairwise interaction model in which the nearest neighbor distances are fixed solely by the lowest energy configuration of a two-atom interaction model. In such a case the surface configuration is a precise extension of the bulk lattice. If such a crystal is cut reversibly to obtain two surfaces, the work required will correspond to the sum of the surface energies of the two newly formed surfaces. Since the surface configuration is precisely identical to the configuration these atoms had in the bulk, no surface force is needed and the surface stresses are identically zero. In real crystals, however, some surface stresses will always be present, although some relaxation could occur if defects such as dislocations are present. It can be easily shown that these surface stresses are components of a second-order tensor, a result which further emphasizes the difference between surface stress and surface energy in solids since the latter is a scalar quantity.

Since surface energy depends on the configuration of atoms at the surface, the key point that determines whether surface stress is equal to surface energy is whether the atomic configuration on the surface is sensibly altered by the stretching process. The equality of surface stress and surface energy depends on the mobility of atoms and thus on the relaxation time required for surface atoms to assume their undistorted configuration by the atomic migration process. The relaxation time for liquids is usually less than the duration of stretch so that surface energy and surface stress are identical. In solids where long-range periodicity is interrupted by the presence of non-coherent boundaries, the relaxation time is finite so that such boundaries can be easily created or destroyed by the atomic diffusion process if the rate of extension is less than the reciprocal of the relaxation time. In such cases surface energy and surface stress are identical and such boundaries behave as liquid films. This condition is used by Udin, Shaler, and Wulff [1952] in devising a zero creep experiment to measure the surface tension of a vertically suspended, polycrystalline wire. Similarly, the equilibrated solid-liquid interface in metallic systems will tend to maintain the equilibrium configuration of atoms on the interface so that the difference between the surface energy and surface stress may be negligible. Thus, experimental studies on solids which are based on equilibrium shapes of interfaces are preferable to the dynamic experiments in which the time of relaxation is much larger than the rate of motion of the interface. For this reason the grain boundary grooving method for determining the solid-liquid interface energies will be emphasized.

7.5.2 Anisotropy of Surface Energy

For crystalline solids, surface energy will vary with the crystallographic orientation of the surface. Conceptually, the origin of the orientation dependence can be readily visualized by considering the pairwise interaction model at zero Kelvin in which only nearest neighbor bonds are considered. In this case, if the bond energy is constant, the reversible work required to create a unit area of the surface will be proportional to the total number of bonds broken per unit area of the surface. Since the number of atoms per unit area and the number of bonds per atom with the atoms in the parallel plane will be different for different surface orientations, the surface energy will be a function of orientation. The variation in surface energy with orientation is generally represented by a polar diagram, which is known as a γ-plot or Wulff plot, as shown in Figure 7-4. Several important implications that arise from the variation in surface energy with orientation will now be discussed.

First, surface energy minima are observed for low index orientations giving rise to cusps (discontinuous $d\gamma/d\theta$). As the orientation is changed slightly, the surface energy also changes so that a surface which is not at the cusp orientation has a tendency to rotate in the direction of lower surface energy orientation. Thus, surface energy γ tends to contract the surface and $\partial\gamma/\partial\theta$, called the torque term, tends to rotate the surface. These torque terms are quite important in solids and should be taken into account when equilibrium between different phases is considered. Theoretically, cusps should be present at all rational orientations at zero Kelvin, although many of these cusps disappear as the temperature increases due to the surface entropy effects, and only low index orientations show cusps at higher temperatures. In low entropy of fusion materials, for which the solid-liquid interface is rough, cusps at even low index orientations also disappear, as observed experimentally for pure succinonitrile (Huang and Glicksman 1981) and pure pivalic acid [Singh and Glicksman 1989].

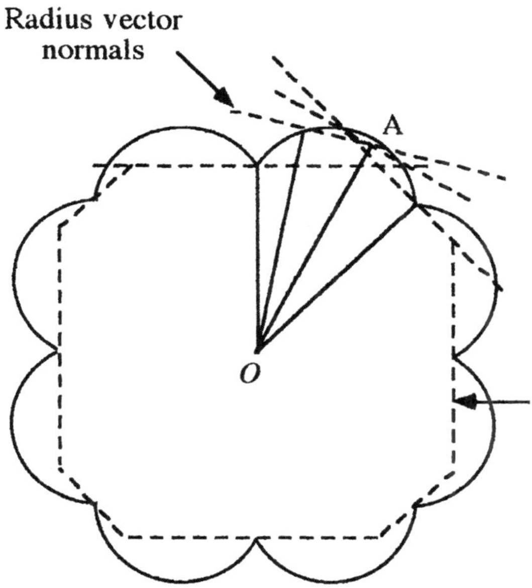

Figure 7-4 A schematic polar plot of interface energy versus orientation, and the equilibrium shape of a crystal

For anisotropic surface energy, the condition for the equilibrium shape of a given phase, Eq. (7-3) needs to be modified as follows to take into account the variation in surface energy with orientation:

$$\Omega_S = \int_S \delta(\gamma dA) = \text{minimum} \tag{7-24}$$

The equilibrium shape of the crystal can be determined from the γ-plot through a Wulff construction. In this case one considers vectors from the origin to the surface of the γ-plot, and then draws planes that are perpendicular to the vectors at the point of intersection on the γ-plot. The inner envelope of all these planes gives the equilibrium shape of the crystal, as shown in Figure 7-4. Note that the distance from the origin (or point of symmetry in the crystal) to the surface of the crystal will be proportional to the surface energy of that orientation so that the equilibrium shape of the crystal will give a measure of relative surface energies for different prominent orientations. Once the γ-plot is known, the equilibrium shape is unique. The inverse is not true in that it is not always possible to draw a unique γ-plot from the equilibrium shape of the crystal unless some specific assumptions are made about the form of the γ-plot or if all orientations are present in the equilibrium shape of the crystal. The relative values of surface energies can thus be determined experimentally from the equilibrium shape of the crystal.

Another important consideration for anisotropic surface energy is that the reversible surface work is now given by

$$d\Omega_S = \gamma\, dA + A\, d\gamma(\theta)\,. \tag{7-25}$$

Substituting this result in Eq. (7-6) modifies the Gibbs-Thomson relationship, Eq. (7-8), as follows [Trivedi 1975]:

$$\Delta P = \left[\gamma + \frac{\delta^2\gamma}{\delta\theta_1^2}\right] K_1 + \left[\gamma + \frac{\delta^2\gamma}{\delta\theta_2^2}\right] K_2 \tag{7-26}$$

where the subscripts 1 and 2 on curvature and orientation refer to the two principal curvatures. As an example, the equilibrium melting point of a one-component system, given by Eq. (7-16) for isotropic interface energy, will now be given by the following equation for the anisotropic interface energy:

$$\Delta T = \left[\frac{1}{\Delta S_V}\right]\left[\left(\gamma + \frac{\delta^2\gamma}{\delta\theta_1^2}\right)K_1 + \left(\gamma + \frac{\delta^2\gamma}{\delta\theta_2^2}\right)K_2\right] \tag{7-27}$$

Thus, the melting point will be a function of curvature and orientation of the interface. The above equation is valid only for the condition $\sigma = \gamma$. When surface energy differs from surface tension, the problem is more complicated, and the precise modification of the Gibbs-Thomson equation is not yet established.

7.5.3 Equilibrium Condition at a Triple Point

When three phases are in equilibrium, the surface forces must balance at the triple point junction to maintain equilibrium. The equilibrium condition for isotropic interface energies is given by the balance of surface tension (or surface energy when $\sigma = \gamma$) (also shown in Figure 7-5):

$$\sum_{i}^{3} \gamma_i \bar{t}_i = 0 \tag{7-28}$$

where $\bar{t}_i$ is a unit vector in the ith interface perpendicular and directed away from the line of intersection. For a liquid drop on a solid surface, Figure 7-5b, this equilibrium condition leads to the Young equation:

$$\gamma_{SV} = \gamma_{LS} + \gamma_{LV} \cos\theta \tag{7-29}$$

where appropriate interface energies and the angle θ are defined in Figure 7-5b.

When the interfacial energies are anisotropic, the torque terms also need to be taken into account, which gives a more general equilibrium condition as:

$$\sum_{i=1}^{3} \left[\gamma_i \bar{t}_i + \left(\frac{\delta\gamma}{\delta\theta}\right)_i \bar{n}_i \right] = 0 \tag{7-30}$$

where $\bar{n}_i$ is a unit vector perpendicular to both $\bar{t}_i$ and the line of intersection.

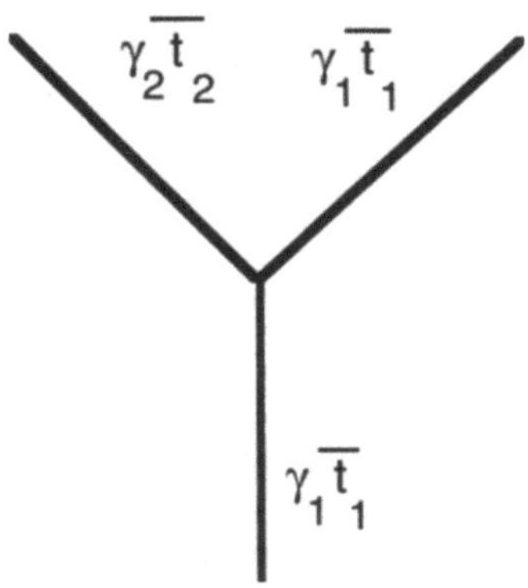

Figure 7-5a Equilibrium conditions at a triple point junction

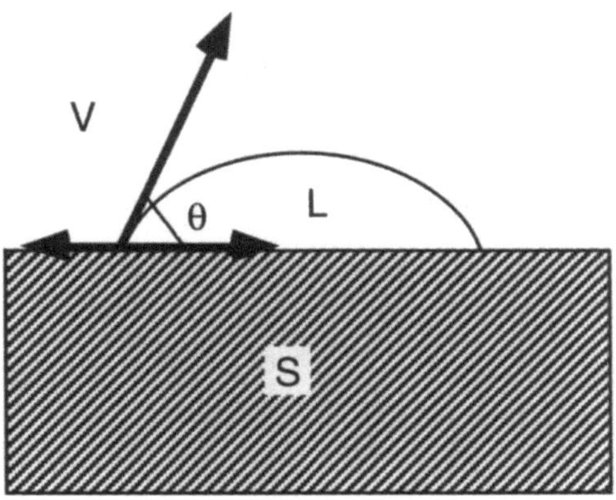

Figure 7-5b Equilibrium conditions for a liquid drop on a substrate

7.6 EXPERIMENTAL TECHNIQUES

So far, various equilibrium criteria for interfaces have been discussed. These expressions are commonly used to obtain interfacial energy values. Therefore, the major techniques that are commonly used to measure the values of interfacial energies will be briefly described [Murr 1975]. These techniques will be divided into three categories: solid-vapor, liquid-vapor and liquid-solid interface energies. For each of the above categories, several techniques have been developed; however, the discussion of all these techniques is beyond the scope of this chapter. Only those techniques that are commonly used for metallic or ceramic systems of importance in soldering shall be focused on.

Some of the techniques for metallic systems are based on the dynamical changes of interface configurations or on mass transport phenomena. Such models are often developed for idealized conditions that may not be satisfied in a given experiment. For example, interfacial energy of a solid-liquid interface is obtained from the maximum undercooling that can be achieved for the nucleation of a solid. From this undercooling the value of surface energy is obtained by assuming homogeneous nucleation [Turnbull 1950]. Experimental studies have shown that it is difficult to satisfy homogeneous nucleation conditions, and larger undercoolings have been observed when the size of the liquid droplet is reduced [Perepezko 1984]. Furthermore, when one of the phases is a solid, these dynamical processes may not be accurate since the distinction must be made between the surface energy and surface stress for the proper evaluation of interfacial energy. Therefore, these techniques will not be considered in this chapter.

7.6.1 Solid-Vapor Interface Energy

The major technique for the measurement of solid-vapor interface is the zero creep technique that is used to obtain an average value of surface energy for all orientations. For examining the orientation-dependence of surface energy, the common techniques are the equilibrium shape method and the twin boundary grooving method.

Zero Creep Method

When a fine wire or thin foil is heated at a constant temperature, it contracts to minimize the surface energy. For a thin foil or a fine wire, the surface-to-volume ratio is large so that the shrinkage force is larger than the static weight of the sample. The shrinkage occurs due to the Nabarro-Herring creep mechanism in which atomic motion from grain boundaries to the surface occurs. Different weights are then applied to the sample and the weight at which zero creep occurs is determined.

The experimental procedure for wires consists of taking a few fine wires, less than 0.1 mm in diameter, to which different weights are attached by spot welding at the bottom ends of these wires. In these thin wires the grain boundaries align themselves normal to the wire axis so that a bamboo-structure is formed, as shown in Figure 7-6a. Two fiducial marks are made circumferentially at about 20mm interval by scratching the wires. Let L_o be the initial distance between the fiducial marks. The samples are then placed in a vacuum or controlled atmosphere chamber and heated to a desired high temperature. The wires are annealed over a long period of time, typically varying from 2 to 10 days. The distance between the fiducial marks, L, is then measured as a function of time, t. The strain rate, $\dot{\varepsilon}_x$, is then given by:

$$\dot{\varepsilon}_x = \frac{L - L_o}{tL_o} \quad . \tag{7-31}$$

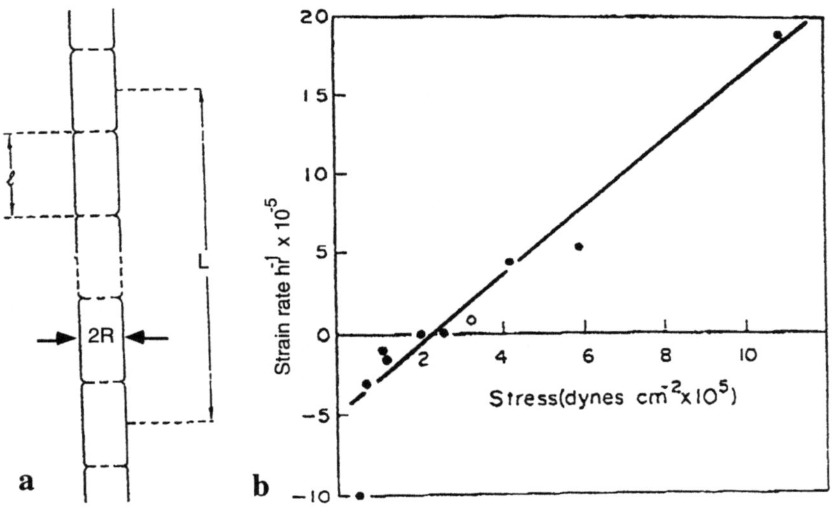

Figure 7-6 (a) Zero-creep experimental set up and (b) the relevant plot for the analysis of the results

The strain rate is then plotted against the effective weight for different wires. The effective weight is obtained by cutting the wire at the middle of the fiducial mark, and weighing the wire and the applied weight together. The strain rate versus weight is plotted, as shown in Figure 7-6b, and the weight corresponding to zero strain rate is determined.

At zero creep condition the stretching work done by the weight is balanced by the surface and grain boundary work that tends to shrink the wire. For an infinitesimal displacement δL between the fiducial marks, which causes a change, dA, in area of each grain boundary within the marks, the zero creep condition gives

$$W_o \, \delta L = \pi \, r \, \gamma_{sv} \, \delta L + n \, \gamma_{gb} \, \delta A \tag{7-32}$$

where W_o is the weight corresponding to the zero creep condition, r is the radius of the wire and n is the number of grain boundaries within the distance L, n = L/l, where l is the average length of the grain. γ_{sv} and γ_{gb} are the surface and grain boundary energies, respectively. Since $n(\delta A/\delta L) = -(\pi r^2/l)dL$, one obtains

$$W_o = \pi \, r \, [\gamma_{sv} - \gamma_{gb} \, (r/l)] \, . \tag{7-33}$$

At the grain boundaries, thermal grooves will be formed, so that the angle θ at the groove can be measured from scanning electron microscope pictures. The balance of energies at the groove gives: $\gamma_{gb} = 2 \, \gamma_{sv} \cos \theta$. Using this relationship, the surface energy is given by the expression

$$\gamma_{sv} = \frac{W_0}{\pi r \left[1 - 2 \left(\frac{r}{l} \right) \cos \theta \right]} \tag{7-34}$$

The measurements of W_o, r, l, and θ allow the value of the solid-vapor energy to be obtained.

The zero creep method only gives an average value of surface energy over all orientations. Note that the variation in solid-vapor energy with orientation is only about 4–5%. Zero creep method gives quite reproducible results for average energy, and such average energies have been measured for several metals [Murr 1975].

Equilibrium Shape of a Crystal

The measurement of relative energies of different orientations of a surface can be obtained by examining the equilibrium shape of a crystal. In this technique a small crystal is placed on an inert substrate, or a thin film of metal is deposited onto a substrate. The composite is then wrapped in a foil of the same material as the film and heated in an ultra-high vacuum at a temperature just below the melting point of the metal. The film breaks up and forms micron-sized single crystal particles. When the particle is heated for a long time, it achieves its equilibrium shape. This equilibrium shape is then photographed in a scanning electron microscope to measure the distance between parallel surfaces of identical orientation. If h_i is the distance between two surfaces of a specific orientation i, then according to the equilibrium conditions

$$(\gamma_1/h_1) = (\gamma_2/h_2) = \ldots\ldots = (\gamma_i/h_i) = \text{constant}\,. \tag{7-35}$$

Thus by measuring h_i, one can obtain the ratio of surface energies of different orientations.

One of the major problems of this technique is that surface energy is significantly altered by adsorption of impurities on the surface. Thus, an ultra-high vacuum and high purity materials are required. A similar technique is the formation of a bubble in the solid, equilibrating this bubble, and measuring h_i. Since gas pressure inside a bubble can be very high, some stress field will be generated in the solid; the distinction must be made between the surface stress and surface energy to obtain reliable values of relative surface energies.

Twin Boundaries

This technique is used for FCC materials which readily form annealing twins. When coherent twin boundaries intersect a surface, a groove and a ridge are formed when the sample is heated in an ultra-high vacuum, as shown in Figure 7-7. Mykura [1961] analyzed the equilibrium conditions at the triple point, at a groove and a ridge, and obtained a pair of equations:

$$\gamma_t = \gamma_A \cos A + \gamma_B \cos B - \left[\frac{\delta\gamma_A}{\delta A}\right] \sin A - \left[\frac{\delta\gamma_B}{\delta B}\right] \sin B \tag{7-36}$$

and

$$\gamma'_t = \gamma_{A'} \cos A' + \gamma_{B'} \cos B' - \left[\frac{\delta\gamma_{A'}}{\delta A'}\right] \sin A' - \left[\frac{\delta\gamma_{B'}}{\delta B'}\right] \sin B' \tag{7-37}$$

γ_t is twin boundary energy, and other surface energies and angles are defined in Figure 7-7. Experimentally it is possible to measure the angles so that the above pair of equations have five unknowns: γ_t, γ_A, γ_B, $(\partial\gamma_A/\partial A)$ and $(\partial\gamma_B/\partial B)$. Thus some assumptions need to be made to calculate the value of surface energy. Winterbottom and Gjostein [1966] have developed a detailed procedure for calculating the values of the ratios $\gamma_t / \gamma(\theta,\phi)$, where θ and ϕ are polar coordinates of a given surface orientation. By using an interpolation formula for $\gamma(\theta,\phi)$, one can obtain a system of equations which become overdetermined when measurements are made on a large number of twin pairs. Using this technique, a complete γ-plot can be obtained.

7.6.2 Liquid-Vapor Interface Energy

The most common techniques for the measurement of liquid-vapor interfacial energy are static methods of the capillary rise method, sessile drop and pendant drop methods, and the dynamic methods of capillary waves, or the levitating drop.

Capillary Rise Method

The basic idea of the capillary rise technique is shown in Figure 7-8. For some cases the liquid in the tube, instead of rising, will be depressed. Since the basic idea for both these cases is the same, only the capillary rise method will be discussed. The capillary tube is of circular cross-section and the variation of the inner diameter should be very small. Typically, the diameter of capillary tube is on the order of 0.5-1.0 mm. The outer container should be sufficiently large, a few centimeters in diameter, so that the capillary rise is negligible. The tube should be of material that readily wets the liquid and does not react with it.

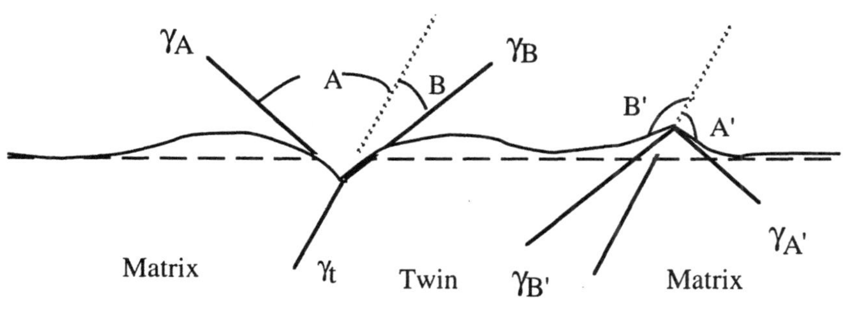

Figure 7-7 Surface profile for a pair of twin boundaries that intersect at a free surface

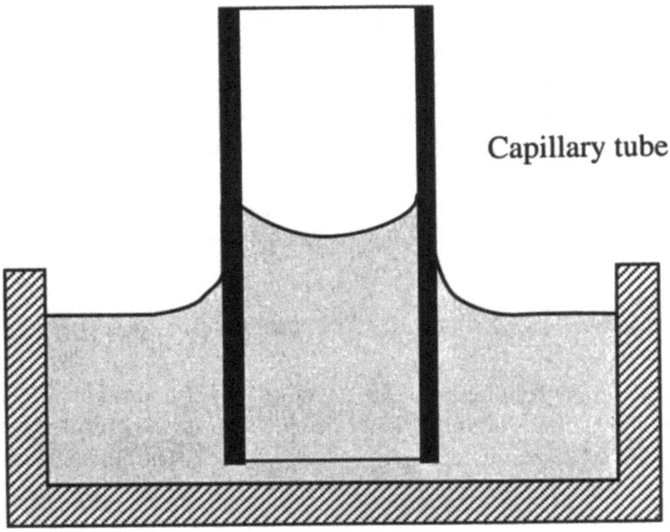

Figure 7-8 The capillary rise method

The pressure difference across the meniscus is given by $\Delta P = \gamma_{lv} K$, where K is the curvature of the meniscus and γ_{lv} is the liquid-vapor interfacial energy. This pressure difference must be equal to the hydrostatic pressure drop in the column of liquid, which is equal to $\Delta\rho gh$, where $\Delta\rho$ is the difference in density between the liquid and the vapor, g is the acceleration due to gravity, and h is the height of the meniscus above a flat liquid interface. Thus, the interfacial energy is obtained as [Adamson 1967]:

$$\gamma_{lv} = \Delta\rho gh / K . \tag{7-38}$$

If the meniscus is spherical in shape, then the calculation of liquid-vapor energy is simple since the curvature is given by $K = 2 \cos \theta / r$, where r is the radius of the tube so that both principal radii of curvature are given by $r/\cos\theta$. The interfacial energy is thus obtained as:

$$\gamma_{lv} = \Delta\rho gh \, r/(2 \cos \theta) . \tag{7-39}$$

In general the meniscus is not hemispherical, although it will have a shape of a figure of revolution. In this case the two principal radii of curvature will be different. A general expression for curvature of a body of revolution can be used to solve the problem. If y is the height of any point on the meniscus with respect to the flat interface in the container, and x is the coordinate in the radial direction, then one obtains the relationship:

$$\gamma_{lv} = \frac{\Delta\rho g h}{\left[\dfrac{y''}{(1+y'^2)^{3/2}}\right] + \left[\dfrac{y'}{(1+y'^2)^{1/2}}\right]} \tag{7-40}$$

where $y' = (dy/dx)$ and $y'' = pdp/dy$.

Note that the interfacial energy can also be related to weight of the liquid, W, in the tube above the flat interface, which gives

$$W = 2\pi r \gamma_{lv} \cos\theta \tag{7-41}$$

This expression is equivalent to the balance of upward surface tension force at the meniscus-wall interface with the gravitational force. If a quartz capillary tube is used, then the shape of the meniscus and the contact angle can be measured from a micrograph, and the weight of the liquid can be calculated from the volume of the liquid in the tube and the liquid density data.

Sessile Drop Method

In the sessile drop method, a liquid droplet is placed on a substrate in a furnace that is maintained at a constant temperature under inert atmosphere or vacuum, Figure 7-9a. The shape of the droplet will tend to be spherical due to surface tension effects, whereas a distortion from spherical shape will occur because of gravity effects. By balancing the surface tension and gravity forces, the shape of the droplet can be determined. Conversely, if the shape of the droplet is determined experimentally, the value of the surface energy can be obtained. The shape of the droplet is governed by the equation:

$$\Delta\rho g y = \gamma_{lv} K(y) \tag{7-42}$$

where y is the vertical coordinate with origin at the top of the droplet, as shown in Figure 7-9a, and the curvature is a function of y. If the droplet is large, then only one radius of curvature needs to be taken into account, i.e. the radius in the plane of drawing. In this case the shape of the bubble is given by

$$\Delta\rho g y = \gamma_{lv}\left[\frac{y''}{(1+y'^2)^{3/2}}\right] \tag{7-43}$$

The integration of this equation from y = 0 to y = h, where h is the distance from the top of the droplet to the equatorial plane where the slope is infinity, gives

$$\gamma_{lv} = \frac{\Delta\rho g h^2}{2} \tag{7-44}$$

Since the shape of the droplet is considered only from the top to the equatorial plane, the result is independent of the contact angle. If the bubble is small, both principal radii of curvature must be considered, as in Eq. (7-40).

The sessile drop technique is the most commonly used technique for measuring the liquid-vapor surface energies of metallic systems. The shape of the droplet is obtained from X-ray shadow-casts or direct viewing through appropriate ports. Certain experimental problems could arise from wetting irregularities that may cause the shape of the droplet to depart from rotational symmetry. A modification of the sessile drop technique, shown in Figure 7-9b, is to replace the substrate with a container that has an accurately chamfered rim which improves symmetry.

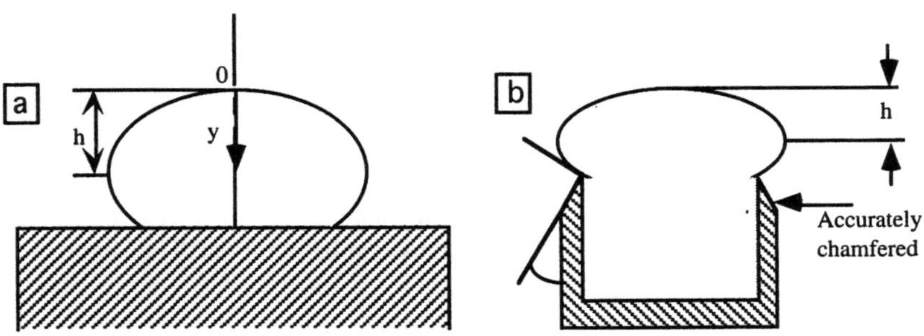

Figure 7-9 Sessile drop method for a drop (a) on a substrate and (b) in a container

Pendant Drop Method

In the pendant drop method, a liquid droplet is allowed to form and hang at the bottom of a tube, as shown in Figure 7-10. The shape of the droplet will again be governed by the balance of surface tension and gravitation forces. In order to obtain simple expression, the shape equation is considered from the bottom of the pendant to the equatorial plane [Andereas et al. 1938], so that the result is obtained in terms of d_s, the diameter of equatorial plane, and d_e, the diameter that is equal to the distance from the bottom, as shown in Figure 7-10. The solid-vapor interface energy is then given by

$$\gamma_{lv} = J\Delta\rho g h^2 \tag{7-45}$$

where J is a function of d_s/d_e, which is numerically calculated and tabulated [Fordham 1948].

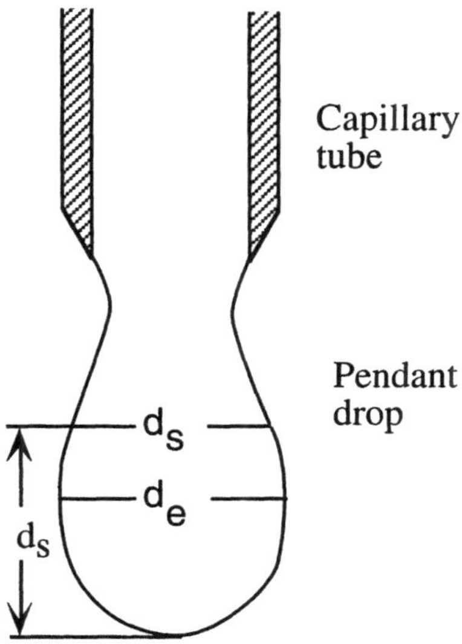

Figure 7-10 Shape of a pendant drop

Capillary Waves

The surface tension effect influences the wavelength of ripples on a surface according to the following relationship, derived by Lord Kelvin [1871]:

$$\gamma_{lv} = \left[\frac{\lambda^3\rho}{2\pi\tau^3}\right] - \left[\frac{g\lambda^2\rho}{4\pi^2}\right] \tag{7-46}$$

where λ is the wavelength and τ is the period of the wave. A detailed mathematical analysis and experimental techniques are reviewed by Lucassen and Hansen [1966], and Mann and Hansen [1963]. A relatively recent development is the levitating drop method [Frazer et al. 1971; Keene et al. 1986]. A small sample of an electrically conducting material is melted and levitated in the apparatus shown schematically in Figure 7-11a. Small perturbations in the system lead the drop to oscillate at its natural frequency ω_n. Provided that the amplitude of the oscillation is small, the drop shape may be considered spherical and the relationship between the surface energy and mass m, is given by the Rayleigh equation [Lord Rayleigh 1879]:

$$\gamma = \frac{3}{8}\pi m \omega_n^2 \tag{7-47}$$

In practice, the drops become aspherical and additional modes are present. Figure 7-11b shows a typical frequency spectrum in which there are five major peaks. The Rayleigh frequency ω_n is given by the central peak, which is flanked by two satellite peaks which are possibly due to asphericity and rotation of the drop. This method clearly has an advantage in that the sample is not contaminated by a container.

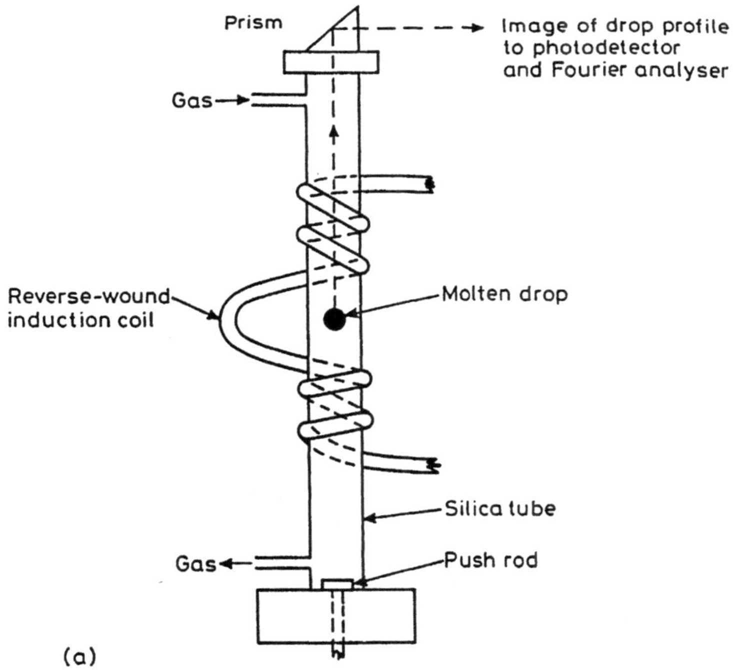

Figure 7-11a Liquid drop levitation technique.

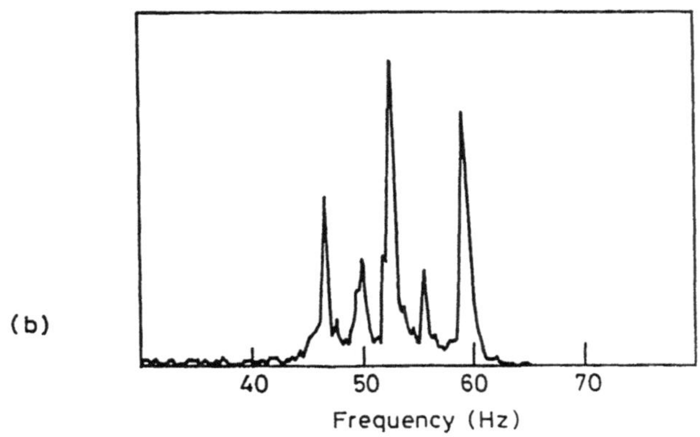

Figure 7-11b A typical frequency spectrum from the liquid drop levitation technique in which there are five major peaks from the ocillations of the droplet

7.6.3 Solid-Liquid Interface Energy

One of the most common techniques for measuring solid-liquid interface energy is the method of grain boundary grooving in a temperature gradient. In this technique, the solid-liquid interface is equilibrated with a grain boundary in a temperature gradient, and interface energy is obtained from the measurements of equilibrium shape of the groove profile [Jones 1974; Jones and Chadwick 1970, 1971, 1978; Shaefer et al. 1975; Hardy 1977; Nash and Glicksman 1971; Bolling and Tiller 1960; Gündüz and Hunt 1985].

The basic experimental set-up is shown schematically in Figure 7-12. A temperature gradient is established radially by using a heating wire along the axis of the cylindrical sample and using a water cooled bath on the outside of the sample. Detailed dimensions of the apparatus and the heating wire are discussed by Gündüz and Hunt [1985]. Once equilibrium conditions are established, the sample is quenched by turning off the current in the wire. It is critical that the temperature variations along the cylinder be minimized, typically ± 0.2 K/cm. Also, the radial temperature gradient needs to be maintained constant so that the fluctuation in temperatures, and thus in the interface position, is negligible.

The quenched sample is first cut transversely into small lengths, typically about 25 mm, which contain few grain boundaries. Each section is then cut longitudinally and ground down to the axial plane, as shown in Figure 7-13. The sample is now polished, etched, and photographed. When the cusps are symmetrical and the grain boundary is perpendicular to the axis of the sample, the groove profile can be analyzed from the photographs. When the grain boundary is not perpendicular to the sample axis, the sample is longitudinally cut close to the boundary along a plane that is perpendicular to the initial longitudinal and transverse sections, as shown in Figure 7-13. The orientation of the boundary is determined from these three mutually perpendicular sections, and appropriate corrections to the shape of the grooves are made, as described by Gündüz and Hunt [1985]. Typical groove profiles obtained in the Sn-Pb system are shown in Figure 7-14.

Figure 7-15 shows an equilibrium groove profile. If the conductivities in the two phases are equal, the temperature of the profile varies with y according to the relationship

$$T(y) = T_E - Gy \tag{7-48}$$

where G is the temperature gradient and T_E is the equilibrium temperature corresponding to a flat interface. At equilibrium, the interface temperature variation must be balanced by the Gibbs-Thompson relationship given by Eq. (7-16), i.e.

$$T(y) = T_E - (\gamma_{S-L}/\Delta S)/r \tag{7-49}$$

where r is the radius of curvature. Equating Eqs. (7-47) and (7-48) gives

$$Gy = (\gamma_{S-L}/\Delta S)/r\,. \tag{7-50}$$

Since it is easier to measure the angle θ, as shown in Figure 7-15, one can integrate the above equation with respect to y from y = 0 to y = y, to give

$$\frac{1}{2}Gy^2 = \left[\frac{\gamma_{S-L}}{\Delta S}\right][1 - \sin\theta] \tag{7-51}$$

The above equation allows the interface energy to be determined. ΔS is the entropy of melting per unit volume, which for an alloy is given by [Gündüz and Hunt 1985]:

$$\Delta S = \frac{RT_e(C_S - C_L)}{[mV_S C_L(1 - C_L)]} \tag{7-52}$$

where m is the slope of the liquidus and V_S is the gm. molecular volume. Note that a correction to the value of G is required when the thermal conductivities in the solid and in the liquid are not equal [Nash and Glicksman 1971; Gündüz and Hunt 1985].

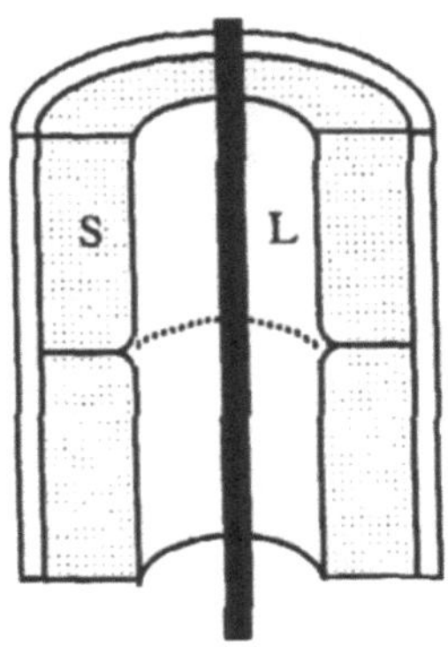

Figure 7-12 Experimental set-up for the measurement of solid:liquid interfacial energy by the technique of grain boundary groove in a temperature gradient

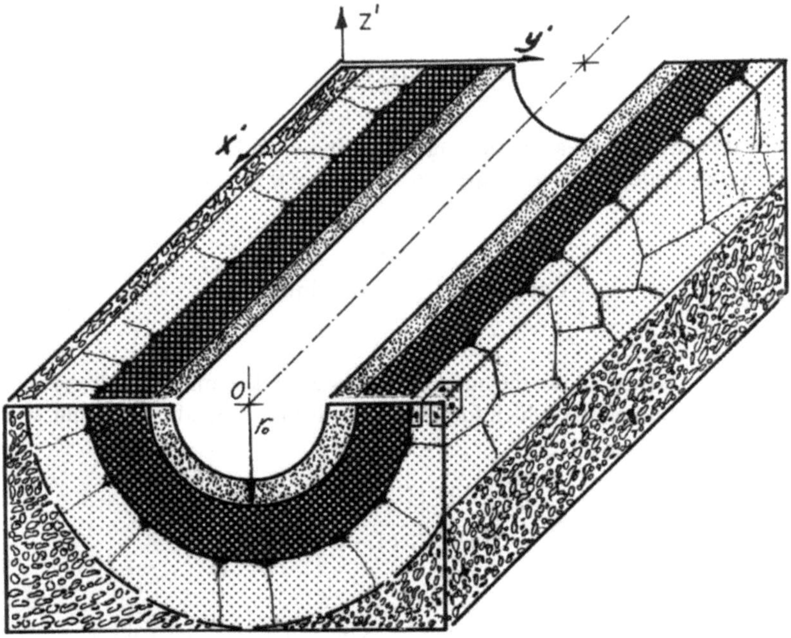

Figure 7-13 Relevant cross-sections showing the three-surfaces needed for the analysis of groove profiles

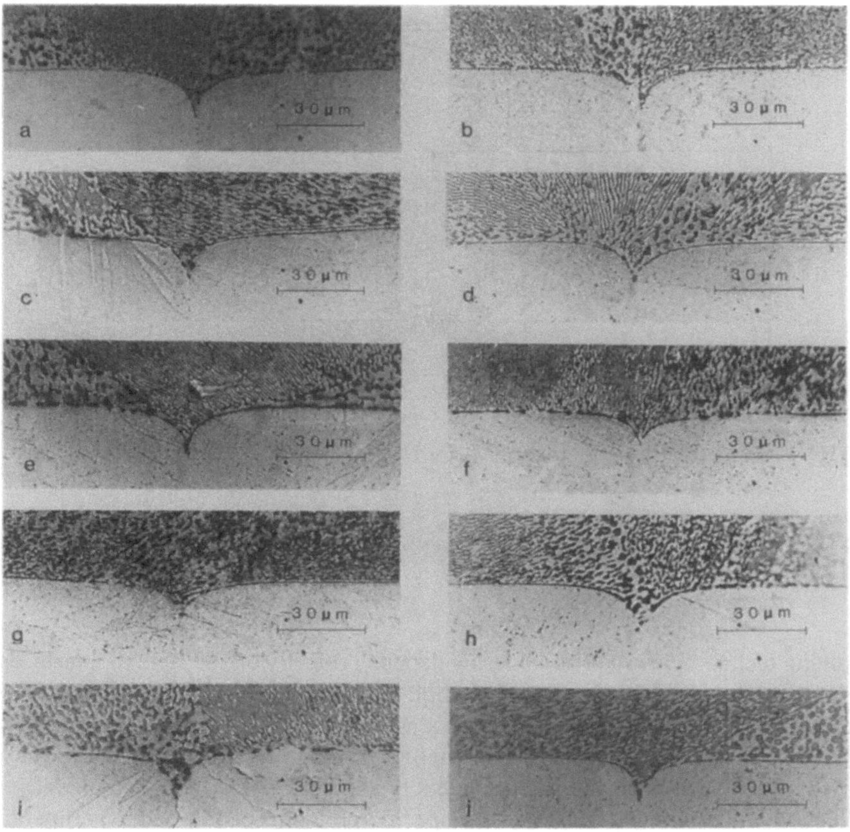

Figure 7-14 Typical grain boundary groove profiles in the Pb-Sn system [Gündüz and Hunt 1985]

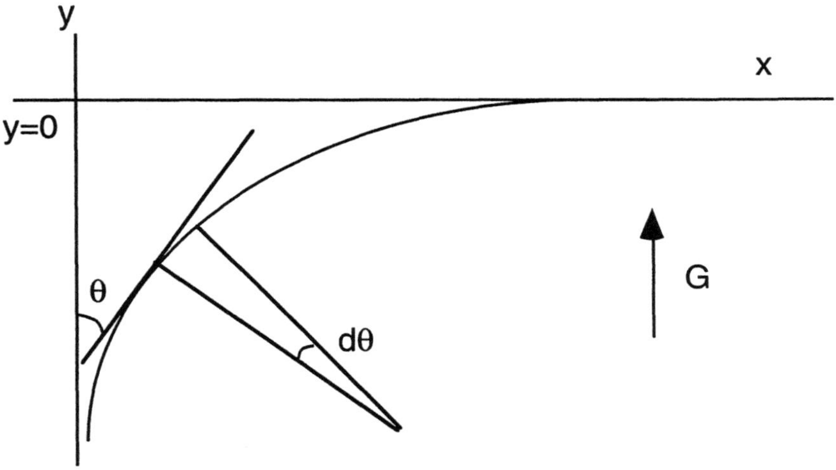

Figure 7-15 A grain boundary groove profile showing the coordinate y and θ [Gündüz and Hunt 1985]

7.7 CONCLUSIONS

Joining technologies such as soldering depend critically on the state of interfaces. The wetting behavior of solid-vapor, liquid-vapor, and solid-liquid interfaces is largely determined by their chemical constituency while interfaces involving a solid phase are also affected by crystalline anisotropy and stress state. These effects have been reviewed in the context of classical surface thermodynamics to illustrate the usefulness as well as the specific problems of selected experimental techniques for the measurement of these interfacial energies. It was shown that the Gibbs dividing surface provides a framework for the precise definition of surface energy. The conditions for equilibrium at a curved interface and the Gibbs-Thomson equation were derived for their usefulness in the measurement of solid-liquid interfacial energy. The experimental details of this measurement were also presented. For interfaces involving a solid phase the importance of crystalline anisotropy and distinction of surface energy and surface stress was emphasized. Several useful techniques for the measurement of liquid-vapor surface energies were described and yet other techniques can be found (6) for this purpose. Several methods for measuring solid-vapor surface energy were also discussed but these must be used with the understanding that slow solid state diffusion kinetics may affect results. Not as plentiful are techniques for measuring solid-liquid surface energy. However, the technique discussed, grain boundary grooving in a thermal gradient is considered to yield reliable surface energy data.

7.8 ACKNOWLEDGMENTS

This work was carried out in part at Ames Laboratory, which is operated for the U.S.Department of Energy by Iowa State University under contract no. W-7405-ENG-82. This work was supported in part by the Office of Basic Energy Sciences, Division of Materials Sciences.

REFERENCES

Adamson, A. W. 1967. *Physical Chemistry of Surfaces*. New York: Interscience Publishers.

Andereas, J. M., E. A. Hauser, and W. B. Tucker. 1938. *J. Phys. Chem.* 42:1001.

Bolling, G. F., and W. A. Tiller. 1960. *J. Appl. Phys.* 31:1345.

Fordham, S. 1948. *Proc. Roy. Soc. Ser.* A. 194:1.

Frazer, M. E., W. K. Lu, A. E. Hamelec, and R. Murarka. 1971. *Metall. Trans.* 2:817.

Gibbs, J. W. 1961. *The Scientific Papers of J. Willard Gibbs*. Vol. 1. New York: Dover Publications, Inc.

Gündüz, M., and J. D. Hunt. 1985. *Acta Metall.* 33:1651.

Hardy, S. C. 1977. *Phil. Mag.* 35:471.

Herring, C. 1951. *Physics of Powder Metallurgy*. Edited by W. E. Kingston. New York: McGraw-Hill. p. 143.

Huang, S. -C., and M. E. Glicksman. 1981. *Acta Metall.* 29:717.

Jones, D. R. H. 1974. *J. Metals. Sci.* 9:1.

Jones, D. R. H., and G. A. Chadwick. 1970. *Phil. Mag.* 22:291.

---. 1971. *J. Crystal Growth.* 11:260.

---. 1978. *Phil. Mag.* 27:569.

Keene, B. J., K. C. Mills, A. Kasama, A. McLean, and W. Miller. 1986. *Metall. Trans.* 17B:159.

Kelvin, Lord. 1871. *Phil. Mag.* 42:368.

Landau, L. D., and E. H. Lifshitz. 1958. *Statistical Physics.* London: Pergamon Press.

Lucassen, J., and R. S. Hansen. 1966. *J. Colloid. Inter. Sci.* 22:32.

Mann, J. A. Jr., and R. S. Hansen. 1963. *J. Colloid. Sci.* 18:805.

Mullins, W. W. 1963. In: *Metal Surfaces: Structure, Energetics and Kinetics.* Edited by W. D. Robertson and N. A. Gjostein. Metals Park, OH.: American Soc. for Metals. p. 17.

Murr, L. E. 1975. *Interfacial Phenomena in Metals and Alloys.* Reading, MA.: Addison-Wesley.

Mykura, H. 1961. *Acta Metall.* 9:570.

Nash. G. E., and M. E. Glicksman. 1971. *Phil. Mag.* 24:577.

Perepezko, J. H. 1984. *Mat. Sci. Eng.* 65:125.

Rayleigh, Lord. 1879. *Proc. Roy. Soc.* 29:71.

Shaefer, R. J., M. E. Glicksman, and J. D. Ayers. 1975. *Phil. Mag.* 32:725.

Shuttleworth, R. 1950. *Proc. Phys. Soc.* A63:444.

Singh, N. B., and M. E. Glicksman. 1989. *J. Crystal Growth.* 98:573.

Trivedi, R. 1975. In: Lectures in the *Theory of Phase Transformations.* Edited by H. I. Aaronson. Warrendale, PA.: Metallurgical Soc.

Turnbull, D. 1950. *J. Chem. Phys.* 18:768.

Udin, H., A. J. Shaler, and J. Wulff. 1949. *Trans. AIME.* 185:186.

Winterbottom, W. L., and N. A. Gjostein. 1966. *Acta Metall.* 14:1033.

8

Advanced Soldering Processes

J. L. Jellison, J. Golden, D. R. Frear, F. M. Hosking, D. M. Keicher, F. G. Yost

8.1 IMPETUS FOR CHANGE

Although solder technology's roots are buried in antiquity, solder processing innovations are occurring at a faster pace than at any time in history. There are three primary drivers for this: rapidly changing design requirements, increasingly stringent process and product reliability requirements, and environmental issues. Improving solder processes to meet the needs of electronic manufacturing, while addressing a heightened awareness of environmental issues, including compliance to governmental regulations, is a great challenge to technologists supporting solder processes and will require the best efforts of multidisciplinary teams.

Indirectly, the primary impetus for improved solder processes is the quest for increasingly higher density packaging. As the density of interconnects and conductor lines increase and the size of devices and electrical terminations decreases, the capability of solder processes must be improved. Also, high intensity heat sources are needed to overcome the problems encountered in attempting to solder next to heat sinks and heat pipes with conventional heat sources. Design flexibility is likely to become a more important issue. Examples are the mounting of devices on non-planar surfaces, greater reliance on the mechanical properties of solder joints, and greater use of multiple solder alloys in the same assembly to achieve design requirements.

The high density and complexity of many electronic packages have heightened the need for reliable interconnect processes. In some cases rework of solder joints is impractical. Rework is nearly always a costly approach to product reliability. The cost of manufacturing is certain to remain a major factor.

Federal, state, and local government regulations are mandating the elimination of most of solvents that have been used in solder flux removal. The use of chlorinated solvents and chlorofluorocarbon solvents will be very restricted by the late 1990s. Also, "labeling" regulations, in effect as of April 1993, will reduce customer acceptance of products manufactured with the use of chlorofluorocarbon solvents. Consequently, soldering processes that are compatible with substitute solvents will be needed. Moreover, some of the alternate solvents under consideration pose new problems such as incompatibility with device materials used in electronic packaging and low flash points, which are a safety hazard.

Finally, regulations to reduce the use of lead in electronic manufacturing are probable. Consequently, we can expect that processes in the future must perform reliably with lead-free solders.

8.1.1 What Would the Ideal Soldering Process be Like?

Based on the above discussion of demands on advanced solder processes, we can begin to speculate about what the ideal soldering process would be like. First, complete control of the spatial and temporal distribution of energy would be desirable. That is, we would like to put heat into the solder and substrate where and when we want to. For high density packaging of heat-sensitive devices, localization of heat clearly can become an issue. Conceptually, manipulation of the wetting and flow of solder by establishing dynamic temperature gradients provides a way of limiting solder to spatially defined areas and controlling final geometry of solder joints. Also, localization of heat facilitates step-soldering, i.e. the fabrication of assemblies with solders of different melting temperatures. Progress in the use of lasers and microwaves to provide control of heat during soldering will be discussed.

Second, the ideal solder process of the future would only utilize chemicals and materials that are environmentally friendly. Since most of the undesirable chemicals are used presently for flux removal, fluxless soldering or soldering with no-clean fluxes is appealing. Use of no-clean fluxes is an attractive near-term approach, but significant improvements in no-clean flux formulations probably will be needed to attain at least equally good solderability as has been achieved with more conventional mildly activated rosin and activated rosin fluxes. Some of the fluxless soldering approaches that will be discussed show promise of providing improved solderability compared to conventional processes. Especially interesting is the potential for fluxless soldering of more difficult-to-solder materials such as nickel, iron, and some of their alloys. Elimination of fluxes may also facilitate process automation. Flux residues tend to clog, gum up, and impede the smooth operation of complex moving mechanisms on automated solder assembly equipment.

The third major element of the solder process of the future will be the ability to precisely control solder placement. Current processes primarily operate on the basis that solder will be distributed to all surfaces that can be wet by the solder. To accommodate point heat sources and to achieve accurate distribution of solder in high density packages, advanced techniques for placing the solder in the electronic assembly are needed. Various methods of spraying, pouring, and spreading liquid solder and solid-liquid solder onto substrates have been conceived. Also, pick-and-place techniques for solid solder preforms have been developed. Development of new technologies for the accurate placement of solder is much needed. Some applications will probably require placement of two or more solder alloys within the same assembly.

Finally, the solder process of the future must support highly automated, agile manufacturing to provide design flexibility, achieve high quality solder joints through control of all important process factors, and control both set-up and in-process costs. The mass soldering processes in use today support high throughputs of hardware, but do not provide significant independent control of process factors, nor are they highly adaptive to changes in product design. The challenge of advanced processes will be the achievement of spatial control of both energy and the placement of solder without sacrificing throughput. Integration of the solder process with other assembly operations appears to be a particularly attractive approach to meeting this challenge. For example, if at the time a device has just been placed in an assembly, solder attachment was also completed, two operations could be combined into one. Thus the temporal and spatial control of the process required to produce high quality solder joints could be exercised for that individual device without compromising the quality of the rest of the assembly. This approach has been demonstrated in the case of pick-and-placed surface mount devices.

8.1.2 Possible Enabling Elements of Advanced Soldering Technology

Soldering processes and the associated processing steps for preparing parts and assemblies for soldering are undergoing rapid evolution. Significant changes are being conceived and developed involving preparation of surfaces, promotion of wetting, and heating of solder joints. Even radically new solder joint geometries are evolving, such as solder bumps. The advancements in these enabling technologies are as important as the development of solder assembly processes themselves.

8.2 ALTERNATIVE APPROACHES TO PROMOTE WETTING

The metallized layers wet by solder to form the interconnection between the various levels of an electronic package are typically heavily contaminated with metal oxides, carbon compounds, and other materials. Typical solderable metallizations are copper, nickel, or nickel plated with gold. Contamination on the metallizations occurs during extended exposure to the manufacturing environment. A metal surface contaminated by these materials cannot be wet by solder. However, once this surface contamination is removed, the solder wets the metallization and forms a metallurgically sound solder joint which will both hold the various electronic components in place and pass electrical signals.

Historically, oxides have been removed from metallized surfaces during solder processes by the application of liquid fluxes. Flux (from the Latin, meaning "to flow") is applied to a surface to assist in wetting by the solder. A flux consists of active agents dissolved or dispensed in a liquid vehicle. Typical vehicles are alcohol or water to which varying concentrations of acids or salts are added as activators. The function of the activators is to reduce base metal oxides. The acids most commonly found in fluxes are: abietic acid (also present in the resin itself), adipic acid, and hydrochloric acid. Zinc chloride is a salt commonly found in inorganic fluxes. The purpose of a flux is four-fold: 1) remove the oxide from the metallization, 2) remove the oxide on the molten solder to reduce the surface tension and enhance flow, 3) inhibit subsequent oxidation of the clean metal surfaces during soldering, and 4) assist in the transfer of heat to the joint during soldering. To remove contamination and create a sound metallurgical joint, only functions #1 and #3 are critical. For a more in-depth review of liquid fluxes, see Chapter 3 of this volume.

Fluxes can be applied by a variety of techniques. For electronic applications, the fluxes are brushed or sprayed onto the assembly. Often the flux has a low viscosity so that it can readily flow into gaps where solder joints will form and coat the entire assembly.

After a solder joint is formed, a flux residue remains. The residue consists of the carrier (resin) that has not evaporated, acid or salt deposits, and the removed oxides. The residue can be deleterious to the long-term reliability of an electronic package if it is not removed. The resin can absorb water and become an ionic conductor which could result in electrical shorting and corrosion. The residual activator can, over a period of time, corrode the soldered components and cause electrical opens. Furthermore, the flux residue will also be present underneath the components where inspection is difficult and the residue is difficult to remove.

Current practice is to remove flux residue using solvents that the flux and its by-products are soluble in. The most common solvent is the chlorinated fluorocarbon (CFC), CFC-113. The solvent is applied by spraying or dipping in a bath. This solvent works well in removing flux residue but has recently been found to be environmentally hazardous. Molina and Rowland [1974] found that CFCs deplete the stratospheric ozone layer.

International concern over the depletion of the ozone layer led to the enactment of the Montreal Protocol. This protocol, initially signed in 1987 by 24 countries, restricts the production and consumption of ozone-depleting chemicals. The protocol was later found not to be sufficiently restrictive, so the London amendment to the Montreal Protocol was signed in 1990 by 92 countries. Under this new amendment, CFCs will now be completely phased out by the year 2000 in developed countries and 10 years later in the entire world. In the U. S., the Clean Air Act was amended to contain provisions pertaining to the Montreal Protocol and stratospheric ozone protection. Excise taxes are applied on chemicals that deplete the ozone layer. By 1995 a use-tax on CFC-113 procurement will be $3.10/lb and increases by $0.45/lb every year thereafter [U.S. Dept. of treasury n.d.] In a typical surface mount soldering process on a printed circuit board, an average of 0.5 lbs of CFC-113 are consumed per board.

Therefore, there are strong economic and environmental incentives to develop either alternative cleaning processes that do not use ozone-depleting chemicals or, optimally, a soldering process that uses no flux and eliminates the need for hazardous solvents (fluxless soldering). Whatever the process, it is necessary that the solder joints be metallurgically sound and have at least the same cleanliness as those made with traditional flux and solvent cleaning procedures. Solvent substitution processes are being actively explored and are summarized in depth elsewhere [Oborny 1988; Oborny et al. 1990a]. The following is a summary of alternatives to traditional flux soldering for electronic applications. The areas covered in this section are "no-clean" fluxes and fluxless soldering. The fluxless soldering processes covered are the following: soldering under controlled atmospheres (including both acid vapor soldering and inert cover gases), laser ablative cleaning and soldering, laser soldering, ultrasonic soldering, and plasma cleaning and soldering.

8.2.1 No-Clean Fluxes

Fluxes are typically comprised of activators, resins, and solvents. The residual activators and resins that remain after soldering are called solids. These residues can cause reliability problems due to corrosion, electromigration, or electrical shorting if they are not fully removed after the soldering process. The problems with using solvents to remove the residues have been discussed above. One possible solution to the solvent problem is to use a flux with a decreased amount of solids; thus, the residues will not be sufficiently harmful so as to require removal. Such a flux has been termed "no-clean" flux or low solids flux. The elimination of post solder cleaning has the additional advantage of reduced operating costs. Cleaning equipment is no longer needed and the time for processing is reduced. Figures have been reported that cleaning processes account for two-thirds of the capital cost of a soldering line and 40% of the consumables (in the form of solvents) [Warwick and Wallace 1990]. More importantly, environmental concerns over the use of environmentally hazardous solvents to remove the flux are also eliminated.

A typical low solids flux consists of from 2 - 5 wt% solids content. The solids content is reduced by increasing the amount of solvent (isopropyl alcohol) and reducing the resin and activator (such as abietic acid). During the soldering operation the solvent evaporates, the activator attacks surface oxides and the vehicle encapsulates the activator residues. Activator systems have been specifically chosen for low solids fluxes to be relatively inert at electronic assembly operating conditions to minimize long term corrosivity concerns [Rubin and Warwick 1990]. The vehicle is either a synthetic resin, rosin, or modified rosin. Most low solids fluxes are halide-free and contain only organic acid types of activators. The low solids fluxes that contain rosin also contain halides. Therefore, a synthetic resin is a preferable choice for a low solids flux [Sholley 1989]. Due to the large amount of alcohol present in the low solids flux, other solvents or foaming agents are added to promote foaming so the flux can be applied to printed wiring boards for manufacturing. However, an excessive amount of foaming agents can deleteriously increase the corrosivity of the flux [Guth 1991]. The low solids flux comes in both bulk as well as a constituent of solder paste for wave soldering and reflow applications, respectively.

The optimal no-clean flux would leave non-corrosive and colorless residues. An advantage is that the low solids flux exhibits lower viscosity, which allows it to fill small joint volumes found in surface mount and through-hole assemblies.

The properties of no-clean fluxes have been summarized as follows [Guth 1991]:

- The concentration of flux needed for a given soldering operation must be optimized to produce adequate wetting yet minimize residues (too much low solids flux residue has the same result as a rosin-based "high solids" flux);
- Specific gravity is important for determining the amount of flux needed for a given application. The specific gravity of the flux is difficult to control due to the high vapor pressure of the alcohol present in the formulation;
- Water absorption is pronounced in a low solids flux because alcohol is hydroscopic. Excessive water in the flux will defeat the fluxing action and cause the metallization to oxidize;
- The low solids flux is difficult to foam (an important method of applying flux for wave soldering applications) due to the large alcohol content; and

- The process window for making a quality solder assembly is made more narrow by the demands placed on the manufacturing process by the low solids flux.

Despite these drawbacks, the benefits of low solids flux could be enormous.

8.2.2 Controlled Atmosphere Soldering

Controlled atmosphere soldering is exactly what the name suggests; the ambient air is replaced with an atmosphere that is either inert or acts to reduce contaminants present on the solderable surface (acid-vapor environment, atomic hydrogen, ionic hydrogen). Both of these types of controlled atmospheres are described in more depth below.

Inert Atmosphere Soldering

Inert atmosphere soldering can be used in wave soldering and pretinning operations or in conjunction with low solids flux applications. The areas to be soldered are heated in the absence of oxygen and therefore do not oxidize at processing temperatures. This is especially critical for low solids flux applications where the flux, because of the decreased volume of activators, has a lower activity for removing oxides. Decreasing the amount of oxide that forms during processing increases the possibility that the low solids flux will be adequate for the processing of quality solder joints. The increase in solderability with decreasing oxygen content has been modeled and is attributed to a decrease in surface oxidation rate [Jaeger and Lee 1992]. Another benefit of the use of an inert atmosphere is that oxide skin formation on the molten solder bath is reduced so that the tendency to form bridges is also reduced, resulting in a general improvement in solder quality [Klein-Wassink 1989].

Using an inert atmosphere also creates the possibility of soldering without flux. The three general areas where fluxless soldering in an inert atmosphere can be performed are pretinned surfaces, surfaces plated with noble metals (e.g. gold, platinum, palladium, silver), and etched and cleaned metals (e.g. copper) [Nowotarski and Konsowski 1990]. Pretinned surfaces are the easiest to solder without flux. When the surface is immersed into molten solder in an inert environment, the molten solder breaks up the surface oxide on the pretinned surface. In the inert environment, the resultant molten solder joint does not oxidize, thereby minimizing solder bridging and icicle formation. Noble metal surfaces can be soldered without flux in an inert environment because the oxide formation on these surfaces in minimal. Etched and cleaned metals, like copper, can be soldered without flux in an inert environment, but the soldering operation must occur shortly after cleaning or oxides will reform on the surface, thereby inhibiting wetting.

Typically, the inerting atmosphere is nitrogen. However, work by Esquivel and Chavez [1992] indicates that CO_2 works better than nitrogen because the larger molecular weight of the carbonic gas displaces the oxygen more efficiently.

The equipment used for soldering in an inert environment could be a solder dip pot, a reflow furnace, or wave soldering equipment. For the solder dip pot, a continuous flow of inert gas blankets the surface of the molten solder, creating an inert environment [Nowotarski and Konsowski 1990]. For the reflow furnace and wave soldering, enclosure hoods confine inert gases to the work area [Travato 1991; Hwang 1990].

Soldering in an inert environment in conjunction with no-clean fluxes can be a powerful method to eliminate a cleaning step after soldering. The method works best when the oxides present on the surfaces to be joined are minimal or nonexistent. Other means of removing the oxide without a cleaning step are necessary for the situations where a more stubborn oxide surface exists.

Acid Vapor Fluxless Soldering

Activated acid vapor fluxless soldering is a process in which liquid flux is replaced by a dilute solution of a vaporized acid in an inert carrier gas. By a strict definition, the acid vapor is a flux in that its function is to reduce surface metal oxides in order to enhance solderability. However, the acid vapor leaves little residue that requires cleaning, as long as acid concentrations are low.

The process of acid vapor soldering involves passing the carrier gas and acid vapor over the metallized surface to be joined. The acid chemically attacks the surface oxide as the metal is heated, removing the oxide from the surface. In this manner the surface is "activated" for wetting by the solder. The solder is introduced while the acid vapor flows across the joints; the protective atmosphere prevents reoxidation of the metallized surface during solder wetting. The process normally involves the use of solder preforms but it can be applied to wave soldering.

Recently, a number of studies have been performed to investigate the reactions involved in the formation of fluxlessly soldered acid-vapor solder joints [Frear et al. 1992; Frear and Keicher 1992; Keicher et al. 1992]. Wetting has been evaluated as a function of acid type and acid-vapor concentration under controlled experimental conditions. Although acid-vapor fluxless soldering has been proposed for commercial applications, the knowledge base of the reaction is not yet well understood.

Three requirements must be met to perform acid-vapor fluxless soldering:

- The acid must be able to reduce the relevant surface oxides in a vaporized form at soldering temperatures (~220°C) (*manufacturing concern*);
- The acid vapors must react quickly, on the order of a few seconds (*manufacturing concern*); and
- The vapors must be reactive enough to reduce metal surface oxides, but without causing corrosion on any part of the soldered assembly (*reliability concern*).

The acid vapors that best satisfy these requirements are shown in Table 8-1, below, along with their relevant physical properties. The forming gas, formic, and acetic acids are the most promising because of their low boiling point, which allows them to be in a gas form during solder processing.

To evaluate acid-vapor fluxless soldering, sessile drop tests have been performed using 60Sn-40Pb solder on Au/Ni and copper metallizations [Frear et al. 1992; Frear and Keicher 1992; Keicher et al. 1992].

The use of forming gas, or even pure hydrogen, has been proposed as a fluxless soldering atmosphere. However, no wetting was observed in tests with forming gas. Therefore, forming gas is not a suitable option for *in-situ* fluxless soldering. At 220°C hydrogen has a low reaction rate for reducing metal oxides and is consequently not suitable for soldering. At higher temperatures (>350°C) the hydrogen reduces metal oxides rapidly, but these temperatures could damage electronic components. However, if the metallized surfaces were cleaned by some other method prior to soldering, the forming gas could act as a blanket that displaces oxygen in the chamber and prevents oxidation during soldering.

Table 8-1 Characteristics of Reducing Agents Used for Acid Vapor Fluxless Soldering

Reducing Agent	Concentration in Volume %	Melting Point (°C)	Boiling Point (°C)
Forming Gas (H_2)	4% H_2 in argon	–259	–253
Acetic Acid (CH_3COOH)	0.5 – 2.5% in nitrogen	16.6	117.9
Formic Acid (HCOOH)	1.5 – 7% in nitrogen	8.4	100.7

Formic and acetic acid vapors wet the Au/Ni and copper metallizations well. Experimental sessile drop tests have been performed using a variety of acid-vapor concentration of both acetic and formic acids. The data from the sessile drop test that quantify the effectiveness of the acid-vapor atmosphere best are the time-to-wetting. Time-to-wetting tests are a measure of how fast surface contamination (e.g. surface oxides) can be removed with the assumption that no molten solder spreading can occur until all the surface contamination is eliminated. In an acetic acid vapor environment (above 1.5% by volume) metal acetates form, which increases the time to wetting. Concentrations of acetic acid below 1.5% are insufficient because the acid vapors do not remove the contaminants rapidly enough (times-to-wetting greater than 10 seconds). The optimum concentration of formic acid is 4% on copper surfaces and 1.6% on Au/Ni. Below these concentrations, the metal oxides are not removed rapidly. The wetting angle of 60Sn-40Pb solder using both acid vapors at their optimum concentrations resulted in wetting angles that are adequate for electronic solder joints (10 – 20°).

Although the use of acid vapors for fluxless soldering is still in the experimental stage, the preliminary results are promising for their use on electronic applications. There are commercial applications that use a combination of nitrogen and formic acid vapor to create an oxygen-free environment [Nowotarski and Konsowski 1990; Elliot 1991; Raleigh and Melton 1991; Walsh 1991]. The equipment typically consists of an enclosed wave soldering system where, through a series of doors and preferential vacuum pumps, a dilute solution of formic acid in nitrogen replaces the air atmosphere. A belt pulls the assemblies to be soldered through the system and across the solder wave. However, the commercially available systems are not a truly fluxless soldering operation because an abietic liquid flux is applied to the surface of the boards by an ultrasonic atomizer.

Atomic and Ionic Hydrogen as Reactive Atmospheres

Although thermodynamic data show the potential for reducing metallic oxides in hydrogen or vacuum, the rate of such reactions at typical soldering temperatures, 200–300°C, is very low. A reactive atmosphere is consequently required if fluxless soldering is the objective. Ionic or atomic gases have the greatest potential for effectively reducing surface oxides. For example, thermodynamic data reveal that atomic and ionic hydrogen at 250°C have a greater copper oxide reduction potential than molecular hydrogen. The non-molecular gas species can be generated by hot filament, microwave, or cathodic plasma processing. The result is a "clean," oxide-free surface that can be fluxlessly soldered. Heating can be done directly with the process gas or by means of a secondary heat source (e.g. infrared or laser heating).

Cursory fluxless wetting experiments have been conducted [Hosking et al. 1992] with hot filament and plasma generated atomic/ionic hydrogen on oxygen-free, high conductivity copper substrates. The "reactive" atmospheres chemically reduce the surface oxide and permit direct wetting of the exposed base metal by Sn-Pb solder. The hot filament process has also been demonstrated on printed wiring board coupons. Further work is necessary to develop the technique into a commercial process. The effect of the reactive gas on electronic packaging materials must be clearly defined. More effective techniques to deliver heat to the soldering area are needed since the filament and plasma processes are neither easily controlled nor efficient at the lower temperatures which are used to solder.

Fluxless Laser Soldering in a Controlled Atmosphere

Laser soldering normally involves the melting of a preplaced, or fed, solder preform with a directed CO_2 or Nd:YAG laser beam [Jellison and Keicher 1988; Lea 1989]. Rapid heating and cooling of the solder joint produces a very fine microstructure with reduced intermetallic formation. The process is highly automated and offers improved process control. Higher strength solder joints are possible since the localized heating of the laser allows the use of solder alloys with higher melting temperatures without damaging nearby materials. Solder mixing with other solder alloy compositions of nearby solder joints can also be avoided because of localized melting.

A new development in laser soldering is the incorporation of controlled atmospheres with laser heating to give a fluxless soldering process [Hosking and Keicher 1992]. The process was originally developed to eliminate entrapped, corrosive flux residues while making closure joints for an electronic package, Figure 8-1. It is being extended to soldering discrete leaded devices to printed wiring boards. The study used a 100 watt, continuous wave (CW) Nd:YAG laser to produce fluxless laser soldered joints. Experiments were conducted to show the feasibility of laser soldering Ni-Au electrodeposited Kovar™ closure joints. Kovar™ (nominally 53Fe-29Ni-17Co-0.5Mn, wt%) is generally very difficult to directly wet with solder unless very aggressive, corrosive fluxes are applied. Since most active fluxes are not permitted during microelectronic soldering of high reliability components, Kovar™ is typically electroplated with a 5.0 μm (200 μin.) layer of nickel followed by a 1.2 μm (50 μin.) overplating of gold to enhance its solderability. The Ni-Au metallization is critical to the success of fluxless soldering. The noble gold surface prevents oxidation of the underlying nickel layer before and during soldering.

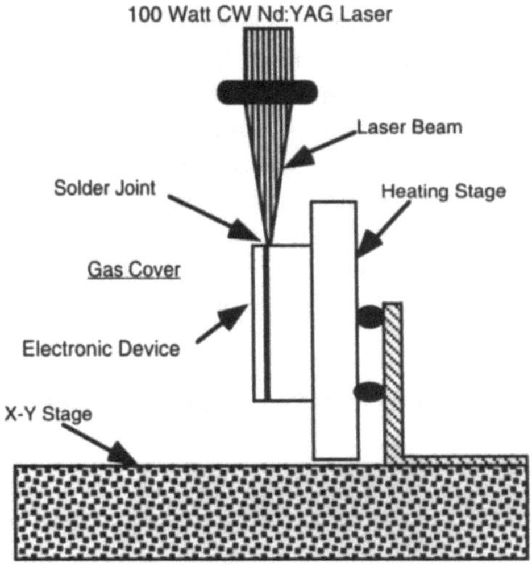

Figure 8-1 Process schematic for laser soldering of electronic module closure joints

A slightly reducing, controlled atmosphere (3 – 5 vol% hydrogen in argon) was introduced over the fixtured parts to protect the base and solder alloy surfaces during heating and promote solder wetting by preventing the oxidation in air of the underlying nickel layer. Thermodynamic data suggest that the higher temperatures of laser soldering (> 300°C) in an inert or reducing atmosphere can significantly enhance the reduction of most metallic oxides (e.g. copper, nickel, iron, tin, and lead) [Bredzs and Tennenhouse 1970; Milner 1958].

The reflective properties of gold inhibit the absorption of the laser energy and reduce the heating efficiency of the process. The laser beam should be consequently directed on the solder preform for best soldering. For 1.06 μm light, gold has an absorption factor of 0.012. The Sn-Pb eutectic solder alloy, however, has an absorption factor of 0.28, a nominal absorption increase of twenty.

Ni-Au plated Kovar™ solder joints were visually inspected to assess the quality of solder wetting/spreading. Good solder joints were generally obtained at higher laser power settings (> 65 W) with Sn-Pb and In-Pb-Ag solder alloys. Laser spot size diameter and part travel speed were varied from 0.38 – 0.88 mm and 2 – 17 mm/s, respectively. Poor wetting generally occurred at lower power levels with very irregular to negligible solder flow. Figure 8-2 shows a Ni-Au plated Kovar™ and 63Sn-37Pb (wt%) solder joint that was made at 100 W with a 0.88 mm laser beam spot size and 2.0 mm/s travel speed. The joint exhibited good wetting with a very fine microstructure with small Au-Sn and Ni-Sn intermetallic precipitates dispersed throughout the joint. Extensive intermetallic growth, however, can be a problem during multiple-pass laser soldering or reflowing with extended dwell times since higher heat inputs and melting of the base metal can occur. The subsequent joint microstructure is significantly altered and can detrimentally affect its mechanical properties.

Future work in this area should include the investigation of fluxless soldering of more difficult-to-wet base metals such as copper, nickel, and iron alloys. The effect of atmosphere, laser parameters, solder alloy composition, and base metal surfaces on wetting are critical variables to the success of the process. Heat management will also be an important element. As electronic circuit boards move toward more advanced mixed technology with very fine surface mount pitches and the increasing restrictions of environmental regulations narrow the manufacturing window, fluxless laser soldering will become a more viable process.

8.2.3 Laser Ablative Cleaning and Soldering

An attractive aspect of the use of lasers for the soldering of electronic assemblies is that the laser cannot only be used as a source of heat and cause reduction of oxides as described above, but it can also physically remove the oxides by ablation. Laser ablative fluxless soldering (LAFS) involves the use of very short pulses of high peak power laser radiation to remove metal oxides and other contaminants from joining surfaces [Peebles et al. 1991]. The laser ablative cleaning step can induce wetting of copper, nickel, Kovar, and aluminum substrates by tin-based solders. Laser ablative cleaning and the subsequent joining process must be performed under a noble gas environment in a glove box or other type of controlled atmosphere chamber to prevent oxidation of the cleaned surfaces that are heated for soldering. A Q-switched pulsed laser beam is mechanically scanned across each joining surface. The very short (~10 nsec) pulses of laser radiation rapidly heat and vaporize very thin layers (100 – 5000 Å) of surface material. The vaporized material rapidly condenses on contact with the cooler ambient gases above the substrate surface to form submicron solid particles. These particles are carried away from the ablated surface by a flowing stream of the inert gas. All particles are removed from the inert gas stream by filtration. Joint formation can be accomplished by solder tinning and reflow in the same inert gas environment.

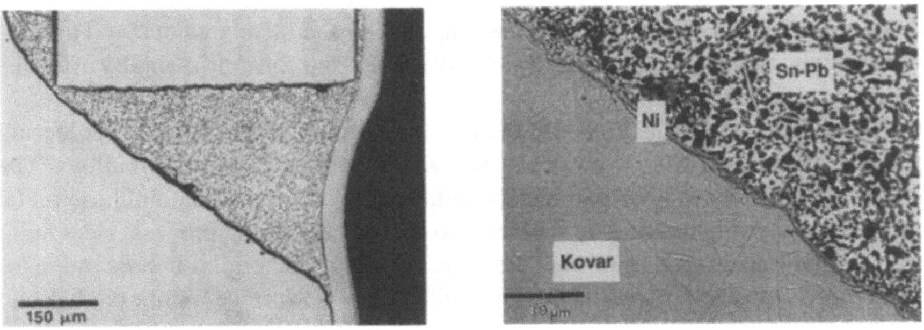

Figure 8-2 Interfacial optical image of a Ni-Au plated Kovar™ and 63Sn-37Pb solder joint made with a CW Nd:YAG laser at 100 W, 0.88 mm spot size, and 2.0 mm/s travel speed

LAFS exhibits some very attractive process characteristics. No reactive chemicals of any type are required in the formation of a solder joint. As a result, there is no possibility that LAFS will leave residue on the surface that could accelerate corrosion or any other form of material degradation. The incident laser energy can be adjusted to allow selective removal of the metal oxide film without significantly overetching the metal substrate. Etching terminates at the interface between the oxide film and the substrate due to the high surface reflectivity of the metal [Peebles et al. 1990]. Because etching is limited to the oxide film, the waste generation rate of the laser ablative cleaning process is extremely small. For an oxide film formed on a copper surface in air at room temperature (typically 100 Å thick), the waste generation rate for the ablation step in this fluxless soldering process will be roughly 60 mg of metal oxide per square meter of surface cleaned.

Experimental results show that oxygen-free high conductivity (OFHC) copper can be wet very well by 60Sn-40Pb (wt%) solder following removal of the surface oxide and other contamination by laser ablation [Bredzs and Tennenhouse 1970]. The wettability by 60Sn-40Pb solder of OFHC copper surfaces cleaned under argon or helium gas by laser ablation equals, or exceeds, the wettability obtained on this substrate using a common RMA flux in air. For copper substrates cleaned under argon gas, a cross section of the solder-copper interface reveals the presence of a Cu_6Sn_5 intermetallic layer between the copper and the solder. No voids between the copper and the solder are observed. The presence of the Cu_6Sn_5 intermetallic layer at the interface indicates that a good metallurgical bond has been formed between the solder and the substrate. Copper substrates cleaned by laser ablation under helium gas show a highly convoluted surface topography. The increased apparent surface wettability induced by laser ablation of the copper substrates under helium gas is the result of enhanced spreading of the solder over the laser-roughened surfaces as a result of capillary forces.

Of course, the same difficulties described earlier in this chapter for laser soldering still apply (i.e. line of sight to joint) to LAFS, but results indicate that the process may be very suitable for certain applications. LAFS is an appropriate fluxless soldering process that may have commercial applications.

8.2.4 Fluxless Ultrasonic Pretinning

Ultrasonic processing provides an alternative to soldering with a flux. The technique utilizes acoustic energy which propagates through the solder bath and removes oxides and other contaminants from metallic surfaces during tinning. A magnetostrictive or piezoelectric device produces the acoustic pulse which is transmitted by a probe or horn into the molten solder. This energy is then coupled to the workpiece and yields a surface which is easily solderable.

Ultrasonic soldering is not a new technology. First mention of ultrasonic soldering dates back to pre-World-War-II Germany and includes a German patent for an ultrasonic soldering iron [Antonevich 1976]. Ultrasonic soldering was later developed to join aluminum and stainless steel, which is generally very difficult to do without the use of very active and corrosive fluxes [Humcke 1976; Schuster and Chilko 1975]. Compact ultrasonic soldering pots are currently used to coat (or "tin") a variety of parts ranging from ultra thin stainless steel, copper alloy tubing, and glass-to-metal seals on microelectronic devices [Fuchs 1990]. These ultrasonic systems typically operate at 20 kHz frequency and up to several hundreds of watts of electrical power. Their use in the electronics industry has been limited to the tinning of conductor leads on passive devices. Active devices present a potential problem of damage to ball bond interconnects during ultrasonic coupling and require a case-by-case evaluation to determine ultrasonic applicability.

Engineering studies in the field of ultrasonic soldering, especially tinning, have largely dealt with its application for a specific product. Relatively little work has been directed toward gathering the more fundamental properties of solder wetting and joint formation by ultrasonic surface activation. Recent work [Vianco and Hosking 1992] has focused on quantifying the fluxless wetting properties of simple substrate geometries with a "point source" ultrasonic horn. "Point source" implies that the horn size is smaller than the geometry of the substrate. Results from these early studies are being extended to the fluxless tinning of prototype microelectronic devices and boards. That research is being done with a commercial ultrasonic soldering machine. The American Technology, Inc. system has dual horns which are attached to 20 kHz piezoelectric elements which operate at a power level of 0 – 300 W.

Preliminary point source experiments have been conducted on oxygen-free, high conductivity copper (OFHC copper) in a molten bath of pure tin (T_M = 232°C). Test temperatures were varied between 245°C and 300°C. Samples were preheated in the hot tin bath for two minutes. Ultrasonic activation was applied after the preheat treatment. The experimental setup and key process parameters are shown schematically in Figure 8-3. The ultrasonic probe was excited by a piezoelectric source with a resonant frequency of 20 kHz and an operating power of 20 – 70 W. Power level stability was ± 2.5 W. Tests were conducted with each sample perpendicularly positioned to the axial centerline of the horn (ø = 90°). The "front" surface of the test samples faced the horn. The "back" surface was opposite of the horn. The horn and sample parameters were: A – the horn and sample separation; B – the vertical position; C – the immersion position; and the lateral position of the horn in front of the sample (at the centerline of the copper substrate). Fluxless solder wettability tests were initially performed with: A – 3.3 mm; B – 14.2 mm; C – 19.1 mm; 25 – 60 W power settings; 5, 15, and 30 s power-on times.

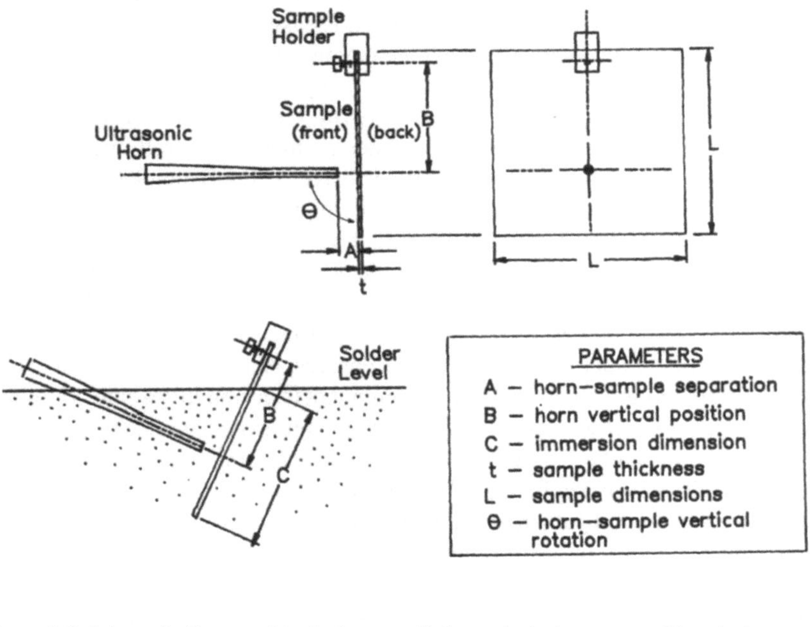

Figure 8-3 Schematic diagram of the "point source" ultrasonic tinning system with typical process parameters

Visual inspection of ultrasonically tinned substrates revealed tin films on both the front and back surfaces of the copper samples. Wetting of the front surface extended well beyond the immediate vicinity of the ultrasonic horn. These observations suggest that wetting is not caused entirely by direct mechanical erosion or cavitation from the horn. Rather, the copper oxide film is also removed by means of acoustic waves propagating within the substrate. This observation implies that, although surfaces unexposed to the horn can be wetted by tin, the ability to deposit a tin film will depend on the base material properties and geometry of the substrate.

Erosion occurred only along the front surface which faced the probe. Erosion was inhibited by the free surface at the tin bath and copper interface. The wetted surfaces were coated in some regions with a thin (2.5 μm) layer of tin, while other areas exhibited a "donut-like" morphology (Figure 8-4). Cross sections through these "donut" regions reveal an annular, wetted region that surrounds and is surrounded by the non-wetted substrate. These features were observed in areas where no erosion occurred. Further work needs to be done to better understand the fundamental mechanism behind this "donut" wetting process.

Image analysis is also being done to quantify the fraction of immersed surface area which the tin wets. A recent experiment [Schuster and Chilko 1975] subjected five samples to ultrasonic activation at 55 – 60 W for 15 s at 282°C. The resultant tin coatings exhibited some experimental scatter (7%, or one standard deviation of the percent area wetted) on both front and back surfaces. Although this data give only a first-order approximation of the experimental error, the low value gives confidence in the reproducibility of the wetting results.

The wetted surface-area percent of tin on copper as a function of applied power is shown in Figure 8-5 for substrates dipped at 245°C with a power-on time of 30 s. The back surfaces showed a smaller wetted area than the front surfaces at power levels between 30 and 50W. Wetting was more dispersed on the back surfaces. The difference in wetting suggests that the propagation of acoustic energy within the substrate is variably dependent on the power level. Visual observations of coated samples also show that while erosion extended over a greater surface area with increased power, the depth of the eroded surface appears to be unaffected by the power level.

The effects of power-on time and tin bath temperature are shown in Figure 8-6. The area fraction of wetting increased as the power-on time is increased for temperatures of 245°C and 276°C. Wetting was similar for each time period at 296°C. At 15 s, the variation in percent area wetted between the different temperatures was significantly reduced. As the time period and temperature were increased, less variability in wetting was observed. However, a qualitative visual evaluation of the samples revealed that the degree of erosion and, to a lesser extent, the depth of erosion on the front surface, increased as the power-on time and bath temperature were increased. Erosion was negligible on the back side for all test conditions.

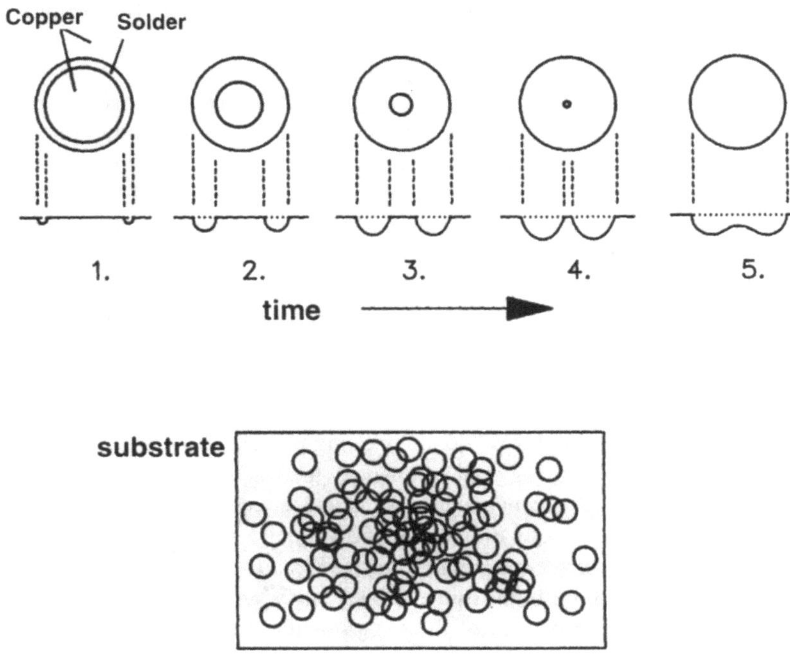

Figure 8-4 Schematic of the "donut" morphology exhibited by optimally coupled flat copper coupons exposed to ultrasonic tinning

The above fluxless wetting results demonstrate that power level, power-on time, and tin bath temperature can be optimized to uniformly coat OFHC copper substrates without severely eroding the workpiece. Acoustic transmission within the substrate permits wetting of tin on surfaces not directly exposed to the ultrasonic horn. Further work is necessary to determine the effect of substrate thickness on ultrasonic tinning. Thinner substrates appear to be more susceptible to erosion on their front face than thicker geometries. The "donut" morphology also becomes smaller and less defined as thickness is decreased.

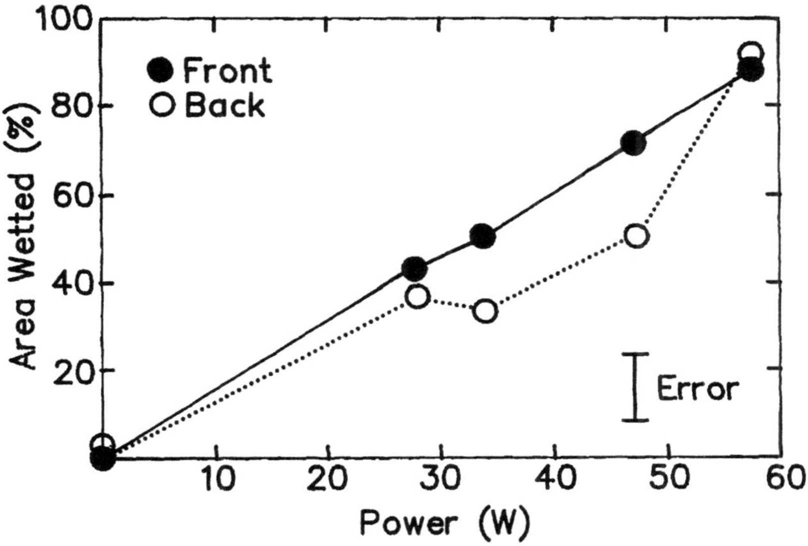

Figure 8-5 Area wetted as a function of ultrasonic power level for 1.6 mm-thick OFHC copper by tin at 245°C and a power-on time of 30 s

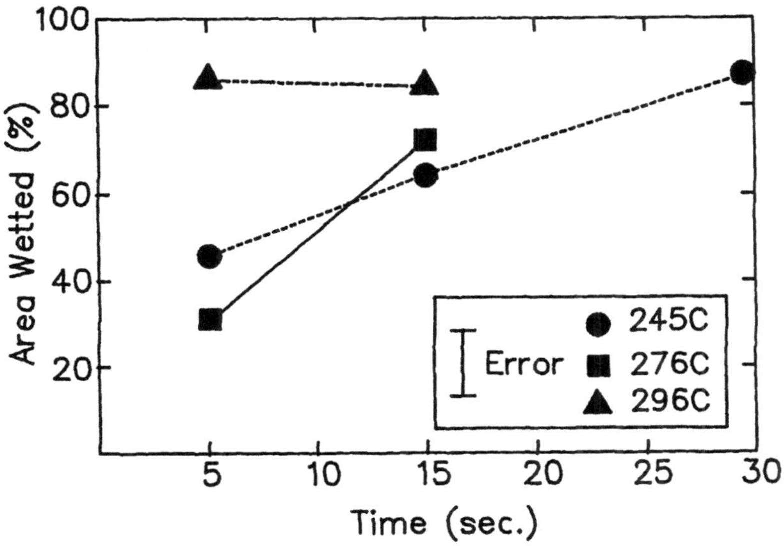

Figure 8-6 Area wetted on the front surface of a 1.6 mm-thick OFHC copper sample as a function of power-on time and tin bath temperature at a power level of 55 – 60 W

The above work with the commercial ultrasonic soldering system has been expanded to investigate other simple geometries. There are plans to demonstrate solder tinning on leaded electronic devices and larger structures such as conductive lines on printed wiring boards and structural pipes. Additional studies are being proposed to also evaluate difficult-to-wet base materials (e.g. ceramics, aluminum, nickel alloys, ferrous alloys) and other solder alloys besides the traditional Sn-Pb eutectic or near-eutectic alloy which is being heavily used throughout the electronics industry. With the push to find Pb-free solder alloy replacements, it is critical that these new alloys be included in the ultrasonic soldering study. The end result will be the development of an advanced environmentally compatible soldering technology which will meet electronics manufacturing specifications.

8.3 ALTERNATE HEAT SOURCES

8.3.1 Laser Heat Sources

Use of lasers in soldering has received considerable attention since the late 1970s. Actual production experience with laser soldering is limited. As electronic devices become increasingly smaller and packaging densities become increasingly higher, there is a growing realization that hand soldering and mass soldering techniques place limits on electronic assemblies. Laser soldering has been investigated as a possible alternative to conventional soldering techniques, and some of the advantages of laser soldering have become recognized. The ability to precisely manipulate a small diameter, chemically inert heat source by optical means has been the principal motivation in the investigation of laser soldering. The intrinsic high irradiance of lasers generally has not been utilized in laser soldering, since power densities seldom exceed 10^3 W/mm^2. The reasons for selecting laser soldering over conventional soldering processes relate to the fact that lasers can accurately deliver small quantities of energy to the precise locations to be soldered within dimensions required of microminiature devices. Laser soldering is not yet competitive with current mass soldering processes and has only been used in special applications, especially in the repair of high cost assemblies.

Types of Lasers

Both gas and solid state lasers have been used as heat sources for soldering. Gas lasers are primary represented by CO_2 lasers [Lea 1989; Suenaga et al. 1988; Whitehead and Polijanczuk n.d.; Hall and Whitehead 1987; Hartmann et al. 1991; Stow 1984; Lish 1986], although other gas lasers, such as argon lasers and copper vapor lasers, are potential alternates. Ruby lasers and Nd:glass lasers were among the first solid state lasers to find application to metal processing. However, Nd:YAG lasers have become the dominate solid state lasers and have been evaluated extensively for laser soldering [see Lea 1987, 1989; Suenaga et al. 1988; Whitehead and Polijanczuk n.d.; Lish 1986; Zakel et al. 1992; Miller 1988; Okino et al. 1986; Mizutani 1987; Miura 1988a,b; Burnett 1991].

The appropriateness of a laser for a soldering application is primarily differentiated on the basis of the laser's characteristic wavelength. The characteristic wavelengths for laser soldering fall within 0.4 – 10.6 μm. In general, laser energy of shorter wavelengths is absorbed better by metals, especially those that are good conductors such as copper and gold, whereas organic materials tend to absorb longer wavelengths [Lish 1986; Miller 1988; Greenstein 1989]. Some organic materials may actually transmit laser beams of shorter wavelength (0.4 – 1 μm). For applications where direct absorption of the laser energy by metallic leads or conductors is desired, lasers of shorter wavelengths have the advantage.

However, CO_2 lasers (10.6 μm wavelength) have proved to be very efficient for laser soldering in applications where the laser energy is absorbed by organic solder flux. Shorter wavelengths (0.4 – 1.06 μm) are needed for fluxless soldering since the laser energy is absorbed directly by either the solder or metallic conductors [Jellison and Keicher 1988]. A disadvantage of longer laser wavelengths is the greater tendency for charring substrates containing organic materials that are common in printed wiring boards [Lea 1987; Miller 1988].

There are some operational advantages and disadvantages related to laser wavelength. First, optical components such as lenses and prisms are less expensive and readily available if they can be made of glass or quartz, as is the case for shorter wavelengths in the 0.4 – 1.06 μm range. In the case of CO_2 lasers (10.6 μm wavelength), these optical components must be made of special inorganic crystals such as zinc zelenide. However, the fact that ordinary glass and most plastics greatly attenuate light transmission in the infrared range enhances laser safety when CO_2 lasers are used. Simple glass or plastic shields that are transparent in the visible range block most of the laser energy emitted by CO_2 lasers.

Finally, fiber optic technology has only been developed for shorter wavelengths (0.4 – 1.06 μm range). Quartz fibers are being used in production applications to transmit laser beams of hundreds of watts over distances of meters. This capability can be important to the automation of laser soldering.

Advantages of Lasers

In the ideal soldering process, complete control of the spatial and temporal distribution of energy is desirable. Lasers can come close to this ideal. The following summarizes some of the advantages that have been cited for lasers as heat sources for soldering [Whitehead and Polijanczuk n.d.; Lea 1987; Miller 1988; Miura 1988b]:

1) Localization of controlled amounts of heat makes lasers especially suitable for soldering heat-sensitive devices;
2) Because of the small total amount of heat, residual stresses in soldered assemblies can be minimized;
3) The short duration of melt-freeze cycles typical with laser soldering minimizes the formation of intermetallics;
4) The high irradiances achievable by lasers can be used to overcome heat sinks so that solder joints can be made in otherwise difficult locations;
5) Since laser energy can be applied to any location to which light can be transmitted, more complex, three-dimensional assemblies can be soldered;
6) Because laser soldering does not require hard tooling, agile manufacturing is well supported; design changes can be accommodated through changes in the electronic data base;
7) Since light is inert, lasers are intrinsically environmentally friendly heat sources;
8) Because of the ability to focus laser beams to small dimensions, lasers are ideally suited for soldering dense packages, where devices and conductors are closely spaced;
9) Since each solder joint can be made independently without disturbing others, joints can be made with different solders, and devices can be added after adjacent portions of a substrate have been populated, including the other side of the substrate;
10) Because of the spatial control of heat afforded by lasers, wetting of and flow of the solder can be controlled, including minimization of solder bridging between adjacent conductors;
11) Because the amount of heat at each joint can be accurately controlled during laser soldering, the potential for process reproducibility is high;
12) Laser soldering supports the potential of *in-situ* solder joint inspection to achieve high product reliability; and
13) Laser soldering is highly amenable to automation through the combination of optical manipulation of beams and computer control.

Second Harmonic Generation for Low Average Power Lasers

One intrinsic problem encountered in laser soldering is the wide variation in absorptance of the different materials used in soldered assemblies. Solders and most metal conductors are highly reflective, whereas fluxes and board materials may either readily absorb laser energy or transmit it, depending on laser wave length. Jellison and Keicher [1988] found that laser process control can be improved by using a shorter wavelength. They employed frequency doubling to obtain 0.5 μm light from a Nd:YAG laser. Energy absorption is not only much improved for the metals at this wavelength compared to that for 1.06 μm, but the difference in absorption among the various materials is reduced, which makes heating of the metallization without damaging substrates or charring flux easier to accomplish.

Many of the high-conductivity materials used in soldering applications become increasingly reflective with increased wavelength; see Figure 8-7 [Charschan 1972]. Lasers are available that emit light with wavelengths from the deep UV through the visible and into the far IR region of the optical spectrum. As laser wavelength increases through the visible spectrum and into the near IR region, the absorption of this radiation by materials such as copper and gold is reduced to only a few percent. There is no improvement in material absorption as the wavelength increases further. To enhance the solderability of these materials with lasers several schemes can be used. The parts can be pretinned to improve absorption at the Nd:YAG wavelength; modified joint geometry can be used to conduct heat into the joint region via the solder; the absorption of solder flux at CO_2 wavelength can be used to conduct heat into the joint, or a visible wavelength laser can be used to increase the energy absorbed directly into the reflective materials. It has been shown that a laser wavelength in the blue-green region of the visible spectrum greatly enhances the solderability of these materials [Jellison and Keicher 1988]. With this one requirement there are several choices available. Unfortunately, an additional requirement is that the preferred laser source also operate in a quasi-CW mode of operation. The quasi-CW mode of operation allows the solder joint to be heated continuously and avoids the multiple rapid thermal cycles associated with a short pulse length source. In addition, the higher irradiance typically associated with the short pulse length lasers can cause vaporization of irradiated materials. Since a quasi-CW source is preferred in laser soldering, many of the shorter wavelength lasers are eliminated. Although argon lasers can generate tens of watts at wavelengths of 514.5 and 488 nm, these lasers are very inefficient, costly and require high maintenance. One alternative method for producing shorter wavelengths is to utilize non-linear optical methods on the Nd:YAG laser to generate harmonic wavelengths; in particular, to produce the second harmonic wavelength of the Nd:YAG laser at 0.532 μm.

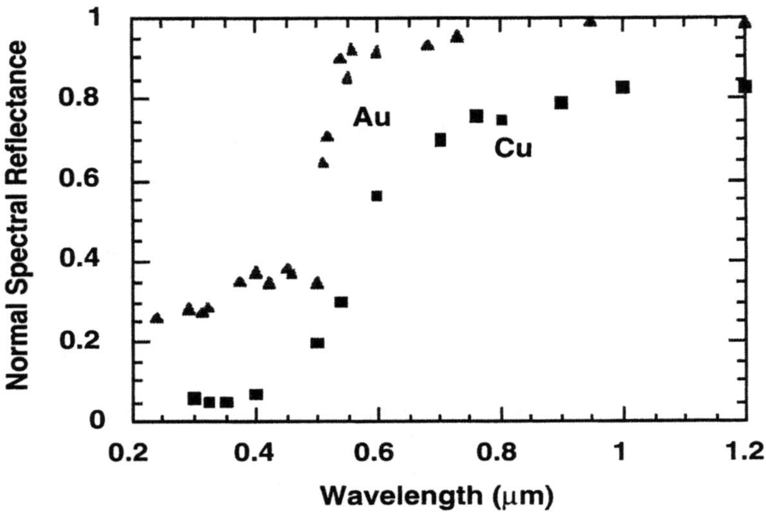

Figure 8-7 Reflectance as a function of wavelength for gold and copper at room temperature [Charschan 1992]

The phenomenon of frequency conversion of light, in particular, second harmonic generation (SHG) of laser light, was first demonstrated more than three decades ago [Franken et al. 1962]. When the laser light propagates through a non-linear optical (NLO) material, the time-varying electric field associated with the laser beam causes time-varying polarization within the crystal to occur. Since NLO crystals lack an inversion symmetry, a directionally dependent restoring force on the electrons exists, leading to an antisymmetric polarization. It is this antisymmetric time-varying polarization that leads to frequency conversion of the original electric field. High efficiency single pass SHG is normally only obtainable with lasers that achieve high peak powers, typically by operating in a pulsed modes. However, for laser soldering, CW mode operations is generally desirable. To overcome the physical limitations requiring high peak power, innovative techniques have been used. These techniques have taken advantage of the power enhancement provided by resonators. The techniques of particular interest are intracavity and pumped cavity SHG. These techniques have allowed efficient SHG to be achieved with low peak power CW lasers.

The most widely used of these techniques is intracavity SHG [Smith 1970]. A diagram depicting an intracavity SHG scheme is shown in Figure 8-8. The cavity is composed of two mirrors, M_1 and M_2, which provide optical feedback through a gain medium, i.e. the Nd:YAG rod. A polarizer is included in the laser cavity to enhance the laser gain preferentially in one polarization plane. Since SHG is polarization-dependent, a polarized beam is required to maximize the available energy for frequency conversion. To perform the parametric process of SHG, the NLO crystal is placed in the laser cavity. By placing the NLO crystal in the laser cavity, the electric field amplitude is increased significantly since the laser output power is a small fraction of the circulating power within the laser cavity. CW power outputs of 9 watts have been demonstrated using this technique [Perkins and Fahlen 1987].

Another method used to efficiently produce SHG from low peak power sources is a pumped cavity SHG technique [Ashkin et al. 1966]. This method, shown in Figure 8-9, also utilizes the power enhancement provided by a resonant cavity. In this configuration, a laser with the desired wavelength for frequency conversion is used to pump a second cavity. Using this technique, an output power of 6.5 W of SH light has been obtained from a pump laser power of 18 W [Yang et al. 1991].

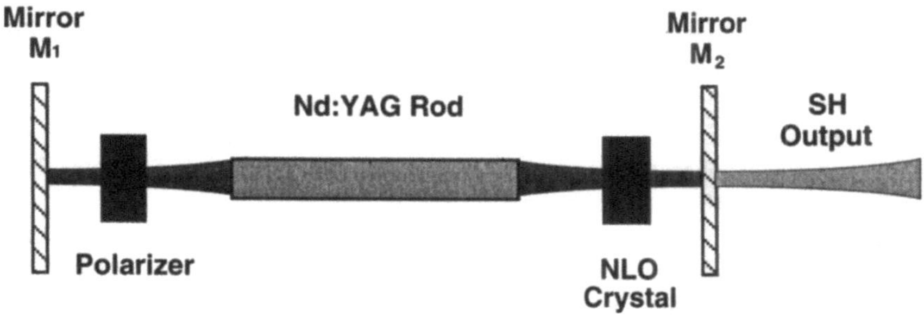

Figure 8-8 Diagrammatic representation of intracavity SHG scheme for Nd:YAG laser

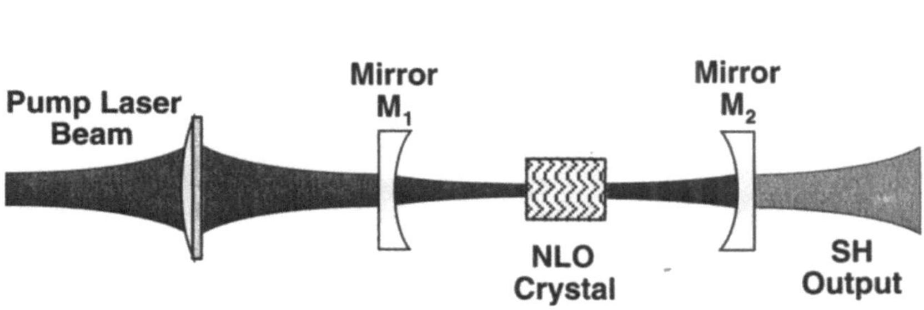

Figure 8-9 Diagrammatic representation of pumped cavity SHG scheme for generic laser input

Both intracavity and pumped cavity techniques which provide efficient SHG have been demonstrated. The intracavity technique provides a simpler technique to achieve a low peak power SHG, and several low peak power pulsed systems are commercially available which utilize this scheme. More recently, the desire for compact, single frequency visible light sources has motivated researchers to investigate the pumped cavity scheme. Although more complicated, the pumped cavity SHG technique provides a more elegant solution to the problem and allows optimization of the pump laser beam characteristics to be independent of the SHG process.

Fiber Optic Beam Delivery for Laser Soldering

In traditional laser soldering systems, the collimated output of the laser is directed by one or more mirrors to an objective lens, which focuses the beam onto the joint to be soldered (Figure 8-10a). In a fiber-delivered laser soldering system (Figure 8-10b), the beam is first focused into one end of an optical fiber. The laser energy is transported down the length of the fiber and is emitted from the other end. The beam is recollimated and then focused onto the joint by the output optics assembly.

The difference in this scheme is the optical fiber. A mechanistic description of the fiber is beyond the scope of this discussion and can be found elsewhere [Smith 1990]. The functional features of the fiber are three-fold: 1) transmission properties, 2) optical properties, and 3) flexibility. Once energy is coupled into the fiber, it is transmitted essentially unattenuated to the other end (typical transmission is > 99.8% per meter). Therefore, long fibers can be effectively used (a 50-meter fiber will lose less than 10% of the energy).

The characteristics (diameter, cone angle) of the exiting beam are determined principally by the core diameter of the fiber, and by the optics used to focus the laser output into the fiber. In particular, they do not change as the output power of the laser is varied. Therefore, within limits (see below) any desired focused spot size can be achieved by properly selecting the focal lengths of the output lenses.

Optical fibers are generally quite flexible (even with protective jackets, bending radii of 8" or less are common). Therefore, they can deliver the beam through tortuous routes (conventional beam delivery requires a clear "line-of-sight" between each pair of optical elements). Further, changes in position or angle of the focused beam can be accomplished by moving only the output assembly.

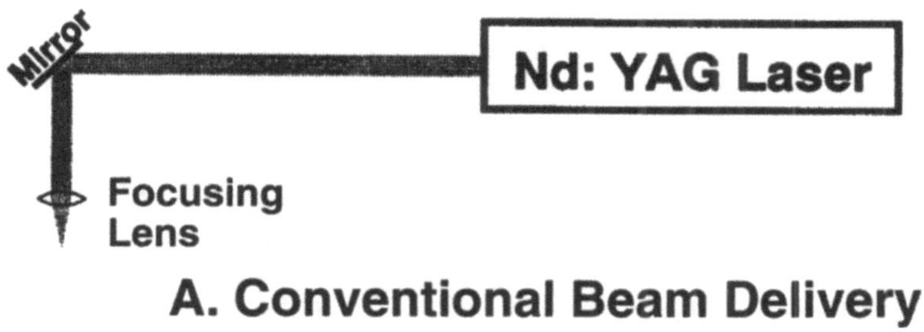

Figure 8-10a Traditional laser soldering system

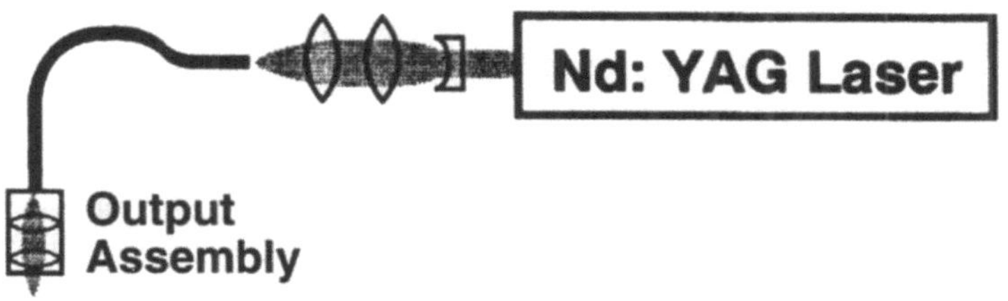

Figure 8-10b Fiber-delivered laser soldering system

Advantages and Applications

Fiber-delivered lasers have several advantages when used for soldering. Foremost is the ability to easily manipulate the output assembly to move and/or change the orientation of the beam with respect to the part. This is especially useful in soldering, where a non-orthogonal beam orientation is often important for a well-formed joint [Miller 1988; refer also to Figure 8-3]. Also, the small size of the output assembly makes it easily adaptable to robotics systems or to integration into placement systems. The latter point will become increasingly important with finer-pitch designs, where soldering at the time of insertion may become mandatory. Several commercial systems already combine a fiber-delivered laser with placement mechanisms [Newbury 1990; Iverson 1992].

The use of a fiber also allows the laser to be placed at a distance from the soldering point, which facilitates integration into an automated assembly line. One such line, for fully automated manufacture of automotive relays, incorporates three fiber-delivered lasers for soldering. Remote location can also protect the laser optics from damage caused by fumes or particulate debris that is common in automated assembly line operations. In extreme cases, the entire laser can be placed in a separate room with controlled environment.

Simultaneous soldering of multiple joints can be easily achieved by energy-sharing devices, which deliver the laser energy through two or more fibers. In addition to increasing throughput, simultaneous soldering of opposite joints can reduce the problems of beam liftoff on SMT devices.

Disadvantages and Limitations

The principal concern of fiber optic laser beam delivery is damage to the optical fiber, especially at the input end, where the laser energy is focused to a small, intense spot. A variety of damage mechanisms can occur [De Hart 1992], especially as a result of misalignment or dust.

Also, the fiber puts a lower limit on the focused spot size, based on its optical properties [Kugler 1990]. With current fiber technology, the minimum focused spot size is about 200 mm (.008 in.). While this is not a problem for most current requirements, it could become a limiting factor as lead pitch shrinks.

Finally, the fiber is not an imaging device, and cannot transmit an image of the part to a coaxial viewing system located at the laser. Coaxial CCTV systems are available which attach to the output assembly; in some cases the added weight and size hinders the flexibility of the system as a whole.

Approaches to Increasing Solder Process Throughput

In view of the many advantages of laser soldering, why hasn't the process gained major acceptance in the manufacture of electronic assemblies? Laser soldering has not yet been implemented in a way that it can compete with more conventional soldering processes in terms of throughput. The strength of laser soldering as a process, i.e. the ability to precisely place energy at the precise point where it is needed, is also its weakness; for typical applications, many joints need to be soldered and the time required to solder thousands of individual joints becomes intractable. A number of approaches to overcoming this disadvantage of laser soldering are beginning to emerge.

Multiple splitting of the beam through fiber optics, sometimes termed multiplexing, has been demonstrated. Multiplexing systems with up to 100 fiber outputs from a single laser have been demonstrated. This is still only a few point heat sources compared to the number of solder joints that must been made on many assemblies. However, high powered CW lasers now becoming available will easily support simultaneous soldering of 1000 or more electronic solder joints. There do not appear to be any fundamental barriers to additional multiplexing to support simultaneous soldering of several joints at a time.

Use of computer-controlled scanning systems [Suenaga et al. 1988] is perhaps the most common approach currently used to making multiple solder joints. Similarly, automation of the process by means of computer-controlled robots has been demonstrated [see *Assembly Automation* 1988]. Combining computer control and scanning with multiplexed fiber optic delivery is a promising approach to producing many joints with good spatial and temporal control of energy.

Another approach to increasing the throughput is to use a cylindrical beam to redistribute the spatial distribution of energy from the laser beam into a linear beam that can simultaneously irradiate a whole row of solder joints at a time [Suenaga et al. 1988]. Okino et al. [1986] developed such a system for soldering fine pitched quad packages to printed wiring boards; it simultaneously soldered two sides of a package and, then, following a 90-degree rotation of the optical assembly, soldered the other two sides.

A conceptually similar approach to the use of a linear beam is the rapid linear scanning of a circular beam along a row of solder joints. Suenaga et al. [1988] implemented this approach by using a galvanometer mirror to oscillate the beam back and forth along a row of leads. Similarly, Hall and Whitehead [1987] used a combination of two scanning mirrors to achieve higher scanning speeds than could be achieved by X-Y or rotary motion of the parts. The application of both approaches to simultaneous irradiating a row of solder joints is illustrated in a cartoon of laser soldering of a double-in-line package (see Figure 8-11).

Another approach for producing multiple beams has been patented by Locke [1987]. Locke conceived the projection of multiple beamlets from a laser by using a mirror with holes arranged to correspond to the desired spatial distribution of energy for the device to be soldered.

Perhaps the ultimate approach of increasing laser soldering throughput is to integrate the process into the rest of the assembly. Rather than using laser soldering as a distinct and separate process, laser soldering could be introduced along with another assembly step such as during the picking and placing of surface mount devices. If combinations of the above methods were used to simultaneously solder all leads of a device at the time they were placed on the circuit, little additional processing would be added since laser soldering could be accomplished within a fraction of a second.

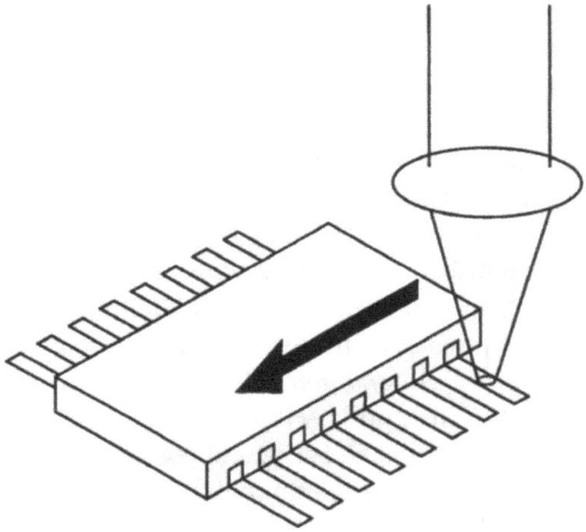

Figure 8-11b Laser soldering of a double-in-line package

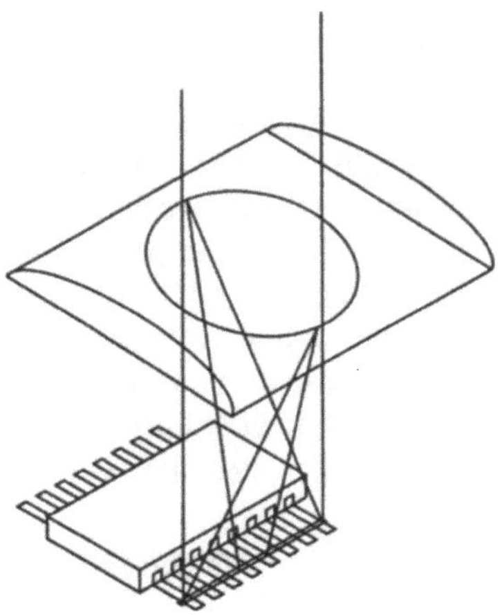

Figure 8-11c

Mirror Scanned Beam

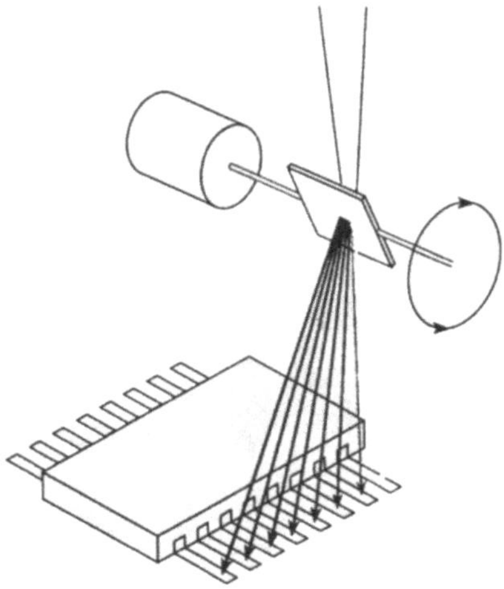

Figure 8-11d

Fiber Optics Multiplexing

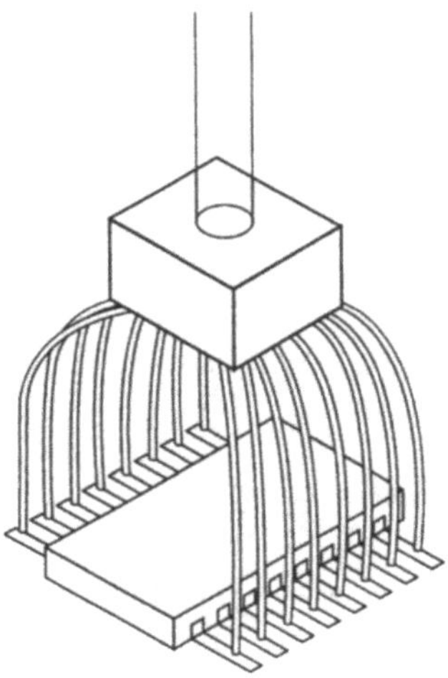

Figure 8-11e

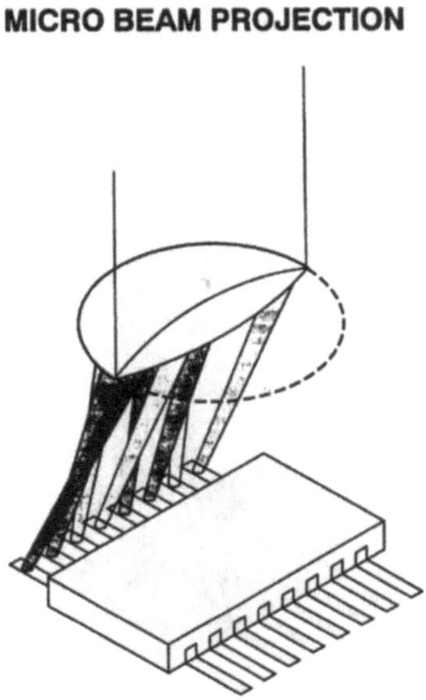

Figure 8-11f

8.3.2 Focused Microwaves

Microwave technology developed in the Ukraine is becoming available to U.S. researchers. Microwave generators, called gyratons, provide up to 35 kW of radiation at frequencies from 35 to 100 GHz focused to a beam with a divergence of approximately one degree. Ukrainian researchers are believed to have applied this technology to the soldering of printed wiring boards [Skyyarevich et al. 1992]. By controlling the particle sizes and properties of the constituents in special solder pastes they report that microwave energy is preferentially coupled to the solder and heating of the metallization and devices is minimal.

8.4 Solder Bump Technology

Most of today's printed wiring boards run at clock rates not exceeding 50 MHz, while evolving integrated circuits are capable of operation at hundreds of MHz. However, to exploit the higher frequencies offered by chips, circuit interconnects must be on the order of 5 cm or less to avoid propagation delays. Printed wiring boards are usually much larger than this. The problem is intensified by the ever increasing input/output (I/O) interconnections per chip, which currently number in the hundreds. These requirements have brought about a need for high density interconnects, compact packaging technologies, and associated thermal management.

An important advance toward these modern technology needs was made almost 25 years ago when L. F. Miller developed the controlled collapse chip connection, or C4 technology [1969]. This technology evolved from earlier work at IBM that became known as "flip chip" or "solder bump" technology. It involves the face-down mounting of an integrated circuit through soldered connections to the substrate. This is an extremely efficient way of locating chips in close proximity to reduce propagation delay and cross talk between substrate conductor lines. The C4 configuration (henceforth called solder bump) is shown schematically in Figure 8-12 with the integrated circuit chip oriented face-down, or flipped. There is one solder bump for each I/O; these can be arranged around the chip perimeter or in an area array. The substrate can be any of a variety of materials and contains a matching pattern of metallization pads that are wettable by molten solder. To assemble, the chip is brought into contact with the substrate and the solder is reflowed. Capillary forces at each solder bump act in unison to achieve precise alignment of chip and substrate. This feature makes the technology particularly useful for aligning optoelectronic devices. A crucial part of this technology is the metallization layer that provides both adhesion to the substrate and a wettable surface for the solder. These layers have been produced by plating and vapor deposition techniques with a metallization scheme such as chromium, copper, or gold. Likewise, the solder may be deposited by plating or vapor deposition and then reflowed. The technology is capable of high joint strength, fatigue life, and manufacturing yields but limited heat dissipation characteristics.

Normally, solder bumps are produced before chips are tested and separated from the wafer– a fact that limits the usefulness of solder bump technology. While several schemes have been advanced to put solder bumps onto chips, the most exciting of these is solder jetting technology [Wallace 1989]. The Laboratory for Manufacturing and Productivity at MIT, The Automation and Robotics Research Institute at the University of Texas at Arlington, and MicroFab Technologies, Inc. in Plano, TX are independently developing jetting technology for electronic applications. The idea is to dispense small (0.127 mm, or 5 mil) drops of molten solder at circuit locations that require it. This is done by pumping molten solder into a chamber which is coupled to a piezoelectric device, as shown in Figure 8-13. When the piezo device is fired, a pressure pulse causes the emission of a single droplet through an orifice and towards a substrate. When the droplet lands, surface tension of the solder is large enough to prevent spatter and spherical shape is maintained. By aiming a stream of such droplets, a solder pattern can be printed at very high speed. Figure 8-14 shows a group of solder bumps that were jetted onto a copper surface in a regular array having a 200 μm pitch. If this jetting technology is used to apply solder bumps to an integrated circuit, some method of providing wetting to the aluminum metallization must be found. As mentioned above, this is currently done with intermediate metallization layers; however, it may be possible to jet solder directly onto bare aluminum if the oxide is stripped just prior to droplet impact.

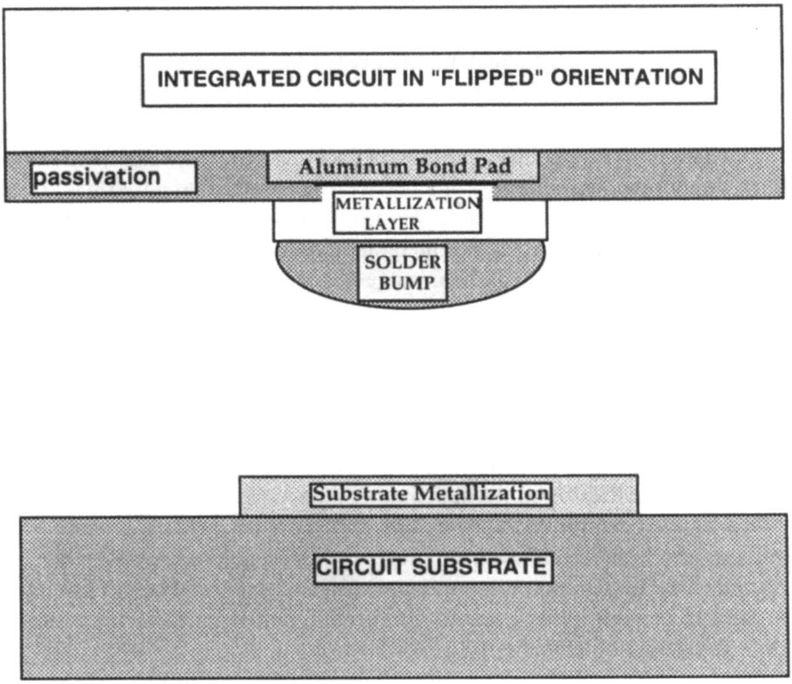

Figure 8-12 Illustration of a solder bump cross section on an integrated circuit and a matching substrate

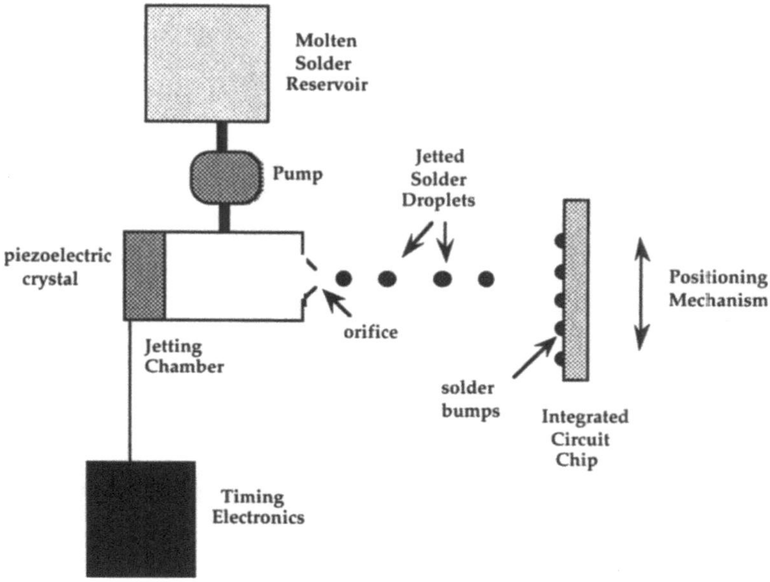

Figure 8-13 Illustration of apparatus necessary for the jetting of solder droplets onto a substrate

Laser ablative cleaning technology for soldering applications has been under development at Sandia National Laboratories for two years [Peebles et al. 1991]. During this period it has been demonstrated that a variety of metal and alloy substrates can be cleaned and soldered without chemical flux of any kind. We have demonstrated that electronic parts, rejected for poor solderability, can be reclaimed through laser ablative cleaning and robotic retinning. In addition, we have demonstrated that these processes can be performed rapidly and automatically, with negligible waste generation. Currently, we are coupling the ablative cleaning potential of a directed laser beam and robotic manipulation and assembly of parts with solder droplet jetting technology. A computer-driven X-Y table will rapidly transport a printed wiring board (or any other circuit substrate or module) beneath the cleaning laser and solder jet such that all appropriate circuit real estate is covered by solder without ever using a chemical flux of any kind. These processes will be controlled so that the time elapsed between cleaning and jetting a particular area is less than that characteristic of reoxidation and/or contamination of the circuit surface in the chosen process ambient.

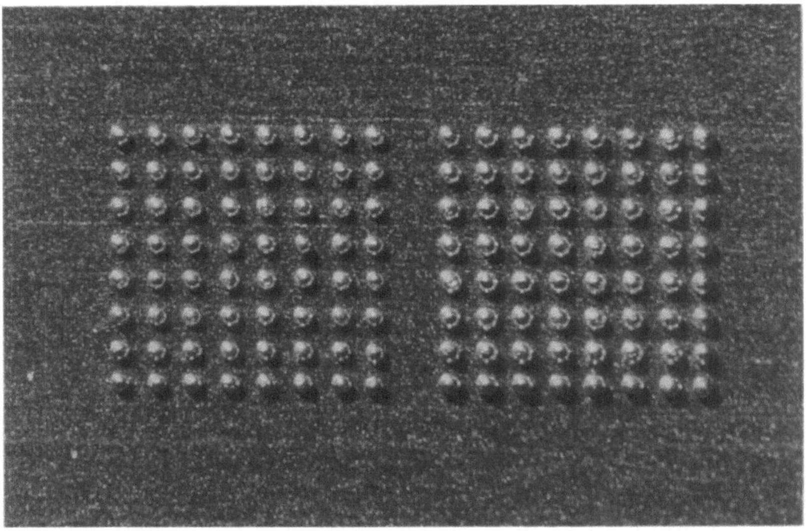

Figure 8-14 Photograph of solder bumps arranged in a square pattern having a 200 μm pitch courtesy microfab technologies, inc.

8.5 FUTURE DIRECTIONS FOR SOLDER PROCESS TECHNOLOGY

Several trends in solder processing are evident. First, electronic packaging is requiring much higher computational throughput and higher clock rates; thus, the pitch and geometry of solder joints is changing. There is an increased reliance on the mechanical properties of solder joints, and solder bridging is potentially a greater problem. Consequently, solder joints must be highly reliable. This is the underlying reason for accurate spatial and temporal control of heat. No doubt, infrared light heat sources will continue to be important, but as technology improves, the application of lasers to soldering will become common. The second trend is the use of "fluxless soldering" methods. Both the desire to reduce environmentally undesirable chemicals from the manufacturing process and the need to improve process robustness is driving the development of new approaches to improve solderability.

The degree to which adaptive control and *in-situ* inspection can be applied to future soldering processes is less clear. Certainly, the goal is to develop process technology that is intrinsically reliable and tolerant to variations in boards and devices. Laser processing appears to be especially amenable to the development of adaptive control methodologies. The use of infrared signals and reflected laser power for *in-situ* inspection and process control has received considerable attention [Fugate and Felty 1987; Vanzetti and Alper 1990; Horneff et al. 1988; Bayba 1989–90].

Finally, as fluxes are eliminated and lasers become an important heat source for solder processing, there will be an opportunity for integrating soldering with other assembly steps. This will allow the designers and manufacturing engineers to concentrate on other aspects of the product.

8.6 ACKNOWLEDGMENTS

The authors gratefully acknowledge the contributions of H. C. Peebles and P. T. Vianco to the research reviewed. Also, we would like to thank P. T. Vianco for an exceptionally helpful review of the manuscript. This work was performed at Sandia National Laboratories under the auspices of the U. S. Department of Energy Contract Number DE-AC04-76DP00789.

REFERENCES

Antonevich, J. 1976. *Welding J.* July 1976. pp.200s-207s.

Ashkin, A., G. D. Boyd, and J. M. Dziedzic. 1966. *IEEE J Quant. Elec. QE.* 2:109.

Assembly Automation. 1988. pp. 195-96.

Bayba, A. J. 1989-1990. *Harry Diamond Laboratories Report.* Nos. 1-17.

Bredzs, N., and C. C. Tennenhouse. 1970. *Welding J.: Welding Research Supplement.* pp. 86-90.

Burnett, T. 1991. *Proc. Technical Program NEPCON West* '91. 2:1372-81.

Charschan, S. S. 1972. *Lasers in Industry.* Toledo, OH.: Laser Institute of America.

De Hart, T. G. 1992. *Photonics Spectra.* Nov. 1992. pp. 107-10.

Elliott, D. A. 1991. *Proc. NEPCON West'91.* Anaheim, CA. pp. 1754-58.

Esquivel, A. E., and E. Chavez. 1992. *Proc. NEPCON West '92.* Anaheim, CA. pp. 219-27.

Franken, P. A., A. E. Hill, C. W. Peters, and G. Weinreich. 1962. *Phys. Rev.* 127:1918.

Frear, D. R., and D. M. Keicher. 1992. *Proc. NEPCON West '92.* Anaheim, CA. pp. 1704-15.

Frear, D. R., F. M. Hosking, D. M. Keicher, and H. C. Peebles. 1992. *Materials for Elec-*

tronic Packaging. Stoneham, MA.: Butterworth-Heinemann.

Fuchs, F. 1990. Proc. *14th Electronic Manuf. Seminar.* NWC. China Lake, CA. pp. 295-307.

Fugate, G. W., and J. R. Felty. 1987. *1st International SAMPE Electronics Conf.* pp. 92-104.

Greenstein, M. 1989. *Applied Optics.* 28(21):4595-603.

Guth, L. A. 1991. *Solder Joint Reliability.* Edited by J. H. Lau. New York: Van Norstrand Reinhold. pp. 143-72.

Hall, D. R., and D. G. Whitehead. 1987. *Proc. 4th International Conf. Lasers in Manuf.* pp. 133-38.

Hartmann, M., H. W. Bergmann, and R. Kupfer. 1991. SPIE Vol. 1598: *Lasers in Microelectronic Manuf.* pp. 175-85.

Horneff, P., H. G. Treusch, D. Knoedler, and W. Moeller. 1988. SPIE Vol. 1022: *Laser Assisted Processing.* pp. 48-51.

Hosking, F. M., and D. M. Keicher. 1992. *Proc. NEPCON West '92*. Anaheim, CA. pp. 1739-48.

Hosking, F. M., D. R. Frear, and P. T. Vianco. 1992. *International J. of Conscious Manufacturing*. 1(1):53-64.

Humcke, R. 1976. *Welding J.* March 1976. pp. 191-94.

Hwang, J. S. 1990. *Printed Circuit Assembly.* 4:30-2, 37-8.

Iverson, W. R. 1992. *Assembly.* June 1992. pp. 26-7.

Jaeger, P. A., and N. -C. Lee. 1992. *Proc. NEPCON West '92.* Anaheim, CA. pp. 394-404.

Jellison, J. L., and D. M. Keicher. 1988. *International Symposium and Exhibition Microjoining '88.* Cambridge, England: The Welding Institute.

Keicher, D. M., C. Hernandez, D. R. Frear, and F. M. Hosking. 1992. *Proc. NEPCON East '92*. Boston, MA.

Klein-Wassink, R. J. 1989. *Soldering in Electronics.* Ayr, Scotland: Electrochemical Publications, Ltd. pp. 502-03.

Kugler, T. R. 1990. *Presentation at ICA-LEO '90.* Boston, MA.

Lea, C. 1987. *Hybrid Circuits.* No. 12. pp. 36-42.

—. 1989. *Sold. and Surf. Mount Technol.* 2:13-21.

Lish, E. F. 1986. *IEEE Electronic Components Conf.* pp. 79-87.

Locke, E. V. 1987. *U. S. Patent, Multi-Lead Laser Soldering Apparatus.* Pat. No. 4,682,001. July 21, 1987.

Miller, C. B. Jr. 1988. *Hybrid Circuit Technology*. July 1988. pp. 27-31.

Miller, L. F. 1969. *IBM J of Research and Dev.* May 1969. pp. 239-50.

Milner, D. R. 1958. *British Welding J.* pp. 90-105.

Miura, H. 1988a. *Electronic Manuf.* Feb. 1988. pp. 43-9.

— 1988b. *Welding International.* No. 5. pp. 467-72.

Mizutani, T., K. Hasumi, M. Nakamura, and Y. Hujigaya. 1987. *IEEE 3rd Electronic Manuf. Technol. Symposium.* pp. 12-15.

Molina, M.J., and F. S. Rowland. 1974. *Nature.* No. 249. pp. 810.

Newbury, Phil. 1990. *Circuits Assembly.* Oct. 1990. pp. 33-5.

Nowotarski, M., and S. G. Konsowski. 1990. *Surf. Mount Technol.* 10:50-3.

Oborny, M. C., E. P. Lopez, and D. R. Frear. 1990. *Proc. International Conf. on Pollution Prevention: Clean Technologies and Clean Parts.* Washington, D. C.

Oborny, M. C., E. P. Lopez, D. E. Peebles, and N. R. Sorenson. 1990. *Proc. International Workshop on Solvent Substitution.* Phoenix, AZ.

Oborny, M. C. 1988. *Proc. 3rd Interagency Cleaning and Contamination Control Seminar.*

Okino, K., K. Ishikawa, and H. Miura. 1986. *IEEE International Electronic Manuf. Technol. Symposium.* pp. 152-57.

Peebles, H. C., N. A. Creager, and D. E. Peebles. 1990. Proc. 1st International Workshop on Solvent Substitution. Phoenix, AZ. p. 1.

Peebles, H. C., D. M. Keicher, F. M. Hosking, P. F. Hlava, and N. A. Creager. 1991. *Proc. Laser Mater. Processing Symposium.* 74:186-203.

Perkins, P. E., and T. S. Fahlen. 1987. *J. Opt. Soc. Am. B.* 4:1066.

Raleigh, C., and C. Melton. 1991. *Proc. Surface Mount International Conf.* pp. 432-8.

Rubin, W., and M. Warwick. 1990. *Surface Mount Tech.* 4:42-6.

Schuster, J., and R. Chilko. 1975. *Welding J.* Oct. 1975. pp. 711-17.

Sholley, C. 1989. *Electron. Manuf.* 35:32-3.

Skyyarevich, V., Detkov, A., Shevelelev, M., Decker, R. 1992, Proc. Mat. Res. Soc. Symp. '92, vol. 269, pp. 163-9.

Smith, R. G. 1970. *IEEE J Quant. Elec. QE.* 6:215.

Smith, W. J. 1990. Editor, *Modern Optical Engineering.* 2nd ed. pp. 265-70.

Stow, J. R. 1984. *Electronic Packaging: Materials and Processes.* pp. 9-14.

Suenaga, N., M. Nakazono, and H. Tsuchiya. 1988. *Welding International.* No. 3. pp. 269-76.

Travato, R. A. Jr. 1991. *Circuits Assembly.* 2:48-52.

U. S. Department of the Treasury. *IRS Publication 510: Ozone Depleting Chemicals.*

Vanzetti, R., and R. I. Alper. 1990. *Proc. Technical Program NEPCON East '90.* pp. 895-902.

Vianco, P. T., and F. M. Hosking. 1992. *Proc. NEPCON West '92*. Anaheim, CA. pp. 1718-29.

Wallace, D. B. 1989. *J. Electronic Packaging*. 3:108-11.

Walsh, T. 1991. *Proc. NEPCON West '91*. Anaheim, CA. pp. 599-622.

Warwick, M., and R. Wallace. 1990. *Electron. Production*. 19:3.

Whitehead, D. G., and A. V. Polijanczuk. *Fundamentals of Phase Change HTD*. 143:47-56.

Yang, S. T., C. C. Pohalski, E. K. Gustafson, R. L. Byer, R. S. Feigelson, R. J. Raymakers, and R. K. Route. 1991. *Opt. Lett.* 16:1493.

Zakel, E., G. Azdasht, and H. Reichl. 1992. *Hybrid Circuits*. No. 27. pp. 7-13.

9

Reliability-Related Solder Joint Inspection

Don Lewis Millard

9.1 ABSTRACT

This chapter provides a historical perspective of solder joint inspection and its present role in the establishment of manufacturing techniques, processes, and controls that produce very highly reliable electronic assemblies. The scope of this chapter is limited to solder joint inspection and, therefore, does not address electrical testing technologies. The criteria necessary for evaluating the reliability of currently produced solder joints are presented and compared with the data necessary for compliance with modern specifications. A review of potential inspection technologies and methods is provided, along with sample inspection data acquired from Rensselaer Polytechnic Institute's effort in the Air Force's MANTECH for Advanced Data/Signal Processing program. A discussion of automation of the inspection process includes an example of a novel technique that was developed to reduce the amount of effort and set-up time necessary to acquire inspection data using x-ray equipment. An assessment of automatic defect detection capabilities of various inspection technologies is provided to indicate how the inspection data can be used to detect manufacturing defects and aid in process control. Considerations of the future requirements for inspection conclude with an illustration of how inspection data can be further used to produce a non-destructive 3-D visualization of the solder joint, which can then be used to aid in the prediction of a design's reliability.

9.2 MOTIVATION

The solder joint is one of the most significant elements associated with the quality of today's fine pitch electronic assemblies. Figure 9-1 shows the trend of increased use of fine pitch surface mount packaging technologies, which has significantly altered the functional requirements of a solder joint. The current definition of a good solder joint has evolved to include mechanical and thermal integrity, along with the historical electrical concerns. This evolution has mandated the development of criteria beyond that of electrical continuity, involving that which assesses the metallurgical, mechanical, and thermal properties of the solder joint. While the functions of inspection are to validate the design while assessing and assuring product quality, increased utilization of fine pitch surface mount technology (SMT) has rendered conventional human-oriented visual evaluation of electronic assemblies difficult, if not impossible. Today's most prevalent method of evaluating electronic assemblies still involves visual inspection, using low power magnification, and is based upon industrial or government specifications. This visual inspection typically applies general procedures that deal more with ambiguities than product specifications. Today's fine pitch electronic assemblies require more sophisticated aids with greater resolution to inspect and automatically control product quality.

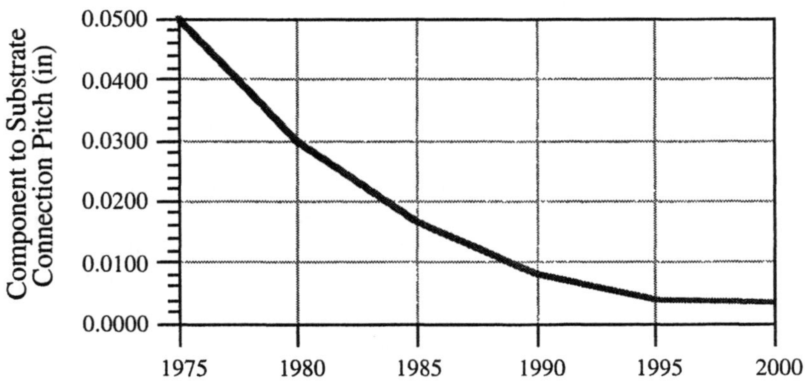

Figure 9-1 Trends in electronics packaging density

Non-destructive metrology tools have long been used to assess the reliability of electronic assemblies, yet only recently have companies brought systems to the marketplace to specifically inspect the finer pitch surface mount technologies produced today. The group of automated inspection tools developed in response to this need include three-dimensional vision, thermal signature, and radiographic systems. Most of these are computer-aided systems that use very complex signal/image processing software.

The role of inspection is also changing with the evolving fine pitch electronic assembly technologies. Its original role was to locate joints that needed rework, yet its present function is increasingly to serve as a process control audit that locates portions of the process that need attention. Inspection is thus helping to shift the emphasis from merely shipping an acceptable product to manufacturing a high quality product the first time.

9.3 INSPECTION CRITERIA

The historical definition of a highly reliable solder joint has been based on the outward appearance of the solder volume. Unfortunately, today's SMT assemblies may have interconnections which are not visible or are obscured by nearby components. Therefore, the historical definition does not address all of the issues necessary to develop a current solder joint defect data base. Conventional databases for defect standards include anomalies that rely upon the capabilities and interpretations of a human inspector. These defects are typically the result of gross differences in the assembly process or can have a direct correlation with diminished electrical integrity. The validity of such cosmetic-oriented, visually detected defect criteria has been under question by electronic systems manufacturers because of the current emphasis on product quality and reliability. Printed wiring board (PWB) defect criteria need to be based on the predicted *in-situ* performance of the solder joint throughout its lifetime, rather than on its relative appearance.

Visual or low power magnification inspection should be used as a means of assessing soldering process control, rather than as a determination of quality. It should be an audit operation that occurs immediately after initial joining of the components to the interconnecting substrate. Data should be taken on exact defect type, defect location, component type, and board type. This data can then be used to determine trends and provide the direction for proper process correction or improvement. More modern specifications, such as MIL-STD-2000, recognize the value of such data and require that it be collected.

Inspection must accurately describe the defect in order to have a positive effect on the product. For example, terminology such as "poor wetting" often occurs in the defect criteria but does not offer sufficient information. Poor wetting should be further broken down into additional categories including dewet land, dewet lead, nonwet land, and nonwet lead. "Insufficient solder" is another commonly misused criteria which is often used to describe anything from a solderability problem to full plated through-hole or via filling.

Surface-mount assemblies require a different set of inspection criteria than plated through-hole technologies, relative to the additional demands on their corresponding solder joints. Electronic assemblies need to be inspected in accordance with *both* electrical and mechanical specifications that additionally allow for potential design variations. Leaded and leadless component configurations pose different joining problems, again requiring specific inspection criteria. The wide variety of device configurations, differing coefficients of thermal expansion (CTEs) for different materials, and choice of processing options considerably complicate the task of quality assessment and assurance.

Investigation of leadless components has produced data that indicate two predominant mechanisms of solder joint failure which result from thermal fatigue [Millard 1990]. In the first mechanism, which occurs early, a crack propagates from the edge of the component outward toward the exterior of the solder fillet as a function of cycling. This typically results from a mismatch in the CTEs of the component and its interconnecting substrate, and denotes the need to adapt the inspection to incorporate an assessment of both the volume and geometry of the corresponding solder joint, as depicted in Figure 9-2. The second mechanism occurs later in the cycling process (given that the CTEs of the board and components are well matched), as indicated in Figure 9-3, and includes the outward propagation of cracking from the interface between the solder fillet and the component (which ultimately occurs from the small difference in CTEs between the component and board). This indicates that an inspection technique needs to be developed to aid in the investigation of the corresponding interface constituency, and to help assess the mechanical reliability of the material therein.

The determination of whether porosity has an effect upon reliability and lifetime performance of a fine pitch assembly has not yet been clearly defined. Large variations in the amount of solder joint porosity can provide process control insight. Conflicting arguments have involved indications that porosity led to failure in some instances, while acting as a stress relief in others. Solder pastes, which are typically used in SMT assembly, may contain a large amount of volatiles that result in a higher percentage of porosity within corresponding solder joints. There is also some indication that the relationship of porosity to solder joint reliability is dependent upon position. A stress-induced crack which strikes a "glancing blow" (as depicted in Figure 9-4) to a large void will tend to propagate out of the fillet region at a higher rate than one that runs toward the center of the void. In some instances, a crack which encounters the center point of a porous region will stop and reinitiate its progression in another direction (as depicted in Figure 9-5) at a subsequent point in time. Inspection criteria must take this into account by setting a severity threshold based upon both void volume (as a percentage of the solder joint) *and* three-dimensional void location data.

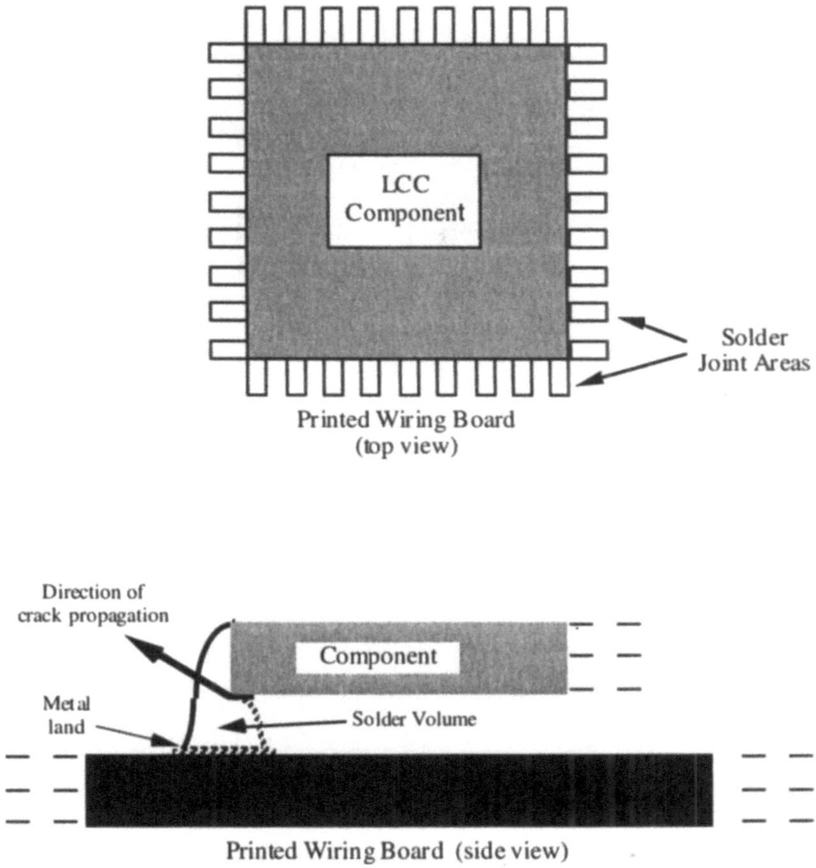

Figure 9-2 Late thermal fatigue failure mechanism

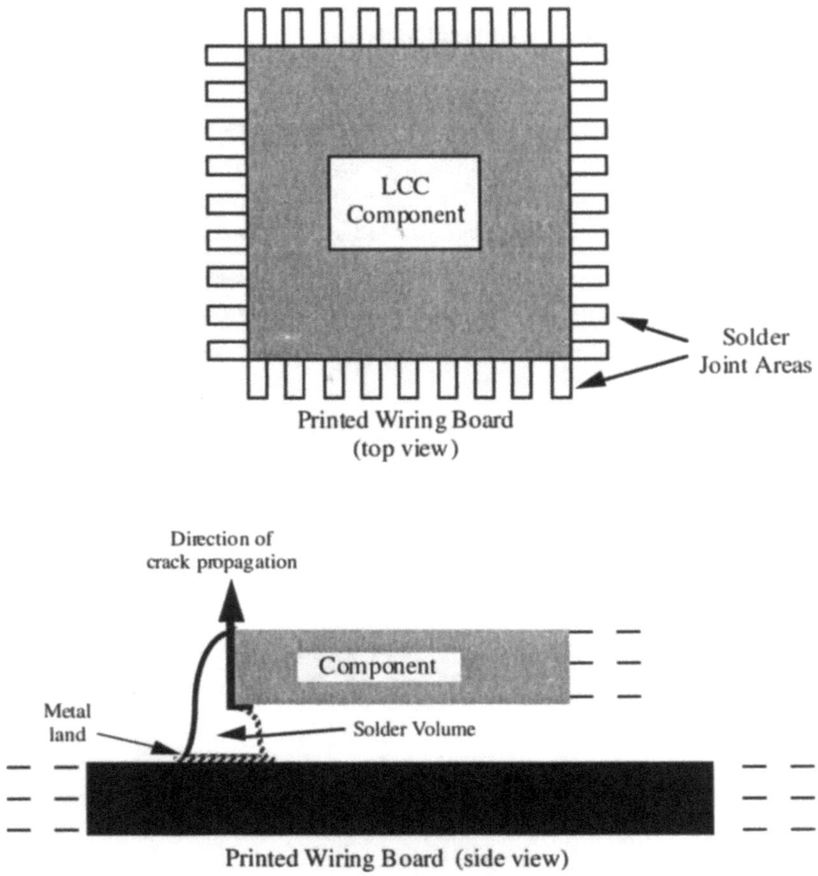

Figure 9-3 Late thermal fatigue failure mechanism. Note the change in the direction of crack propogation.

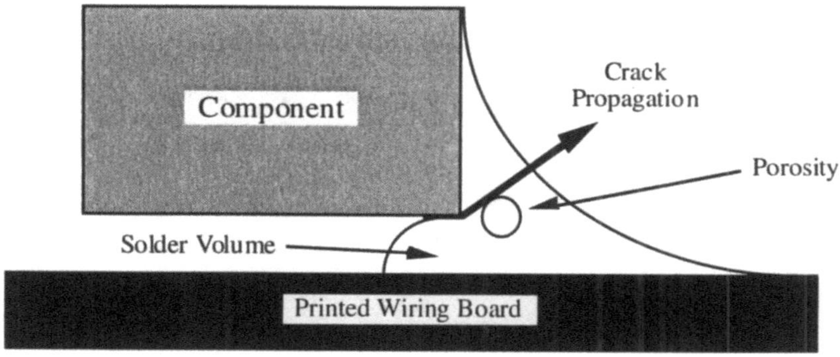

Figure 9-4 Crack propagation with tangential porosity encounter

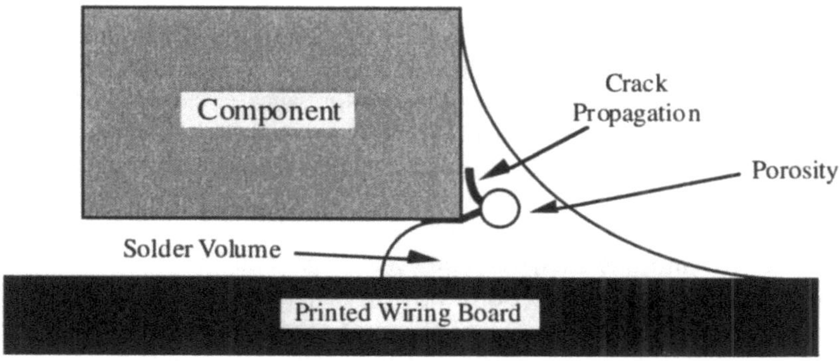

Figure 9-5 Crack propagation with centered porosity encounter

These examples of lifetime-oriented solder joint criteria, along with additional processing-related variables, are a mere subset of potential evaluation parameters that have become inspection necessities in today's manufacture of fine pitch electronic assemblies. Table 9-1 is a list of these and other performance-related solder joint criteria that have been shown to affect the reliability of fine pitch SMT assemblies [Solomon 1989].

Table 9-1 Performance-Related Solder Joint Criteria

Critical Solder Joint Reliability Criteria	
Leadless	**Leaded**
CTE match	Wetted lead
Fillet size	Height
Component tilt	Parallel sides
T_g	Compliant lead
Height	CTE match
Parallel sides	Uniformity
Uniformity	

The present set of criteria contained within current standards predominantly pertains to visual inspection of the fine pitch assembly appearance, rather than to a defect classification based on the *in-situ* performance of the soldered interconnections. Unfortunately, it is difficult to extract and relate failure information from equipment that has already been placed in service. Defect criteria have been derived from thermal and mechanical testing that used dc current to measure electrical failures occurring as a function of the corresponding in-service stressors. The final in-service fine pitch assembly, however, is presently being excited by electrical signal frequencies in excess of fifty megahertz. This excitation potentially introduces an entirely different set of electrical conditions than those historically used to derive defect criteria. The use of fine pitch packaging technologies presents new variables in the manufacturing environment that mandate new defect criteria. Therefore, design and manufacturing engineers now need to consider how these variables affect the solder joints of a fine pitch electronic assembly.

9.4 INSPECTION TECHNOLOGIES

A variety of Non-Destructive Evaluation (NDE) technologies have been employed to investigate solder joint integrity, with varying degrees of success. Recently, two- and three-dimensional reflectance, thermal, and x-ray imaging techniques have been used to measure physical parameters thought to be indicative of solder joint reliability criteria. Equipment that is based upon these techniques can be categorized as reflectance, x-ray, acoustic, or thermal based inspection systems. Many have similar digital-processing capabilities which result in an automated assessment of joint integrity that is based on a previously accumulated data base. This section briefly describes each of these systems.

9.4.1 Visual Systems

Three basic types of reflectance systems are currently applicable to solder inspection. The first system uses conventional lenses and mirrors to aid the human eye in the visual inspection of the solder joint, in a manner much like that of an optical microscope. Inspectors visually examine the assembly by means of an optical system (as depicted in Figure 9-6), which results in about a 2x to 3x magnification of the solder joint. Illumination of the inspection site is very important for accurate evaluation. In practice, this can vary from ambient room lighting to special microscope illuminators. Low power stereo microscopes are also commonly used for inspection of the device-to-substrate interconnections. These will vary in magnification from 5x to 50x. Typically, the finer the pitch, the greater the magnification employed by the inspector. The evaluation is performed by comparing the solder joint under question with a set of criteria. Table 9-2 is a list of typical criteria used to evaluate solder joint quality.

Figure 9-6 Automated optical inspection system schematic

Visual or low power optical inspection evaluates the surface characteristics of a solder joint by the nature of the process. The theory is that the surface characteristics will reflect the metallurgical quality of the solder joint. This was reasonably true in point-to-point hand soldered joints and single-sided printed wiring solder joints. With the advent of the plated through-hole printed wiring board, this became less true, since many defects that occurred were internal and left no external trace. Still, many defects such as land dewetting, nonwet leads, or lack of plated through-hole fill could be evaluated. With the latest fine pitch surface mount technology, visual or low power optical inspection is even less effective. Fine pitch solder joints that are located under a component (such as those associated with a surface mounted pad grid array—SMPGA) cannot be seen, much less evaluated. Therefore, newer automated techniques must be used to evaluate these solder joints.

Table 9-2 List of Typical Solder Joint Inspection Criteria

Typical List of Solder Joint Evaluation Criteria
• Charring, burning, or other damage to insulation
• Spattering of solder on adjacent connections or components
• The presence of icicles that jeopardize board performance
• Voids or holes not in accordance with section 3.6.1 of IPC-S-815B
• Solder which obscures the connection configuration (excess solder)
• Wicker which obscures conductors
• Loose leads or wires
• Cold solder connections
• Rosin-solder (entrapped flux) connections
• Fractured solder connections
• Cuts, nicks, stretching, or scraping which exceeds 10% of the diameter of leads or wires
• Unclean connections: lint, residue, solder, splash, dirt, etc. (if cleaning is prescribed)
• Dewetting
• Insufficient solder

- Clinched-lead overhang which reduces the spacing between conductors below allowable minimum values
- Splicing
- Lands connected by interfacial connections which show evidence of failure to wet the metallic surfaces
- Measling in excess of that permitted in IPC-A-600
- Excessive warping (IPC-A-600)
- Insufficient stress relief as described on the assembly documentation
- Solder that reduces electrical spacing below end product requirements

---

The advantages of this type of optical inspection technique are that it can be fast and relatively inexpensive. Unfortunately, however, the assessment of quality is based entirely on a subjective evaluation of the product. The inspector makes a decision after interpreting the conformance of the joint to prior visual-defect and appearance criteria relative to the rest of the electronic assembly population. The effectiveness of this technique depends on proper training, experience, motivation, and the capabilities of the particular inspector, and can vary day to day. Aided visual inspection typically results in an excessive number of defects marked for touch-up, of which only a small subset are actually unacceptable mechanical/electrical connections. This phenomenon often leads to a greater touch-up rate than an actual defect rate. Fine pitch, high pin count SMT devices are less tolerant of conventional touch-up techniques, which could render a once-acceptable joint incapable of proper electrical operation.

9.4.2 2-D Inspection Systems

Two-dimensional inspection systems produce an image on a television monitor that results from the reflection of incident light on the surface of the assembly. An image of the reflected light is captured by a camera focused on the inspection site and is then digitized into a pixel array by means of a frame grabber. Each pixel in the imaged array has a brightness level, typically assigned a value between 0 and 255, relative to eight bits of gray-scale data. Once the image has been digitized, it can be evaluated in accordance with light information collected by the camera and previously stored. Image quality is a function of the optical resolution of the camera (represented by the number and size of resident pixels), surface illumination technique, and associated digitizing hardware. Desired attributes of the processed image are compared with a solder joint defect data base that is resident in the computer memory. The result of this comparison is then evaluated to determine whether the corresponding image of the solder joint falls within acceptable limits.

There are many advantages of 2-D inspection over optical microscopy. The use of a computer for evaluation reduces the dependence upon human subjectivity (although human subjectivity tells the computer what is good and what is bad), and the digitized image can be stored in memory for further processing. The computer can ultimately feed data directly back to the assembly equipment to automatically control the manufacture of the joint if the cause of the problem is found to be a process error and not a function of a poor design or material deficiency. Variations in gray-scale data of the digitized image, which depend on a combination of the illumination technique and software algorithms, allow height to be inferred. This results in an interpretation that is subjective, at best, for volumetric information. Therefore, it is difficult to obtain precise information from a two-dimensional system that corresponds to component height off the interconnecting substrate and solder joint geometry. Algorithm speed, as with any image-processing technique, depends on proper integration of the hardware and software associated with the overall inspection system.

9.4.3 3-D Reflectance Systems

The third reflectance system typically uses a laser to scan the surface of the assembly, from which the reflected structured light is detected by one or more sensors. The acquired output of the sensor(s) is then processed, using a technique of geometrical triangulation, into a three-dimensional representation of the image, as depicted in Figure 9-7. Two-dimensional representations of the assembly under inspection can also be developed using the intensity of the reflected laser light. The major advantage of three-dimensional techniques is the capability to acquire quantitative data about both the geometry and location of the solder volume associated with its corresponding pad and fine pitch component. These systems are not as dependent on proper illumination as the previously discussed systems and provide the user with vast amounts of data from which to evaluate joint integrity. The defect algorithms applied in these systems use an approach similar to that of the 2-D equipment, although height dimensions are also incorporated within the evaluation/assessment data base. Surface contours can be thoroughly examined by such systems, along with component anomalies. In general, high-resolution 3-D imaging systems appear to be increasingly applicable to fine pitch surface-mount and other emerging technologies such as tape automated bonding (TAB) electronics manufacturing.

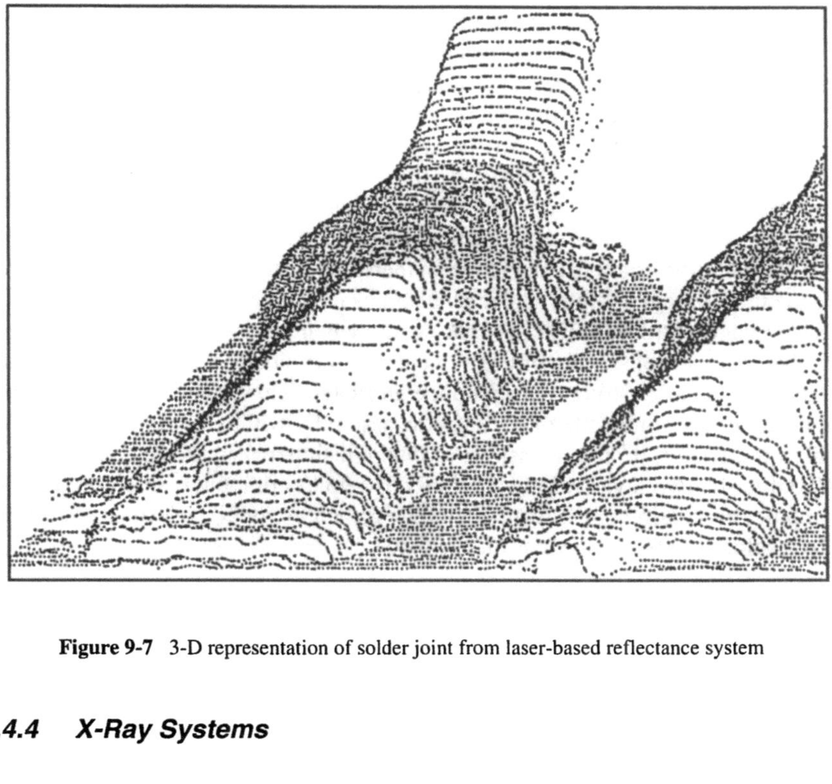

Figure 9-7 3-D representation of solder joint from laser-based reflectance system

9.4.4 *X-Ray Systems*

There are currently two types of systems that use x-rays to evaluate solder joints. In both systems, x-rays generated by a high-voltage source pass through the assembly and bombard a fluorescent screen. An image is reproduced by a video camera focused on the fluorescent screen. The output image from the camera is then digitized and again processed in the manner discussed in the "Reflectance Systems" section.

Transmissive Radiography

Real-time transmissive radiography is a technology in which the source and a detecting screen are fixed, and the assembly is positioned in accordance with the desired field of view. The output of the video camera is a signal in which the degree of x-ray transmission, and therefore material density, is represented by variations in gray scale on a video monitor. The percentage transmission of the x-ray is a function of the material content within the viewing field, which is thus used to infer knowledge about the solder joint and its boundaries. The x-ray energy level is kept below a level that can cause damage to the integrated circuits on the printed wiring board. Porous regions in the solder volume allow greater penetration by the x-ray source, and therefore the image produced is different from that of a joint with a higher percentage of solder. An operator can focus on different aspects of the assembly by adjusting the voltage of the x-ray source and the corresponding penetration. For example, devices and packaging components contain different materials than solder; this difference can be viewed in the resulting image (Figure 9-8). This non-destructive technique can then be used to gain information concerning the interior region of the solder joint and its neighboring structures. A potential disadvantage of transmissive systems occurs when attempting to view an assembly that has multiple layers or components on two sides because the x-rays must penetrate the entire structure to produce an image. This resulting image is based on the all of the composite material of the assembly within the field of view. This makes it difficult, if not impossible, to distinguish between the fine pitch devices affixed to the two layers of the assembly. Complex, fine pitch assembly-specific computer processing of the resultant image is then necessary to extract useful data of the solder joints and devices on each side.

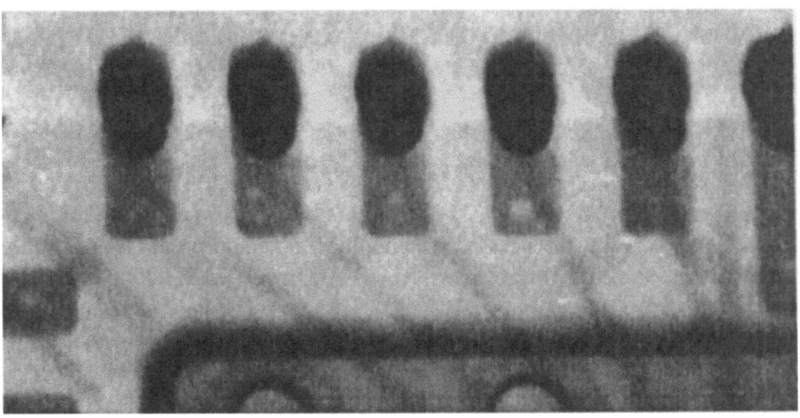

Figure 9-8 Transmission x-ray system image

Laminographic Radiography

Laminographic radiography is an emerging technology that also uses x-rays to image the solder volume and the neighboring structures. This technique uses a source and a fluorescent screen in synchronized motion, so that only one area of an assembly is in focus within the x-ray field at a particular instant. An area is effectively fixed by the source-detector combination operating in conjunction with the positioning system, while the assembly is then translated in the z dimension to observe different layers. This system produces a video output signal that can again be further processed in a manner similar to that of transmissive radiography, although it is also capable of displaying x-ray transmission through objects at various depths within the field of view. This technique offers the distinct advantage of allowing the investigation of two-sided assemblies without the need to generate complex, fine pitch assembly-specific image processing software.

9.4.5 Acoustic Systems

Advances in ultrasonic research have generated the potential for applying acoustic imaging techniques to the evaluation of electronics. Acoustic imaging is a non-destructive inspection technique that uses piezoelectric transducers to acquire signals generated from the interaction of ultrasound waves with a sample under test. Two general methods are used to produce high resolution ultrasonic images: 1) transmission mode, and 2) reflection mode. A transmission mode image results from a lack of signal reflection during ultrasonic wave propagation through a sample, whereas a reflection mode image is formed by the reflected signal.

Scanning acoustic microscopy (SAM) and scanning laser acoustic microscopy (SLAM) are typical methods of producing high resolution ultrasonic images. Scanning acoustic microscopes sharply focus pulses of ultrasound via a surface lens, which is mechanically scanned over the sample (as depicted in Figures 9-9 and 9-10). The reflected ultrasound is converted into an electrical signal by the transducer and displayed on the monitoring system. High resolution surface or lower resolution subsurface tomographic imaging can be obtained by varying the transmitted acoustic frequency. Higher frequencies will display surface and near-surface features, while lower frequencies (with a selective filter) allow inspection at set depths within the sample. The acoustic image can be developed from reflections occurring in a specific time-window of the time of flight of the signal, depending on the acoustic transmission of the material, integrity of the interface, and the polarity of the reflected acoustic signal. Various depths within the sample can be scanned by varying the signal acquisition time-window [Burton and Thaker 1986].

Scanning laser acoustic microscopes typically use a liquid coupling agent such as water to transfer the ultrasonic wave from a piezoelectric transducer to the sample under test. The transmitted ultrasonic wave is attenuated by reflection at regions or discontinuities in material properties as the ultrasound travels through the sample. A laser is scanned over the surface of the sample to detect the surface displacement that is generated by the ultrasonic energy. The displacement is converted into a video signal for further processing, monitoring, and analysis. Scanning laser acoustic microscopes can produce real-time images of defects and discontinuities encountered by the ultrasonic wave as it passes through the sample.

An advantage of acoustic imaging is the potential for imaging and establishing the depth of interior anomalies via signal attenuation/reflection. Material discontinuities such as voids, separations, and delaminations cause total reflection of the ultrasound wave because of the large acoustic impedance of air. Ultrasonic systems can detect these anomalies by monitoring the reflection/transmission of the acoustic pulses. Ultrasonic imaging can resolve physical or material discontinuities when minute contrast differences of the materials or discontinuity dimensions make x-ray inspection impossible. Scanning laser acoustic microscopy can also be used to distinguish a lack of complete bonding or voiding under bonds, since SLAM analysis of bond integrity has been shown to exhibit a strong correlation to destructive wire bond testing. The disadvantage associated with high resolution ultrasonic imaging systems is the requirement that a sample be immersed in an appropriate liquid coupling agent. Despite this apparent shortcoming, these systems can be successfully used to inspect the die attach solder bond integrity and the lid seals associated with integrated circuit components.

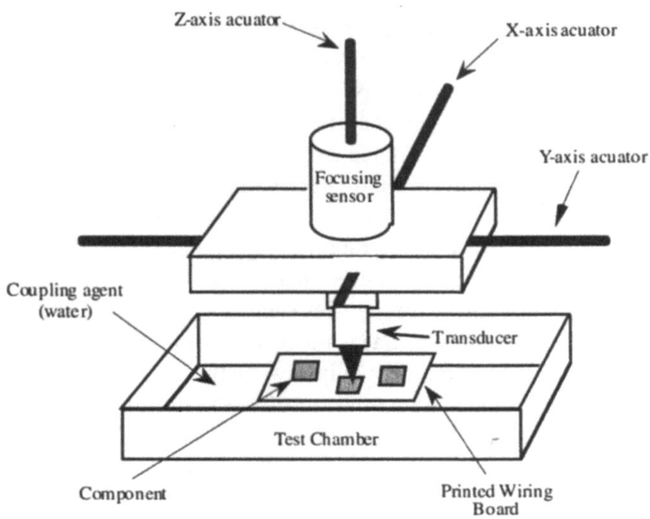

Figure 9-9 Scanning acoustic microscope schematic

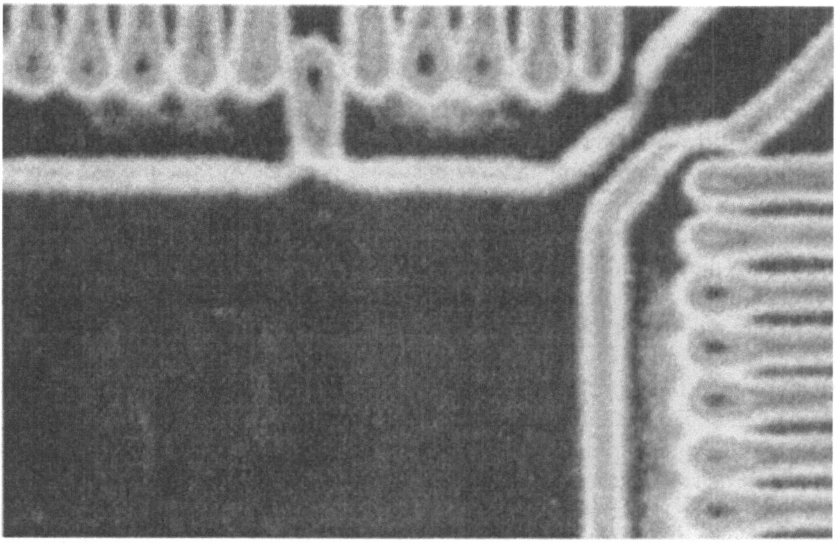

Figure 9-10 Scanning acoustic microscope sample image

9.4.6 Thermal Systems

Two types of systems are currently available that operate on the principle that unlike materials have different thermal properties. Both of these systems actively add heat to a material and track the radiating infrared (IR) energy from the surface of the substance while it is allowed to cool. Material qualities are then inferred from thermal signatures produced by the equipment used to monitor the IR energy incident upon the detector of the system. For example, a larger volume of like material has a greater heat capacity and therefore will exhibit a slower cooling rate. Similarly, a specific amount of material with greater adherence to its substrate appears as an improved heat sink, which produces a shorter period of thermal decay.

The first system (typically referred to as a Vanżetti system) uses a continuous-wave laser to input thermal energy to a solder joint and monitors the cooling profile with an IR detector focused on the solder volume. The output of the detector is a voltage (as a function of time) that is then converted to a digital signal for further computer processing. The computer monitors points along the cooling cycle and compares these new points with previously acquired data stored in memory. Variations between the newly acquired sample points and statistics associated with the stored data are used to assess the integrity of the solder joint. Figure 9-11 illustrates, in graphical form, four data points that correspond to one lead of a surface-mounted component. This system is capable of acquiring data at very high speeds but is highly sensitive to surface effects. Changes in the manufacturing process can readily be detected by the system once a statistically significant sampling of the fine pitch solder joints is stored in the computer memory. Low-volume production facilities that incorporate a high mix of device types in their designs may, however, have difficulty creating the data base necessary to use this system as a means to differentiate among various defect types.

Another technique exposes the entire electronic assembly to a heat source and monitors its thermal properties to assess solder joint integrity. The entire assembly is viewed by an IR camera while it is cooling, with the video output digitized in the same manner as in the 2-D reflectance system. Thermal energy is represented as variations in gray-scale data which are further processed by the computer to assess variations from previously sorted images. These variations can then be referred to an expert system to provide a determination of the defect type present.

As previously mentioned, not all of these technologies may be applicable to the inspection of solder joints associated with a particular packaging style. An Air Force MANTECH study was thus performed to identify which of these technologies may have the greatest success in identifying SMT solder joint defects on a single multilayer printed wiring board, such as that shown in Figure 9-12. The following section summarizes the results of the investigation.

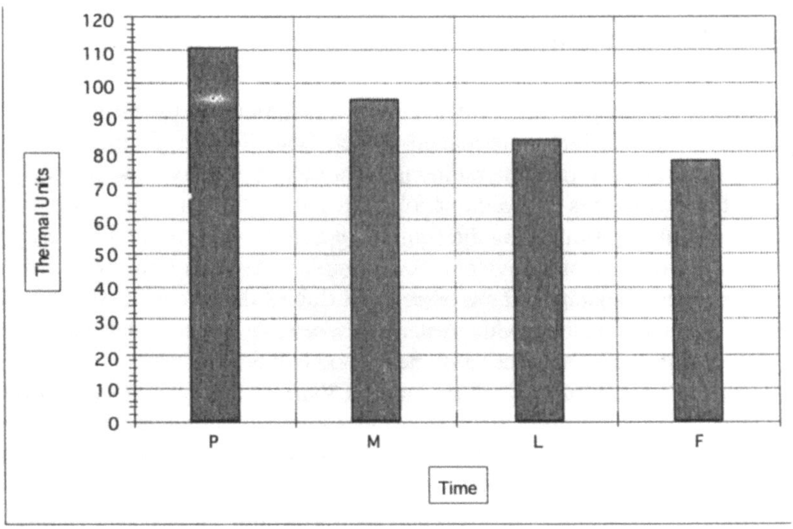

Figure 9-11 Thermal system solder joint inspection data example

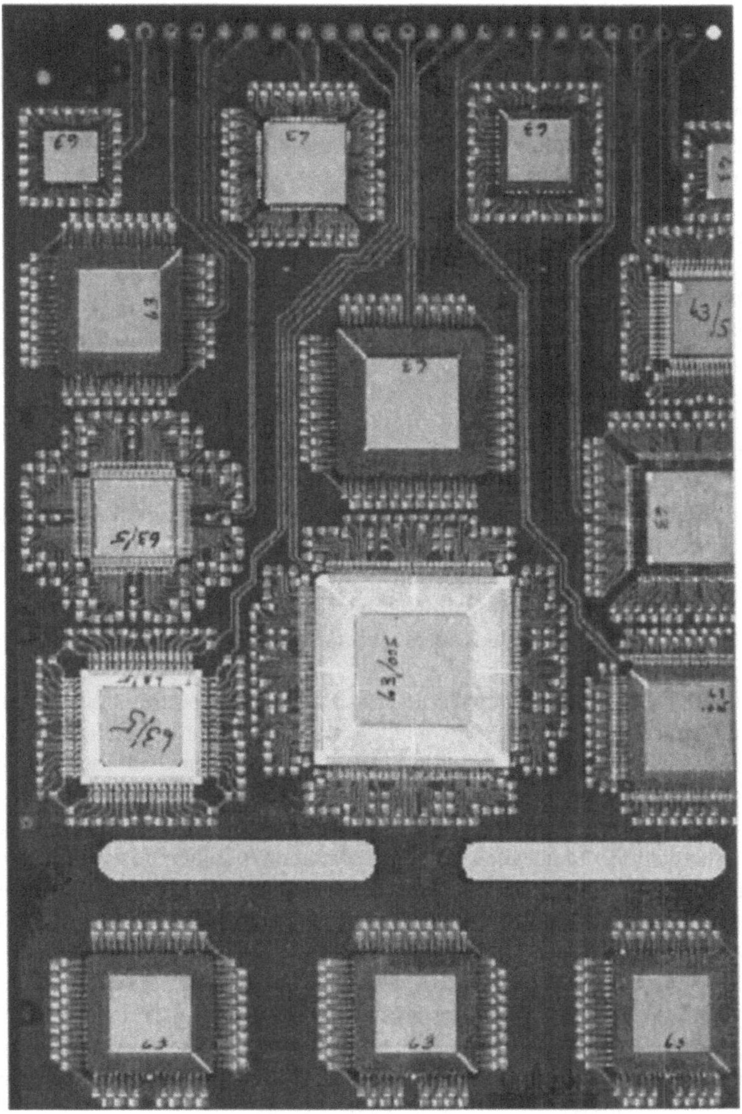

Figure 9-12 MANTECH/ADSP test assembly

9.5 MANTECH/ADSP SOLDER JOINT INSPECTION EXAMPLE

The first phase of the Air Force's MANTECH for Advanced Data/Signal Processing (MANTECH/ADSP) solder joint inspection effort investigated technologies which would be applicable to performance-related solder joint inspection. The following technologies were examined:

- Aided optical inspection
- Machine vision (3-dimensional)
- 3-Dimensional vision
- Scanning electron microscopy
- Acoustic imaging
- Photothermal radiography
- X-ray imaging (transmissive)
- X-ray laminography.

Representative vendors of these technologies were then compiled and contacted to assess their interest in the evaluation of their products' usage as solder joint metrology tools. Data were obtained (using the MANTECH/ADSP test assemblies shown in Figure 9-12) from Nicolet, IRT, Vanzetti Systems, aided optical, and RVSI (prototype-oriented) solder joint inspection systems. The data were analyzed using a spreadsheet program to determine if any inspection data would correlate with the thermal cycling results. Little correlation between the various types of inspection data occurred, nor was there any significant correlation between the various types of inspection data and thermal cycling failure (although 50% of the inspection test vehicles were designed with excellent matching of component/board CTEs). PWBs exhibiting large percentages of porosity, though not necessarily in the fillet region, did not show a strong correlation to failure. Many voids were evident under the LCC components, yet were not found to affect performance.

Phase 2 results of the MANTECH/ADSP program showed a strong correlation between the criteria listed in Table 9-1 and thermal cycling-induced failures. Processing related parameters including fillet size, component tilt and height, component parallelism, and solder joint uniformity were then used as a set of desired inspection criteria to determine which system(s) exhibited potential for further evaluation. Two systems, using 3-D machine vision and x-ray laminography, were designated as technologies which warranted additional study. All of the PWBs assembled in this phase of the program underwent visual inspection, while three assemblies were also evaluated in accordance with the flowchart shown in Figure 9-13.

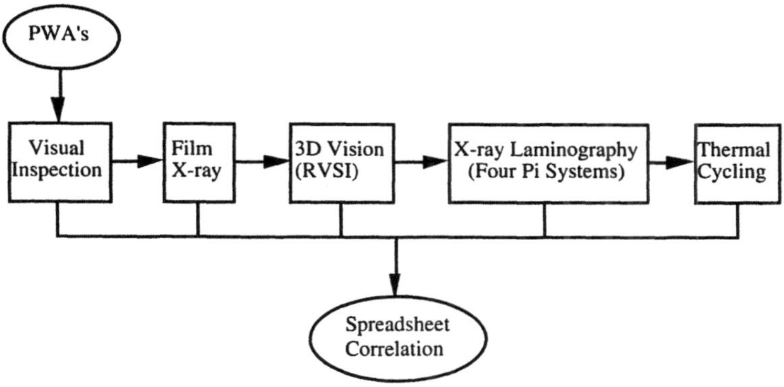

Figure 9-13 MANTECH/ADSP solder joint inspection investigation schematic

9.5.1 MANTECH/ADSP Visual Inspection Results (Phase 3)

The 33 PWBs produced in Phase 3 underwent inspection by three inspectors (A,B,C) using MIL Specifications 454J inspection categories found in Table 9-3. Solder joint volumes were also categorized according to a scale of 0–5 (where 3 is normal) for leadless and 0–3 (where 2 is normal) leaded components. The relative stand-off heights of the four corners of each component were also measured using a coordinate measuring machine (CMM) and documented in a pictorial format shown in Figure 9-14.

A summary of the visual inspection results from the three human/visual inspectors appears in Figure 9-15. The three inspectors reported a total of 5036 defects on the 33 boards, which resulted in an average of 156 defects per assembly. There were a total of 136 electrical continuity failures found as a result of the thermal cycling, which only yielded an average of 4 defects per assembly. Thus, the visual inspection designated an average of 148 more defects per assembly than those which actually led to failure. These 156 defects would have normally been "touched-up" (in conformance with MIL Spec 454J), potentially altering the reliability of solder joints which did not fail under thermal stress.

The thermal cycling results for each component indicated that a total of 136 failures were detected at the completion of the program goal of >2000 thermal cycles. The thermal cycling results also showed that non-uniform height and excessive component tilt have a strong correlation to failure. Insufficient solder volume was also found to have a direct effect upon solder joint reliability in the thermal cycling results. Failures which resulted from the accelerated thermal cycling testing performed in this study may not, however, completely correlate to failures found in the field. The thermal excursions used in this study (–55°C to

+125°C) may greatly differ from the actual environment "seen" by the product. Proper consideration of the operational environment must be included in the review of the associated failure analysis.

Table 9-3 MIL Specification 454J

	Visual Inspection Criteria
Defect #	**Defect Description**
1	Non-soldered connection
2	Bridging
3	Cold solder joint
4	Fractured or disturbed solder joint
5	Insufficient solder
6	Excessive solder, lead not discernible
7	Poor wetting
8	Solder splattering, solder balls
9	Solder pits, pinholes, voids
10	Dewetting of solder connection area(s)
11	Solder not smooth and shiny
12	Flux residue, oils, greases on assembly
13	Contaminants in solder connection
14	Exposed copper
15	Lead too high
16	Poor component/lead registration
17	Component not level
18	Other (specify on IDR)
19	Other (as noted)

Board 11

Component 2 50 mil leadless (68 I/O)

1002 Cycles to Failure
Fail Code: 31

Failed Pin(s): #60

	Pin 59	Pin 60	Pin 61	Notes:
Inspection Calls:	12	12	12	Defect 12 was called on all pins
Pin Volume:	2	2	2	Component Average = 2
Heights:	0.01048	0.00883		
	0.01116	0.01336		

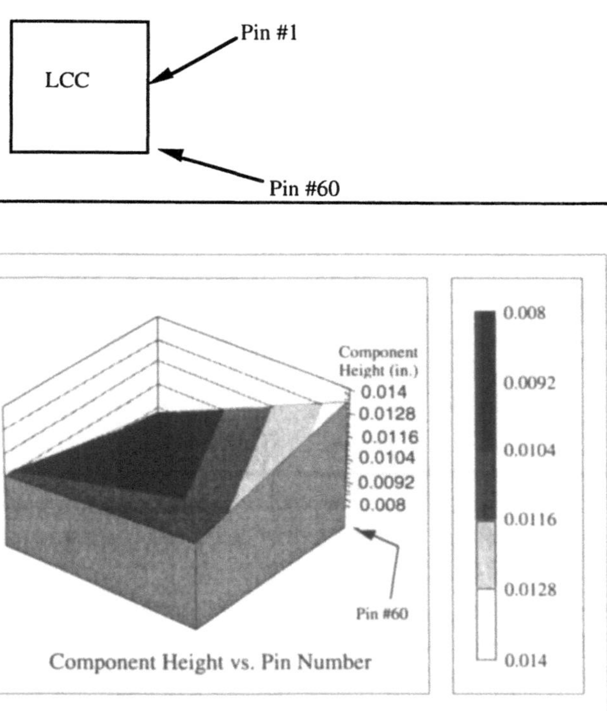

Figure 9-14 MANTECH/ADSP spreadsheet example

Visual Inspection Defect Calls

Total Defects Reported / Number of Inspected Assemblies

Inspector	A	B	C	Total
MM Assembly	50/2	770/4	941/4	1776/10
West. Assembly	330/4	454/4	713/2	1479/10
GE Assembly	441/4	307/3	1033/6	1781/13
TOTALS:	821/10	1531/11	2682/12	5036/33
Avg. #Calls/PWA	82.1	139.2	223.5	153

Figure 9-15 MANTECH/ADSP phase 3 visual inspection results

The finer pitch components did not have significantly more inspector-derived defect calls, which is contrary to what might be expected. Significant differences occurred among the number and types of defects called by the three MIL-qualified inspectors. Automated inspection could potentially eliminate this variability. The large number of inspector-derived "component tilt" (not flush) calls were not found to correlate with the CMM-acquired height data.

9.5.2 MANTECH/ADSP Automated Inspection Results (Phase 3)

Three assemblies produced in the Phase 3 effort were sent to Four Pi Systems (x-ray laminography) and RVSI (3-D machine vision) in order to acquire and store the appropriate inspection signatures prior to thermal cycling. The original intent was to investigate these previously stored signatures for insight as to why a particular joint failed during the thermal cycling. Each of the companies was also asked to run the acquired assembly data through their respective expert systems that are used to accept or reject the corresponding solder joints.

Neither company was able to perform the desired evaluation using their expert systems within the period associated with the schedule. It appeared, as a result of this investigation, that the automated inspection equipment was not yet ready to perform the requested task of relating the acquired data from the three assemblies to MIL Spec 454J solder joint defect criteria in the 1988 time frame.

The raw data acquired using x-ray laminography was further investigated after thermal cycling. A strong correlation was found between the height and tilt of the solder joints associated with the failed component pins. The failed pin in each scenario was either at the point of the greatest tilt or in the vicinity of the highest set of pins on the component in question, in agreement with the resulting CMM measurements. This result indicates that an automated inspection system could potentially predict failure induced by variations in component height which would not otherwise be possible via visual inspection.

9.6 INSPECTION AUTOMATION

The data resulting from each of the inspection systems described above must, in turn, be evaluated in accordance with established defect criteria for a particular product. It is this evaluation process that must be automated in order to reduce the subjectivity presently associated with visual inspection. Most of the inspection systems described incorporate some form of expert system software module to interpret acquired data regarding the presence or absence of defects. Obviously, data that exhibit a higher signal-to-noise ratio (SNR) require less inference from the expert system. These software-driven defect experts are only as reliable as the criteria that are resident within the database with which they have to work.

Table 9-4 exemplifies various fine pitch assembly defects that must be inspected (Millard 1990). These defects serve as a typical data base for the expert system that is part of the inspection equipment. Table 9-4 additionally attempts to assess the potential of the above NDE systems to detect and classify the indicated defect that each system may be potentially capable of detecting, with the proper level of software support. The unsure designation (U) refers to a defect that a vendor has claimed the system could detect, but currently lacks enough data to substantiate. The possibility (P) designation is assigned to a defect that the system may be able to detect. The none (N) designation is assigned to a defect that the system does not appear capable of detecting, while the reliable (R) designation is assigned to a defect that the system can reliably detect.

The final goal of automation is to achieve a hardware/software link between inspection and control of the assembly process. By imbedding inspection into the manufacturing process, defects due to process errors could be both detected and remedied as the solder joint is accomplished. One attempt to create this link uses a laser to manufacture a solder joint under the control of a thermal inspection system [Hymes 1989]. This example is the first of many steps that need to be taken toward the implementation of automated process control.

Table 9-4 Assessment of Automatic Defect Inspection Capabilities of Available Systems

Defect Category			
PWB Defect Group	**Reflectance**	**Thermal**	**X-ray**
Damaged PWB	P	U	P
Solder on PWB	R	U	R
PWB delamination	P	N	P
Artwork defects	R	N	P
Etching defects	P	N	P
Misaligned layers	N	N	P
Incomplete cleaning	R	P	N
Solder mask defects	R	N	N

Solder Defect Group	Reflectance	Thermal	X-ray
Excess solder	R	P	R
Insufficient solder	R	P	R
Unsoldered connection	P	P	P
Improper wetting	R	N	P
Dull Solder	R	P	P
Disturbed solder	R	P	P
Porosity/voids	N	P	R
Solder inclusion	P	N	P
Incomplete flow	P	P	P
Bridging	R	P	R

Component Defect Group	Reflectance	Thermal	X-ray
Component preparation defects	R	P	P
Component registration	R	N	R
Damaged component	P	N	P
Missing component	R	P	R
Wrong component	P	N	N
Part marking	P	N	N
Component height	P	N	P

KEY:
U: Indicates unsure defect detection capability
P: Indicates a possibility for defect detection
R: Indicates reliable defect detection capability
N: Indicates no possibility of defect detection

9.7 ADVANCED SOLDER JOINT INSPECTION TECHNIQUES

Previous efforts have yielded techniques for generating three-dimensional models of solder joints from x-ray data. These 3-D models might be used during process development to measure the effects of varying process variables on solder joint geometry or as input to finite element models of the mechanical response of these joints. Instead of using laminography for end-of-line inspection, models from a sufficient sample size might be used to evaluate solder pastes, fluxes, pad designs, and soldering process parameters. Modeling using an x-ray system is non-destructive, and could provide more data in less time than traditional destructive cross-sectioning. The input to the modeling process would be a collection of laminographic images of a solder joint (from different heights), as shown in Figure 9-16. These eight slices can be processed to define the outlines of the solder volumes in each slice, then be combined into a 3-D computer model, from which the user can choose the desired point of view [Millard et al. 1992]. The output of the model generation process could result in the 3-D example found in Figure 9-17 or a mesh for a finite elements analysis. This data can also be animated so that the 3-D model appears animated on the screen and allows the operator to observe multiple locations. These visualizations can then be saved as digital movies for future reference. The present amount of time associated with the generation of this 3-D modeling data prohibits this technique from being used for 100% inspection; however, recent research efforts have focused on improving both the speed and accuracy associated with the x-ray laminographic based, model generation process. Further work is necessary to improve the resolution of the acquired image data and the speed of the reconstructional algorithm, prior to industry's widespread acceptance of such a tool.

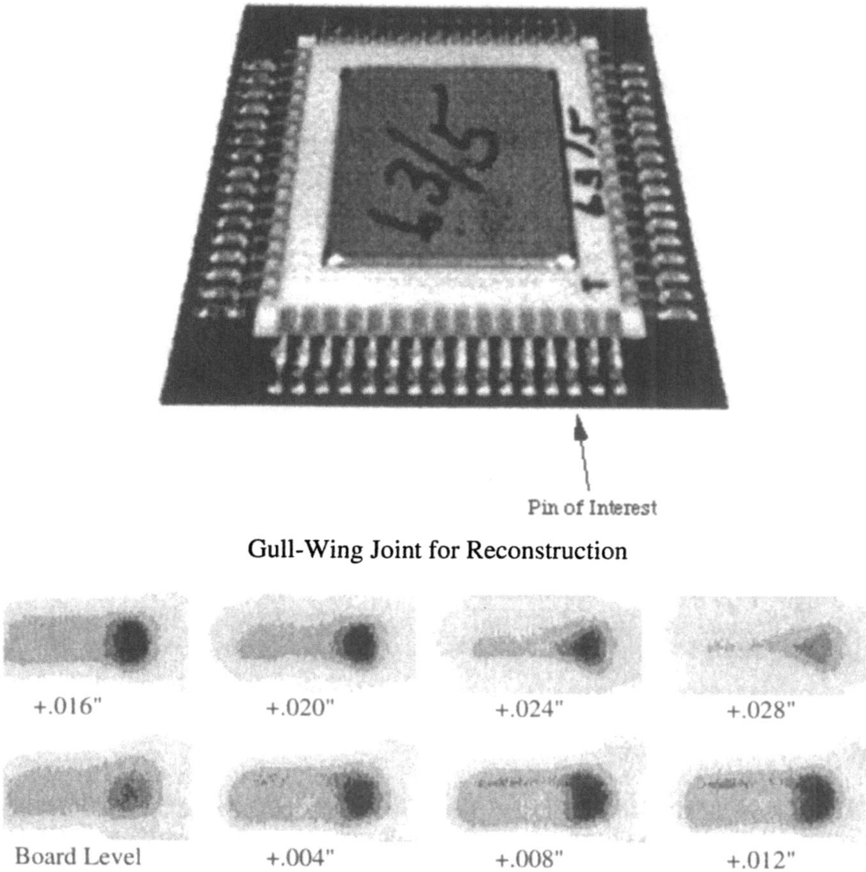

Figure 9-16 Eight x-ray laminographic slices from a gull-wing solder joint taken in equal 0.004" (inch) increments from the board level upward. The slices have been processed so that the variation from white to black correlates with a low to high variation in density.

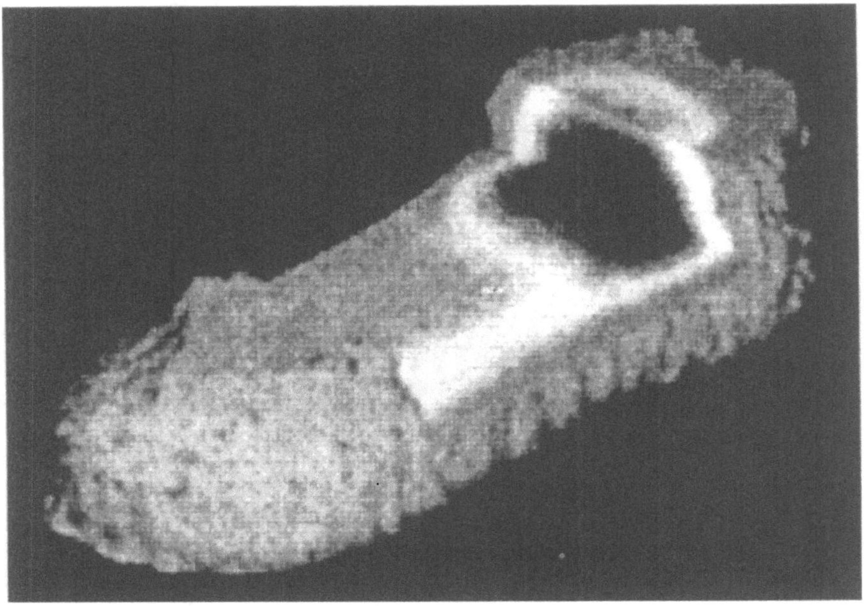

Figure 9-17 Non-destructive 3-d joint visualization using x-ray laminography

9.8 CONCLUSIONS

The goal of automated solder joint inspection is both a challenge and a necessary path for today's manufacturing engineers who are attempting to achieve 100% reliability. Unfortunately, as indicated in the prior sections, there does not appear to be a single inspection technology or system that offers reliable detection of all of the anomalies required to predict the performance of fine pitch assemblies produced today. Therefore, a synergistic approach by both inspection equipment developers and electronics manufacturers is now necessary in order to develop a database that can ensure that a currently manufactured solder joint will reliably maintain its electrical, mechanical, and thermal integrity throughout the lifetime of the product.

REFERENCES

Burton, N., and D. M. Thaker. 1986. *Proceedings of the 1986 ISHM International Symposium on Microelectronics.* International Society for Hybrid Microelectronics. pp. 253-58.

Hymes, L. 1989. "Surface-Mount Soldering." *Electronics Materials Handbook.* ASM International. pp. 697-709.

Millard, D. L. 1990. *MANTECH for Advanced Data/Signal Processing Solder Joint Inspection Final Report.* Manufacturing Technology Directorate. Wright-Patterson AFB/WRDC-TR-89-8025. vol. 5.

Millard, D., D. Knorr, L. Felton, and S. Black. 1992. "Electronics Manufacturing Research at Rensselaer Polytechnic Institute," *Proceedings of the Twelfth IEEE/IEMT Symposium.*

Solomon, H. D. 1989. *Proceedings of the 39th Electronics Components Conference.* pp. 277-92.

10

The Properties of Composite Solders

S. M. L. Sastry, D. R. Frear, G. Kuo, and K. L. Jerina

10.1 INTRODUCTION

Previous chapters of this book have discussed the wetting behavior of near-eutectic Sn-Pb solders. These alloys are the most common in electronic applications and are deserving of the greatest attention. However, there are many applications where the properties of these alloys can be improved. A composite solder, or a solder that contains second-phase dispersions of particles, could improve solder joint properties. The area where composite solders could have the greatest impact is in improving the mechanical properties of the joint. However, the wetting behavior of the solder cannot be compromised for the sake of the mechanical properties –good wetting behavior is a necessity. Therefore, for a composite solder to be useful for electronic packaging applications mechanical properties, wetting and manufacturability must be optimized.

Composites are defined as materials in which distinct individual phases are bonded together in such a manner that the average properties of the composite are determined by the individual properties of each phase. Generally, the composite is constructed in order to gain advantageous characteristics from each of the component materials or to overcome disadvantageous characteristics of each. Materials with a dispersed second phase can also be considered composites because the dispersed phase affects the properties of the solder by contributing to them.

The primary focus of composite solders has been to improve the mechanical behavior of solder joints. Therefore, this chapter contains an overview of the mechanics of solder joints. The first portion of the chapter will discuss the failure mechanisms of solder joints (specifically under conditions of thermomechanical fatigue) and how composite solders can improve solder joint behavior. This is followed by a discussion of the available wetting and solderability properties of composite solders of interest. The chapter concludes with a discussion of the mechanical behavior of composite solder alloys.

10.2 MECHANICS OF SOLDER JOINTS

10.2.1 Origin of Thermomechanical Fatigue in Solder Joints

Solder joints are the electrical and mechanical interconnections that constrain materials having different thermal expansivity in electronic packages. A key issue in the long-term reliability of solder interconnections is joint failure during thermal cycling. The individual components that are soldered together in an electronic package typically have different thermal expansion coefficients. For example, in a surface mount application, a ceramic chip carrier (with a coefficient of thermal expansion of 6×10^{-6}/°C) can be soldered to a polyamide circuit board (with a coefficient of thermal expansion of 15×10^{-6}/°C). When the soldered assembly undergoes a thermal cycle, strain is imposed on the solder joints. These thermally induced cyclic strains result in thermomechanical fatigue and, in some situations, failure. The failure of a single joint could render an entire electronic system inoperable.

The reliability of solder interconnections under conditions of thermomechanical fatigue becomes more critical as new electronic packaging technologies evolve. In the future, larger silicon devices and the rapidly evolving multi-chip module will induce larger strains in the joints. The service environment of the solder joints is also becoming more severe. New power devices are running hotter, increasing the possibility of fatigue failures in personal computer and mainframe applications. In the automotive industry, electronic packages are positioned closer to the engine for faster response and greater cost effectiveness. For increased fuel efficiency, styling is more streamlined and air inlets are decreasing in size. This results in higher operating temperatures under the hood and greater strains in the solder joints. In avionics applications, electronic packages used for sensing, control, and telemetry are exposed to severe environmental temperature swings. The severity and frequency of those environmental cycles create large cyclical strains in the solder joints.

10.2.2 Origin of Vibration and Shock in Solder Joints

Under normal operating conditions, the solder joints in electronic systems can encounter conditions of vibration and shock. These conditions are most severe in avionic applications and automobiles. Shock and vibrations do not normally cause failures in solder joints because, typically, the magnitude of strain is small and more strongly influences the more brittle parts of the package (e.g. the Si or the ceramic packages). However, if the shock is large enough, or the magnitude of the vibration is large, the resultant strains can cause failure.

10.2.3 Failure Mechanisms in Solder Joints

The as-solidified microstructure of near-eutectic Sn-Pb solder alloys consists of a two-phase structure of either lamella or globules of tin- and lead-rich phases. Off-eutectic solders tend to have more lead, or tin, so these solders form proeutectic lead- or tin-rich dendrites. The microstructure is further divided into regions of similarly oriented lamella called eutectic cells or colonies. An optical micrograph that is typical of the near-eutectic Sn-Pb solders is shown in Figure 10-1.

The as-solidified structure of near-eutectic Sn-Pb solders is not stable with respect to aging conditions or mechanical deformation. The phases in the near-eutectic microstructure coarsen during isothermal aging. Mechanically deforming and heating near-eutectic Sn-Pb solders results in recrystallization and the formation of equiaxed tin- and lead-rich phases. It has been shown that this microstructure can behave in a superplastic manner [Grivas 1978]. Schmitt-Thomas and Wege [1986] found 60Sn-40Pb to have the least microstructural stability of all the Sn-Pb solders investigated. They found extensive coarsening and surface roughening under conditions of stress relaxation and creep. Therefore it is expected that conditions of thermomechanical fatigue, which involve both mechanical deformation and time at temperature, could have a great effect on the microstructure and properties of near-eutectic Sn-Pb solder joints.

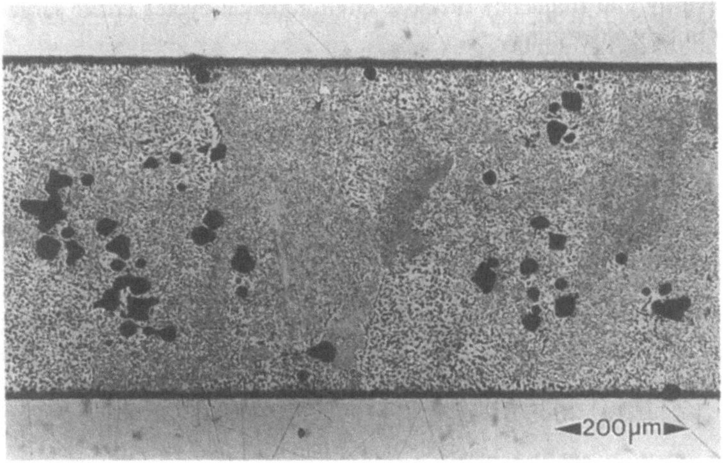

Figure 10-1 Optical micrograph of near-eutectic Sn-Pb solder. The light phase is tin-rich and the dark is lead-rich.

Wild [1972] was the first to examine and metallographically document 60Sn-40Pb solder joints after thermal cycling. The solder joints consisted of pins soldered into plated through-holes that were thermally cycled between –55° and 125°C. Wild found that, after thermal cycling, the surface of the solder joints became "frosty" and he postulated that his surface relief was due to slip displacements in the solder. He found evidence that the microstructure of the solder coarsened in the interior of the joint parallel to the direction of imposed shear strain. Wolverton [1987] thermally cycled LCCs and leaded carriers. He observed that prior to failure the solder joints had bands of coarsened two-phase material parallel to the direction of imposed shear. An example of heterogeneous coarsening is shown in Figure 10-2.

The heterogeneous coarsening of near-eutectic Sn-Pb solders has been studied extensively [Frear, Jones, and Kinsman 1991]. This is an important microstructural feature because the coarsened region is inherently weaker and is known to be the region through which cracks propagate to final failure in thermomechanical fatigue.

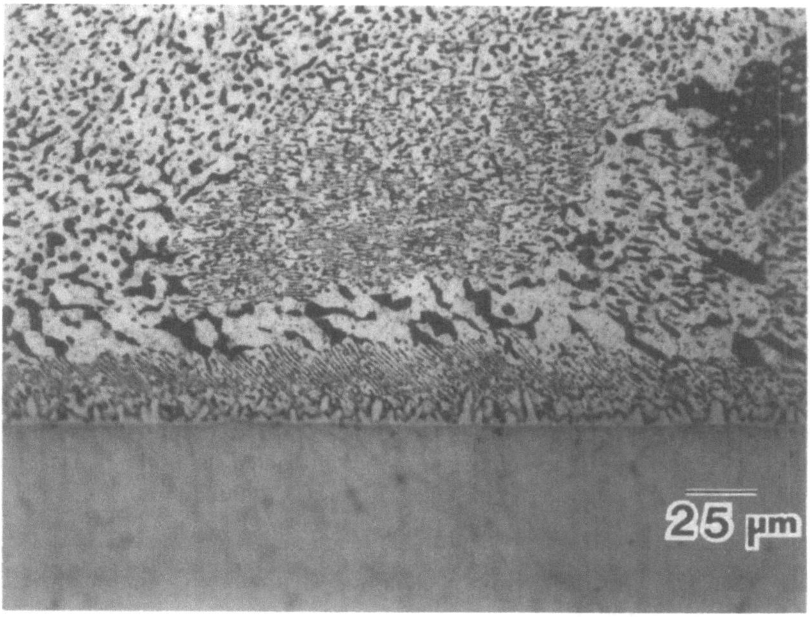

Figure 10-2 Optical micrograph that shows a heterogeneous coarsening band through a solder joint under conditions of thermomechanical fatigue.

The characteristics of the heterogeneous coarsened band change as a function of deformation rate. In thermal shock, the deformation rate is rapid, on the order of $7 \times 10^{-3} s^{-1}$ or faster. The heterogeneous coarsened band that forms is parallel to the direction of imposed shear strain and appears to be independent of any microstructural features. At slower deformation rates (slower than $1 \times 10^{-3} s^{-1}$), present under conditions of thermal cycling, heterogeneous coarsening has a different resultant microstructure [Frear and Jones 1990; Frear, Jones, and Sorensen 1991]. The heterogeneous coarsened band, at a slower deformation rate, occurs primarily at cell boundaries in the joint that are parallel to the direction of imposed shear strain. A mechanism of coarsening at cell boundaries has been proposed [Frear, Jones, and Kinsman 1991; Frear 1989]. After solidification, cell boundaries are slightly more coarsened than the rest of the lamellar solder and are therefore the "weak link" in the microstructure. During the low temperature portion of the thermal cycle, deformation concentrates at the cell boundaries and the cells slide or rotate relative to one another. At the high temperature portion of the cycle, the deformation at the cell boundaries is annealed by recrystallization. The heterogeneous coarsened region develops, having the same appearance as for a near-eutectic Sn-Pb alloy that has been mechanically worked then heated (i.e. recrystallized). Additional deformation in the joint concentrates in the coarsened band because it continues to be the weakest part of the microstructure of the joint.

The heterogeneous coarsened band provides the path through which cracks propagate in near-eutectic Sn-Pb solder joints. Some claim the failure occurs through the lead-rich phase [Wild 1972; Wolverton 1987]. Their hypothesis is that lead is softer than tin and is therefore more easily deformed. However, in these studies the microstructure was characterized well after failure and the mode of initial cracking could not be determined. Some failures were observed to be by interphase separation [Bangs and Beal 1975; Lee and Stone 1990]. Other investigators found cracks initiating and propagating through the tin-rich phase [Frear and Jones 1990; Frear 1989] with some cracks in the lead-rich phase to connect to the adjacent tin. An example of the cracking of tin-rich grains in the heterogeneous coarsened band is shown in Figure 10-3. These intergranular cracks initiate when the large tin grains (which coarsen at the same time as the phases in the heterogeneous bands) can no longer slide and rotate to accommodate the strain. These small cracks coalesce and eventually cause the entire joint to fail.

10.3 ALLOY DESIGN FOR THERMOMECHANICAL FATIGUE RESISTANCE

Understanding of the metallurgical mechanisms that lead to the thermomechanical fatigue failure of solder joints gives direction on how the metallurgy of the solder joints can be modified to improve fatigue resistance. Since the solder most commonly used in microelectronic applications is near-eutectic Sn-Pb, much of alloy design research has concentrated on this alloy.

Given that the failure of near-eutectic Sn-Pb solder joints under conditions of thermomechanical fatigue is by the formation of heterogeneous coarsened bands, fatigue resistance can be improved by metallurgical modifications that either inhibit the formation of the bands or their propagation through the specimen. This can be accomplished by homogenizing the microstructure so that there is no region in which the strain can concentrate. Alternatively, obstacles can be placed in the microstructure that will prevent the heterogeneous coarsened band itself, or cracks in the band, from propagating through the joint. Regardless of the method of improvement, the manufacture (processing and solderability) cannot be substantially changed from that of the base eutectic Sn-Pb alloy so as to not adversely affect producibility.

Figure 10-3 Optical micrograph of cracks that have formed in the heterogeneous coarsened band. These are intergranular cracks through the tin-rich phase.

A number of metallurgical techniques traditionally used to improve fatigue behavior are not suitable for solder alloys. For example, directional solidification or single crystal solder joints could not be produced in a realistic manufacturing process due to the need for frequent reflow of the joints. Precipitation hardening is not effective in eutectic Sn-Pb solder alloys due to the high homologous operating temperatures.

A number of other metallurgical techniques have been presented in the literature that do promise an improvement in the fatigue lifetime of solder alloys. There are four generic metallurgical modifications that can improve the thermomechanical fatigue behavior of eutectic Sn-Pb solders. The first is the formation of a fine, equiaxed, eutectic structure that is superplastic so that deformation is homogeneous rather than heterogeneous. The second method is to make alloy additions that homogenize the microstructure and in turn homogenize the imposed deformation throughout the joint. The third method is to change solder alloy from near-eutectic Sn-Pb to an alloy that does not undergo a heterogeneous deformation process but has wetting characteristics similar to near-eutectic Sn-Pb. The final, and perhaps most promising, method is to employ composite solder alloys that use a second separate phase of material to homogenize the strain distribution or physically hold the joint together when a crack forms in a matrix that has good wetting properties. Each technique is briefly described as follows.

10.3.1 Solder Joints with "Superplastic" Microstructures

Superplasticity is defined as material behavior where extensive elongation (>100%) results without fracture when strain is applied at certain strain rates and stresses. Eutectic Sn-Pb solder exhibits superplasticity after it has been recrystallized into a fine grained two-phase structure. The superplastic microstructure occurs at intermediate strain rates and, very importantly, is homogeneous across the joint. This makes a superplastic microstructure an attractive possibility for thermomechanical fatigue resistance in a solder joint. The homogeneous deformation would limit the possibility of the formation of a heterogeneous coarsened band, thereby increasing fatigue life. Moreover, superplasticity can be thought of as "free" deformation: the fine grains in the microstructure can easily slide and rotate relative to one another, so that damage is not stored in the form of defects that would result in microstructural evolution.

Shine and Fox [1988] analyzed the effect of a superplastic microstructure on solder joints in fatigue and concluded that superplasticity would produce an exceptionally long-life joint. Unfortunately, experimental work has not yet been performed to prove conclusively that a superplastic solder joint does improve fatigue life. Furthermore, superplasticity is limited to certain strain rates and stresses; the deformation during fatigue must occur in this regime.

If superplasticity is beneficial in fatigue, there are two difficulties in attempting to take advantage of superplasticity in a solder joint: 1) how is a superplastic microstructure achieved in an as-solidified joint? and 2) how can that superplastic microstructure be retained throughout the thermomechanical fatigue life? As-solidified joints do not typically contain a structure that is superplastic [Tribula et al. 1989]. However, work by Mei et al. [1990] has shown that in very thin solder joints (0.05 – 0.1 mm) superplasticity can be achieved by means of rapid cooling after solidification.

Once a superplastic microstructure is formed, how can it be retained? Typically, a recrystallized structure will coarsen with time at temperature. Once coarsened, the microstructure will no longer be superplastic (due to the inability of large grains to slide and rotate without opening cracks) and the joints can fail. This remains an open issue, but alloy additions and composite solders that stabilize the microstructure (discussed later) are promising options.

10.3.2 Alloy Additions to Homogenize the Microstructure

The formation of heterogeneous coarsened bands can be suppressed by the creation of a more equiaxed structure. Such a microstructure minimizes strain concentrations and precludes the formation of a continuous heterogeneous coarsened band. The addition of ternary elements to 60Sn-40Pb can cause a more equiaxed structure to form. It is important that these alloy additions be kept to a minimum so as not to be detrimental to processing.

Work has been performed [Tribula 1990; Tribula and Morris 1990] to study the effect of minor alloy additions on the thermomechanical deformation response of near-eutectic Sn-Pb solder. The two alloys that were found to be the most beneficial with respect to inhibiting heterogeneous band formation were 58Sn-40Pb-2In and 58Sn-40Pb-2Cd. For both alloys the structure within the colony is fine, and the colony size is refined (~150 μm) compared with the 60Sn-40Pb (300 μm). Furthermore, the colony boundaries are much thicker and more diffuse in the solders with alloy additions.

The deformation pattern is much more homogeneous and is not concentrated in a single band parallel to the direction of imposed strain for both the indium- and cadmium-containing solders. After the same number of thermomechanical fatigue cycles, the indium- and cadmium-containing solders exhibit distributed coarsening throughout the joint in a homogeneous fashion and no cracks are observed. Work by Nakamura et al. [1986] has also shown that the additions of cadmium and indium have improved the fatigue life of the near-eutectic Sn-Pb solder.

Preliminary work by Tribula et al. [Tribula 1990; Tribula et al. 1990] indicates that the refined colony size allowed the strain to be more evenly distributed through the joint at many colony boundaries and therefore limited the heterogeneous coarsening. Clearly, even minor alloy additions have a strong effect on the microstructure, resulting in improved mechanical behavior.

10.3.3 Other Solder Alloys

The thermomechanical fatigue behavior of joints may be improved by substituting another solder alloy for near-eutectic Sn-Pb. The proposed alloy must have a similar set of good wetting and manufacturability characteristics (relatively low melting temperature and a short melting-temperature range). Many solder alloys may fit this description, but only two have any extensive thermomechanical fatigue test results: 95Pb-5Sn and 40In-40Sn-20Pb.

95Pb-5Sn solder melts at 300°C and has acceptable wetting characteristics. This solder is composed of a single phase lead matrix, with a distribution of β-tin precipitates [Frear, Posthill, and Morris 1989]. Heterogeneous bands have not been observed in this alloy [Frear et al. 1988]. Unfortunately, 95Pb-5Sn has poor thermomechanical fatigue lifetime [Frear et al. 1988]. The solder consists of fairly large lead grains; thermomechanical fatigue strain is accommodated in this alloy by sliding and rotation of the lead grains. If the lead grains are too large, intergranular cracking occurs. Therefore, the single phase matrix solder alloy may not be suitable for thermomechanical fatigue applications unless the strain, or grain size, is kept small.

The 40In-40Sn-20Pb alloy is a low melting-temperature (121°C) ternary solder that has wetting characteristics similar to that of eutectic Sn-Pb. The microstructure consists of a similar colony structure to that found in the eutectic Sn-Pb, with Sn-In and Pb-In phases forming the eutectic. This alloy has a longer thermomechanical fatigue life than eutectic Sn-Pb. The microstructure homogeneously refines, rather than coarsens, during thermomechanical fatigue [Frear 1989]. The failure of this solder occurs by surface damage leading to crack propagation. The improvement in thermomechanical behavior is hypothesized to be due to the presence of indium (segregating to the boundaries in both phases) pinning the grain boundaries, thereby limiting grain growth. These results indicate that a microstructure that refines homogeneously is a great improvement over those that coarsen, especially those that heterogeneously coarsen, because the strain is more easily accommodated by a refined structure.

10.3.4 Composite Solder Alloys

Second-phase particles are common alloying additions to improve the high-temperature strength of metals and alloys. The purpose of particles in solder is to break up the continuous heterogeneous band formation. By interrupting the heterogeneous bands, the strain can be distributed more homogeneously through the joint.

Before describing the strengthening mechanism of composite solder alloys, some background on the general strengthening mechanisms which are well developed and utilized in metallic materials is appropriate. Most of the strengthening methods are based on mechanisms to impede dislocation motion such that higher applied stress is required to cause the metal to deform. The strategy is then either to create a large number of obstacles or to increase the dislocation density. Work hardening, also termed strain hardening, is based on the latter strategy. Obstacles for dislocation movement can be grain boundaries, substitutional or interstitial atoms, coherent precipitates, or non-deformable dispersoids. Based on the fact that tensile strength can be improved by increasing the number of obstacles, grain boundary, solid solution, precipitation, and dispersion strengthening have been developed and utilized in commercial applications for a long time.

Factors which influence strengthening by second-phase particles include the following: the size, shape, number, and distribution of particles; the strength, ductility, and strain hardening behavior of the matrix and the second-phase particles; the crystallographic relationships between the matrix and particles; and the interfacial energy and bonding. The matrix and the second-phase particles each contribute to the overall properties of a composite. If the contributions from each phase are independent, then the properties of the composite are a weighted average of the properties of the individual phases. However, structure-sensitive mechanical properties are also generally influenced by interactions between the phases.

When considering strengthening by second-phase particles, it is important to distinguish between precipitation strengthening, dispersion strengthening, and whisker/particle strengthening because of the fundamental differences in the physics of the strengthening mechanisms and the property modifications produced by the two methods. The strengthening which arises from shearing of dislocations through deformable precipitates is referred to as precipitation hardening. The strengthening which is caused by the bypass of dislocations around non-deformable particles is referred to as dispersion strengthening. The other distinction between shearable precipitates and non-deformable dispersoids is the solubility or the thermal stability at elevated temperature. The second-phase particle in dispersion strengthening systems has very little solubility in the matrix, even at elevated temperatures. Hence, the particles resist growth or overaging to a much greater extent than the precipitates in a precipitation-hardening system.

Dispersion strengthening is produced by a finely dispersed insoluble second phase. The presence of a dispersed phase in an otherwise homogeneous matrix imparts increased strength to the system. Potential improvements in mechanical properties by stable second-phase dispersions include: (1) an increase in room- and elevated-temperature flow-stress, and (2) an increase in creep-resistance and stress-rupture-life. The increase in yield stress of such a dispersion-strengthened system arises from the additional stress necessary for dislocations to bypass the particles before yield can occur. The mechanisms by which dispersoids modify the alloy characteristics include direct dispersoid-dislocation interaction, whereby the dispersoids impede dislocation motion; and indirect effects associated with substructure and grain structure modifications caused by the dispersoids.

Strengthening from dispersoids arises from the additional stress required for bypassing of dispersoids by dislocations. The strength increment in the presence of non-deformable particles is given by the modified Orowan relation (Ashby):

$$\Delta\tau = 1.06\frac{Gb}{2\pi L'}\ln\frac{x}{\gamma_0} \tag{10-1}$$

In Eq. (10-1), G is the shear modulus of the matrix, b is the Burger's vector of the dislocation, x is the dispersoid size, γ_0 is the inner cut-off radius, equal to 4b. L' is the average spacing in the slip plane, which decreases with increasing volume fraction of dispersoids and increases with increasing dispersoid size. Eq. (10-1) is the geometric mean of the bypassing stresses for edge and screw dislocations and includes a statistical factor of 0.85 relating the macroscopic flow stress to the local Orowan stress [Argon 1969]. For effective dispersion strengthening, the particles should be < 0.1 μm.

The theory of dispersion strengthening is mostly based on idealized spherical particles. However, particle shape can be made important primarily by changing the interparticle spacing. At an equal volume fraction, rods and plates strengthen about twice as much as spherical particles [Kelly 1972].

The most important effect of dispersoids on creep deformation arises from dislocation pinning which increases the threshold stress for creep deformation and reduces the creep rate. Dispersoids continue to be effective deformation barriers at elevated temperatures, whereas solid solutions, precipitates, and substructures are ineffective for high temperature strengthening by virtue of being thermally activated barriers. Dispersion-strengthened structural alloys may retain useful strength at temperatures up to 90% of their melting point.

A closely spaced, fine dispersoid distribution provides the additional benefits of inhibition of recrystallization, retention of stored energy, and production of fine dislocation-cell substructures which increase the high-temperature strength and improve the creep resistance. For small particles, the stabilization arises from the drag force they exert on a migrating boundary. The total drag force per unit area of boundary is $3f\gamma/2r$, where f is the volume fraction of particles, γ is energy/ unit area of the grain boundary, and r the average particle size. Hence, if the driving force for the growth of grains is proportional to $2\gamma/D$, where D is the average grain size, the upper limit of grain size in the presence of a dispersion of second-phase particles is given by [Embury 1985]:

$$D = \frac{4r}{3f} \quad (10\text{-}2)$$

The effect of grain size on yield strength is given by the Hall-Petch relationship:

$$\sigma_y = \sigma_o + k\,D^{-n} \quad (10\text{-}3)$$

where σ_y is the yield stress, σ_o is the "friction stress" (representing the overall resistance of the crystal lattice to dislocation movement), D is the grain diameter, and k and n are constants (n ranges from 1 to 2).

Eq. (10-3) implies that grain boundaries act as barriers to dislocation motion, and, as a result, the yield strength increases with decreasing grain size. Dislocation cell structure and subboundaries can also provide a strengthening effect using this same mechanism. In contrast to many other strengthening mechanisms, the contributions to strength provided by separate types of boundaries are not additive. Often the resulting strengthening is determined by the more effective boundary [Courtney 1990]. However, at elevated temperatures, small grain size has an adverse effect on yield strength. Above 0.5 T_{mp}, the steady state creep rate generally increases as the grain size is decreased because a fine-grained material has a larger total grain boundary area available for grain boundary sliding than a coarse-grained material. Hence, for high-temperature applications where creep resistance is desired, a coarse grain structure is preferred. However, coarse-grained structures suffer from intergranular void formation and fracture.

Whereas dispersoids and reinforcements produce significant increases in strength, creep resistance, and high cycle fatigue strength, dispersoids may adversely affect ductility and low-cycle fatigue life. Furthermore, the presence of stable dispersoids in molten solder is expected to alter the flow characteristics, wettability, and solderability of the solders.

If the second-phase reinforcements are >200 nm in size and are separated by distances >500 nm, such systems fall under the category of particulate-/whisker-reinforced composites. The yield stress of such a composite system is given to a first approximation by the weighted average of the properties of the individual components comprising the composites. Depending on the direction of the applied stress, the matrix and the reinforcements may experience the same strain or the same stress. If a tensile force is applied normal to the orientation of the reinforcements, then the stress borne by each phase is the same, but the strains experienced are different. The composite strain will be a volumetric weighted average of the strains of the individual phases:

$$\varepsilon = f\,\varepsilon_d + (1\text{-}f)\,\varepsilon_m \tag{10-4}$$

where ε_d and ε_m are strains in the reinforcement and matrix, respectively; f is the volume fraction of the reinforcements.

The composite modulus is given by

$$E_c = \frac{E_d E_m}{fE_m + 1 - f\,)\,E_d} \tag{10-5}$$

where E_m and E_d are the matrix and reinforcement modulus, respectively.

In most composites the arrangement of the phases with respect to an applied external force is such that the strains in both phases are equal, but the external force is partitioned unequally between the phases:

$$\sigma = f\,\sigma_d + (1-f)\,\sigma_m\,. \tag{10-6}$$

For this condition, the composite modulus is given as

$$E_c = f\,E_d + (1-f)\,E_m \tag{10-7}$$

which is an upper bound of composite modulus.

As in dispersion-strengthened alloys, significant strength and creep resistance improvements have been obtained in whisker-and-particle reinforced composites. Studies to identify the effects of reinforcements on fatigue and fracture toughness have been few and inconclusive.

10.4 METHODS OF PRODUCING COMPOSITE SOLDER ALLOYS

There are four principle methods of producing composite solders. The first three are for the manufacture of dispersion-strengthened alloys and the last category represents a more traditional fiber-reinforced composite.

10.4.1 *Powder Blending*

Composite solders can be made simply by mixing the intermetallic compound Cu_6Sn_5 with eutectic solder in a paste form or a molten state [Wasynczuk 1992; Clough 1992; Marshall 1992; Pinizzotto 1992]. In either case, the intermetallic compounds have to be prepared separately. Cu_6Sn_5 can be produced by supersonic inert-gas spray atomization [Wasynczuk 1992]. The problem with the method is that the intermetallics result in coarse reinforcements (10 - 100 μm).

10.4.2 *Mechanical Alloying*

This method has been explored using Ni_3Sn_4 intermetallics [Betrabet et al. 1991, 1992]. The Ni_3Sn_4 formed by casting is broken down to powders in a high-energy mill for 10 h. Powders less than 20 μm are blended with 60Sn-40Pb solder for 12 h. The blended mixture is mechanically alloyed in a high-energy mill for 6 h. The disadvantage of this method is that the intermetallic can be contaminated by oxygen. An alternative method is to produce intermetallic powders by mechanical alloying and then introduce them directly to molten solder. The mixture is then subjected to ultrasonic stirring and is subsequently quenched. During the stirring operation, a reducing agent is added to the melt to prevent the oxidation of the solder.

10.4.3 In-Situ *Composite Solders by Rapid Solidification*

Dispersion strengthened *in-situ* composite solders of Sn-Pb-Ni and Sn-Pb-Cu alloys containing 0.1 – 1.0 μm dispersoids/reinforcements can be produced by induction heating and inert gas atomization [Sastry et al. 1992]. Composite solder paste can then be made from sieved powders. The advantages of the rapid solidification process are that very fine *in-situ* dispersoids are formed in one process, and no oxygen contamination occurs. This method is very promising and is described in greater detail below.

10.4.4 Fiber-Reinforced Composites

Carbon fiber-reinforced metals are often used to create a composite that combines the beneficial properties of two materials into a single material while minimizing the poorer attributes. An application along these lines has been proposed by Ho and Chung [1991] for copper-coated carbon fibers used for 60Sn-40Pb preforms. The solder preforms are made by coating directional copper-plated fibers with 60Sn-40Pb under pressure. The solder bonds with the copper on the carbon and forms a metallurgical joint. The initial focus of the work was to form a low-expansivity solder by including fibers that had a coefficient of thermal expansion close to zero. The expansivity of the solder was tailored by the orientation and number of fibers. The carbon reinforced solder was tested under conditions of thermomechanical fatigue and compared with 60Sn-40Pb, and was found to have an increased lifetime. The fibers held the solder matrix together even after a heterogeneous coarsened band formed and cracked through the solder joint. In this case, the fibers do not retard or prevent the metallurgical mechanisms that lead to thermomechanical fatigue failure of the solder. However, the fibers of the composite hold the joint together.

The problem with this technique is the difficulty in manufacturing composite solder joints in electronic assemblies. To be effective, the carbon fibers must maintain their directionality during the reflow process. However, there may be applications where this manufacturability issue is not as serious, and a composite solder would work well. For example, die-attached reflow of a semiconductor to a ceramic carrier can be performed with composite solder preforms while maintaining fiber directionality.

10.5 *IN-SITU* COMPOSITE SOLDERS BY RAPID SOLIDIFICATION

Potential dispersoid-and/or reinforcement-forming elements in Sn-Pb alloys are: copper (to form compounds of the type Cu_6Sn_5, Cu_3Sn, and $CuSn_3$; a metastable phase), copper and yttrium in combination (to form Cu_6Y compounds in Sn-Pb matrix), nickel (to form Ni_3Sn_4), titanium (to form Ti_3Sn), and niobium (to form Nb_3Sn). Phase diagrams indicate that large volume-fractions of the above compounds in Sn-Pb matrix can be produced by rapid solidification processing.

Slow cooling rates and insufficient melt undercoolings render the currently used ingot metallurgy and mechanical alloying methods unsuitable for producing a uniform dispersion of fine insoluble particles in a solder matrix — an essential requirement for producing creep-and fatigue-resistant composite solders. Additions of large amounts of dispersoid-forming elements to conventionally treated cast alloys result in small volume-fractions of coarse, equilibrium, constituent particles, which are ineffective for dispersion strengthening. Rapid solidification processing is effective in producing dispersion-strengthened alloys. The solubility of potential dispersoid-forming elements in metal matrices can be increased significantly by increasing solidification rates to above 10^4K/s, with large undercoolings. The solid solutions can be subsequently annealed at suitable temperature to obtain fine incoherent dispersoids. High cooling rates also result in a decreased dendrite arm spacing which facilitates dissolution of normally slightly soluble constituents; thereby effecting a more efficient use of solute elements.

Traditionally, other elements are added to the tin-lead alloy or used with tin and lead separately in different compositions for various reasons of strength and other specific physical properties [Johnson and Kevra 1989]. These other elements include antimony (Sb), bismuth (Bi), cadmium (Cd), indium (In), and silver (Ag). Tin-lead alloys containing up to 3% Sb have slightly greater strength than the binary alloys and can be used in less-exacting applications for economic reasons. One of the most widely used antimony-containing alloys is the 95Sn-5Sb, which provides superior creep strength and good thermal fatigue resistance.

Bismuth is added primarily to lower the melting point of the alloy, but it reduces the mechanical strength of the joint and in high concentrations can produce poorly alloyed and brittle joints. The major advantage of indium is its ductility, which enables indium-containing alloys to withstand thermal stress fatigue testing much more successfully than tin-lead solders. Unfortunately, the cost of this metal is high, and for some indium-containing alloys, the soldering characteristics are very poor. Silver is used to impart strength to the alloy or reduce silver scavenging. The solubility of silver in eutectic tin-lead solder is about 5%. Because of the highly toxic nature of cadmium, the future use of cadmium will be increasingly restricted by environmental regulations, if not entirely prohibited.

Under the condition of thermodynamic equilibrium, some alloying elements can react with tin to form intermetallic compounds. Table 10-1 shows the alloy systems that involve tin-containing intermetallic compounds. Additions of the elements listed in Table 10-1 to conventionally cast Sn-Pb solders result in small volume-fractions of coarse, equilibrium, constituent particles which are ineffective for dispersion strengthening and may be detrimental to ductility and toughness. Large amounts of under cooling and high solidification rates during rapid solidification processing promote the formation of large volume-fractions of fine particles suitable for dispersion strengthening. Among these elements, gold, niobium, copper, indium, and nickel seem to be attractive candidates because these elements form tin-rich intermetallic compounds and, with small amounts of alloying additions, large volume-fractions of dispersoids can be produced.

Table 10-1 Intermetallic Compounds of Tin Which Are Potential Candidates for Particle Strengthening in Sn-Pb Solders

Alloy System	Intermetallics	Crystal Structure
Ag-Sn	Ag-Sn	hcp
	Ag_3Sn	orthorhombic
Au-Sn	$AuSn_4$	face-centered orthorhombic
	$AuSn_2$	orthorhombic
	AuSn	hexagonal
	Au-Sn	hcp
Ca-Sn	$CaSn_3$	ordered fcc ($L1_2$)
	CaSn	orthorhombic
	Ca_2Sn	orthorhombic
Nb-Sn	$NbSn_2$	ordered orthorhombic
	Nb_6Sn_5	complex body-centered orthorhombic
	Nb_3Sn	cubic
Cd-Sn	Cd-Sn	hexagonal
Cu-Sn	Cu_3Sn	fcc or pseudo-hexagonal
	Cu_6Sn_5	hexagonal
Fe-Sn	Fe_3Sn	hexagonal
	Fe_3Sn_2	complex monoclinic
	FeSn	hexagonal
	$FeSn_2$	bct
In-Sn	In_3Sn	body-centered pseudo-tetragonal
	$InSn_4$	hexagonal
Mg-Sn	Mg_2Sn	fcc
Mn-Sn	$MnSn_2$	tetragonal
Ni-Sn	Ni_3Sn	cubic or hexagonal

Alloy System	Intermetallics	Crystal Structure
	Ni_3Sn_4	monoclinic
Sb-Sn	SbSn	rhombohedral, nearly cubic
Ti-Sn	Ti_6Sn_5	hexagonal or orthorhombic
	Ti_5Sn_3	hexagonal
	Ti_2Sn	hexagonal
	Ti_3Sn	hexagonal
Zr-Sn	$ZrSn_2$	face-centered orthorhombic
	Zr_3Sn_2	hexagonal
	Zr_4Sn	complex face-centered tetragonal or cubic

10.5.1 Preparation of Rapidly Solidified In-Situ *Composite Solder Powders*

A schematic of the experimental apparatus for the production of rapidly solidified *in-situ* composite solders by induction-melting/gas-atomization (IMGA) is shown in Figure 10-4. A charge of known composition is melted in a graphite crucible by electromagnetic induction and the molten stream is discharged through a graphite tundish. The melt stream is atomized by a circular array of gas jets. Atomization is accomplished by the focused impact of 12 separate gas jets situated at a 2 inch diameter base circle with a half-apex angle of 19.5°, which is generally considered as an optimum for atomizing a melt stream of metals. Helium gas at pressures of 0.8 – 1.5 MPa is used to atomize and rapidly solidify the melt stream into fine, spherical powder particles. The rapidly solidified powders are sieved, and the –80 mesh (<180 μm) powders are used to make test samples. The solder powders so produced can be used to make solder paste and bulk solder specimens.

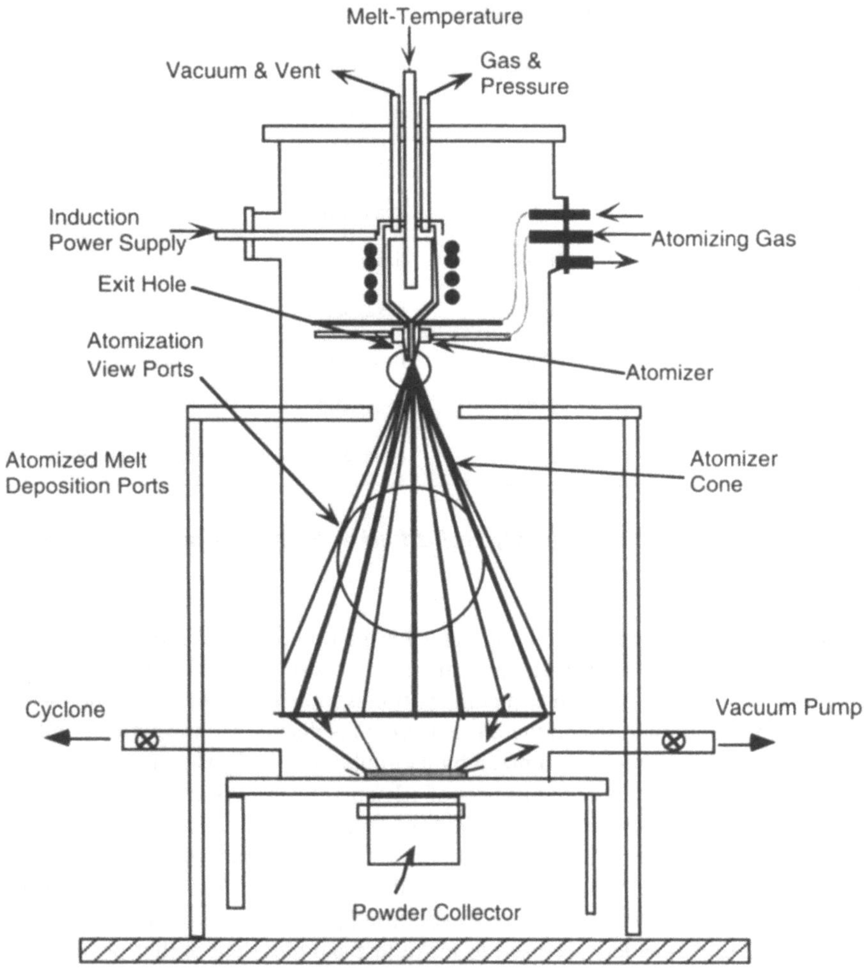

Figure 10-4 Induction-melting/gas-atomizations facility for production of rapidly solidified in-situ composite solder powders

10.5.2 Microstructures of In-Situ *Composite Solders*

Representative microstructures of 63Sn-37Pb and Sn-Pb-X (X = Cu and Ni) *in-situ* composite solders produced by induction-melting/gas-atomization and consolidated by extrusion are shown in Figures 10-5 to 10-8. Rapid solidification produces significant grain refinement in the conventional solder alloy. The microstructures of copper-containing alloys consist of 5–10 μm grains of α and β phases and 0.1–1.0 μm copper-rich Cu_6Sn_5 dispersoids (Figure10-7). With increasing amounts of copper, the grain size decreases, and the size and volume-fraction of copper-rich dispersoids increase. Upon reflow, the grain size increases and significant dispersoid coarsening results.

(a)

(b)

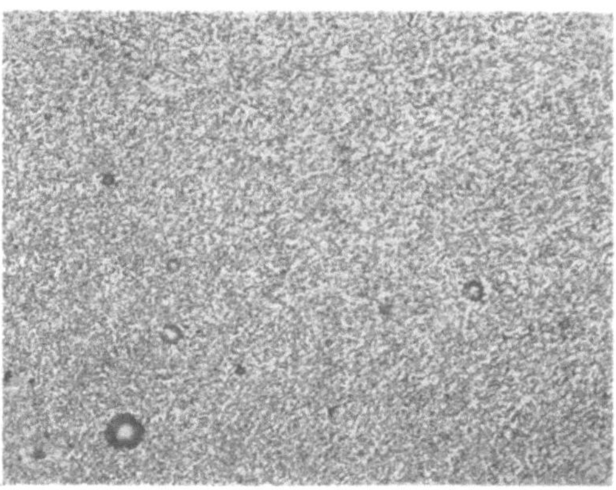

Figure 10-5 Microstructures of rapidly solidified 63Sn-32Pb-5Ni alloy a) as extruded and b) reflowed 63Sn-32Pb-5Ni alloy

The nickel-containing alloys exhibit much finer grains (typically 1–5 μm) and 50-500 nm Ni_3Sn_4 dispersoids. Reflow results in limited grain and dispersoid coarsening with the extent of coarsening significantly smaller than in copper-containing alloys (Figure 10-8). In the rapidly solidified and extruded copper-containing alloys, the Cu_6Sn_5 dispersoids are spherical and 0.2–1.0 μm in diameter. Upon reflow, the dispersoids extensively coarsen with the smallest dimension of the plates increasing to greater than 1 μm.

(a)

(b)

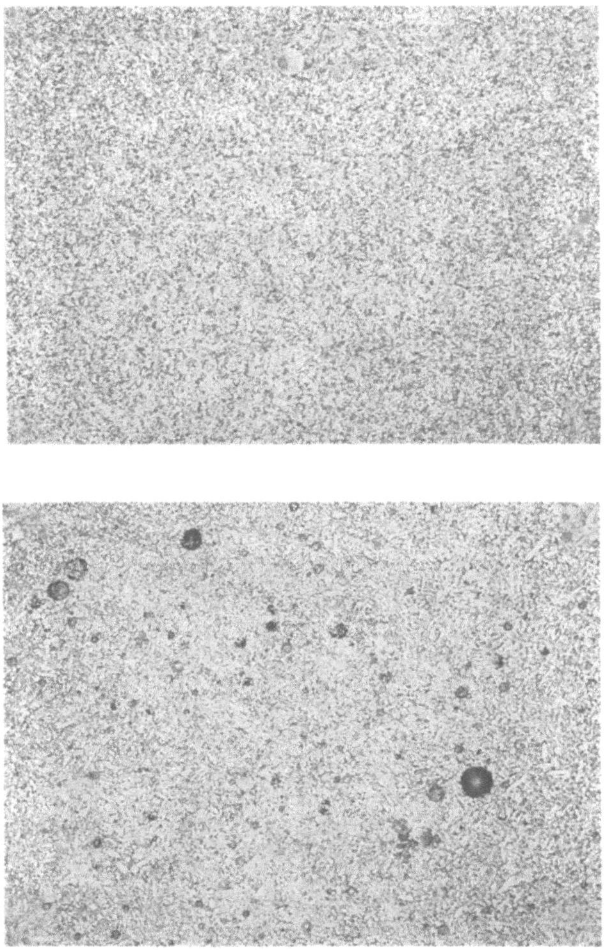

Figure 10-6 Microstructures of rapidly solidified 63Sn-32Pb-5Cu alloy a) as extruded and b) reflowed 63Sn-32Pb-5Cu alloy

The size and morphology of dispersoids in rapidly solidified and extruded nickel-containing alloys are similar to that found in copper-containing alloys. Although some dispersoid coarsening occurs upon reflow of nickel-containing solders, the dispersoids remain spherical and are <1 µm — significantly finer than in copper-containing alloys.

(a)

(b)

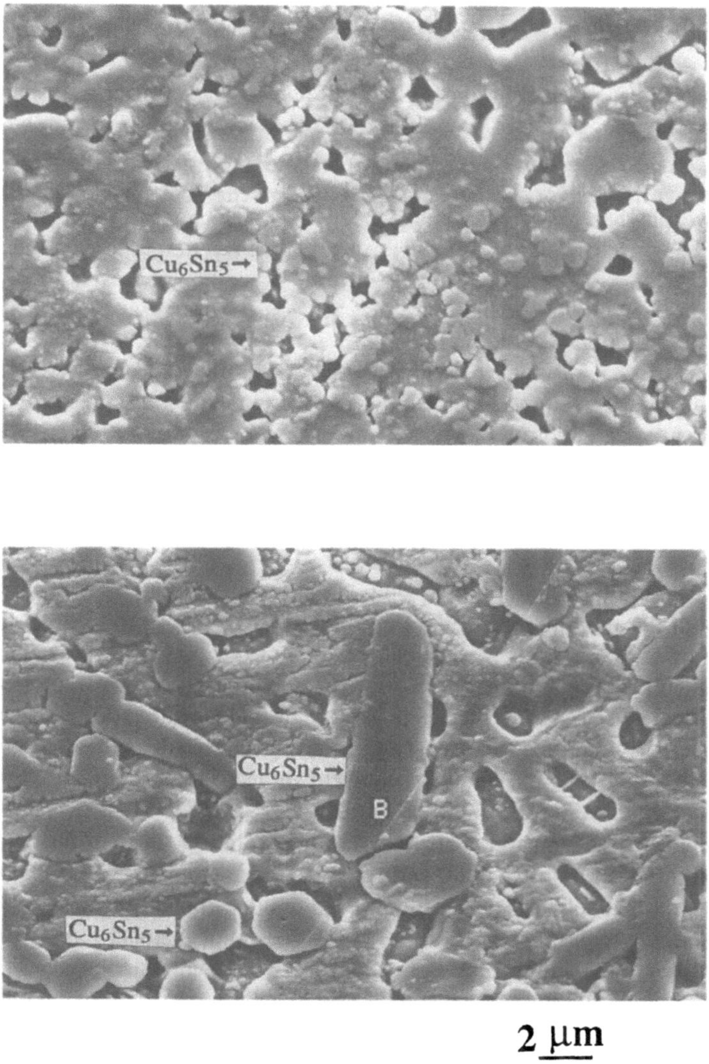

Figure 10-7 Scanning electron micrographs of dispersoids in 63Sn-29.5Pb-7.5Cu: a) rapidly solidified and extruded, and b) after reflow

10.6 PROPERTIES OF COMPOSITE SOLDER ALLOYS

This section will first discuss the wetting behavior of composite solder alloys and their usefulness in joints for electronic systems. The second half of this section will describe the mechanical properties (experimental) of composite solders.

10.6.1 Wetting and Solderability of Composite Solder Alloys

Mechanical behavior should not dominate the selection process of solder alloys to the exclusion of wetting properties and manufacturability. The optimum solution would be a composite solder alloy with a solderability similar to near-eutectic Sn-Pb but with superior mechanical and thermomechanical properties. Unfortunately, the wetting behavior of composite solder alloys has not been explored in great detail. The majority of the work performed to date has looked at only the mechanical properties of the composites in bulk form. However, there is some limited information available in the literature that provides information about the wetting behavior of composite solder alloys.

The wetting behavior of near-eutectic Sn-Pb solders changes when alloying elements, or dispersions, are added to make a composite solder alloy. The composite solders depend upon the good wetting properties of 60Sn-40Pb to form the joint. Under ideal conditions, the added second phase that forms the composite has no effect on the wetting behavior. Unfortunately, this is rarely true. A variety of solder composites have been proposed in this chapter. This section discusses the wetting behavior of each.

The addition of cadmium to solder joints to form a fatigue-resistant alloy causes both the melting temperature and liquid surface tension to drop. The drop in melting temperature is not critical but the lower liquid surface tension can result in solder bridging, or icicle formation, on the circuit board between leads, causing electrical shorts. As the surface tension of the solder liquid drops, the solder has a reduced driving force to form a sphere that can cause adjacent leads to be inappropriately bridged by the solder. The area of spread is reduced by 50% by adding 2% Cd and doubles the wetting time [Ackroyd et al. 1975; MacKay 1981]. The drop in liquid surface tension is attributed to the rapid oxidation of the cadmium. The cadmium oxide causes a dull surface appearance and inhibits flow and spreading of the molten solder. The addition of indium does not strongly affect the wetting properties yet has a similar effect on the thermomechanical fatigue properties as cadmium does. The wetting is slowed somewhat again due to the preferential oxidation of the indium but the area of spread and time to wetting are not greatly diminished with indium compared to 60Sn-40Pb.

(a)

(b)

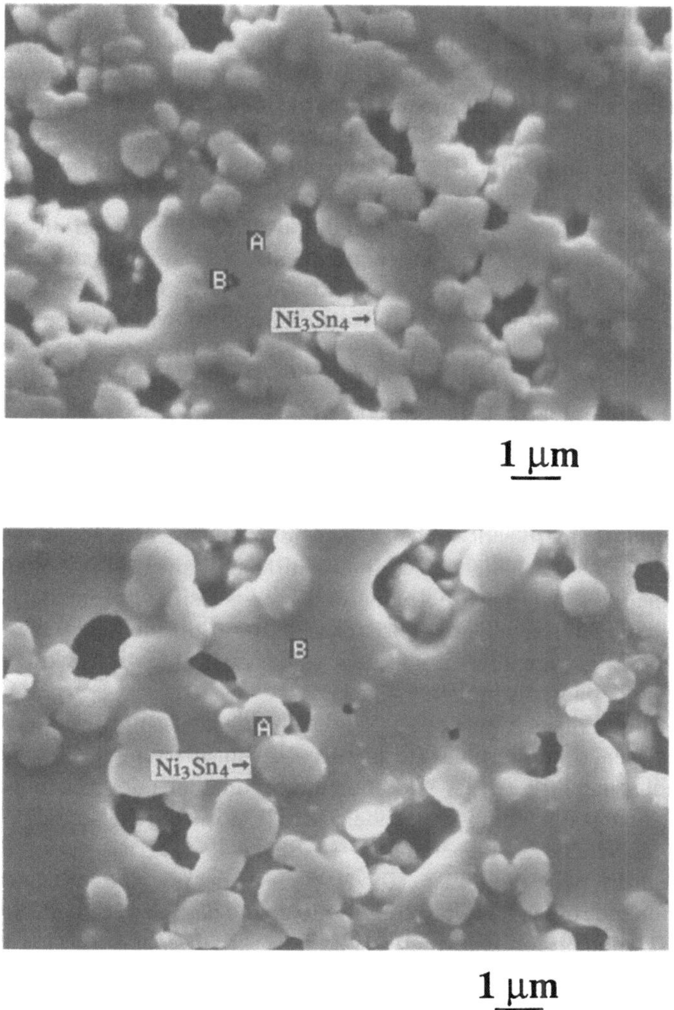

Figure 10-8 Scanning electron micrographs of dispersoids in 63Sn-29.5Pb-7.5Ni: a) rapidly solidified and extruded, and b) after reflow

The influence of a dispersion of intermetallics in solder uniformly changes the surface appearance of the solder joints [MacKay 1981; Ackroyd et al. 1975; Steen and Becker 1986]. The addition of $AuSn_4$, Ni_3Sn_4, and Cu_6Sn_5 dispersions all cause the joint to have a "gritty" surface appearance. This is caused by the intermetallic particles floating to the top of the molten solder (because they are less dense than the Sn-Pb solder [Steen and Becker 1986]) which results in a rough surface upon solidification. The addition of copper to form a dispersion of Cu-Sn intermetallics decreases the spread by 25% [Ackroyd et al. 1975; MacKay 1981]. Wetting balance tests with nickel additions (to form Ni_3Sn_4 intermetallics) decreased the wetting force and consequently increased the wetting time at concentrations beginning at 0.25% Ni [MacKay 1981; Steen and Becker 1986]. The addition of gold results in a similar gritty joint as with the copper additions. However, up to 2.5%, gold has very little effect on the area of spread or other wetting characteristics [MacKay 1981; Steen and Becker 1986]. However, excessive Au-Sn intermetallics have been found to increase the number of voids in the solder joint and are thought to be due to outgassing of volatiles incorporated into the Au during a plating operation [Glazer et al. 1991].

The presence of a small amount of intermetallic dispersions uniformly decreases the wetting properties of solders. The solders have reduced capillary flow and spreading due to an overall decrease in the surface tension of the molten solder. In essence, the intermetallic particles increase the viscosity of the solder. This is exacerbated with intermetallic particles (especially Au-Sn and Cu-Sn intermetallics) because these intermetallics in the bulk solder coarsen at elevated temperatures while the solder alloy is molten.

The addition of inert particles to solders has less of an impact on the wetting properties because the particles are small and do not coarsen. The addition of tungsten particles had little adverse effect on the wetting behavior of a low coefficient of thermal expansion composite solder [IBM 1986]. Betrabet et al. [1991, 1992] have developed a fine dispersion of Ni_3Sn_4 that does not segregate (due to matched density with the solder). The Ni-Sn intermetallic does not coarsen extensively when the solder is molten. Additionally, the particles act as nucleation sites for solidification on cooling which results in a much finer as-solidified microstructure. The wetting properties are purported to be similar to 60Sn-40Pb solder alone and are perhaps due to the small size of the particles not having a large influence on the surface tension of the solder. However, the wetting properties have not been published and the reported results are only qualitative.

The addition of large second-phase dispersants to form a composite solder (e.g. lead dendrites) decreases the wetting behavior of solders. The problem with lead dendrites is that the solidification behavior of lead and tin is non-reciprocal. The lead-rich dendrites are poor nucleation sites for tin but the tin agents are potent sites for lead. The lead dendrites form quickly once below the liquidus temperature, but the tin significantly undercools. Therefore, lead dendrites can float in the molten solder, thereby inhibiting flow and spreading in a manner that is similar to that for intermetallic dispersions.

Carbon fiber-reinforced solders are used as tailoring agents to match coefficients of thermal expansion and improve the thermomechanical fatigue life of solder composites. However, their wetting behavior is extremely limited. The composite solder depends upon the excellent wetting between the copper-coated carbon fiber and the solder to form a contiguous preform. In order to optimize the properties of the composite, the C fibers must be aligned. If the composite is reflowed in an uncontrolled fashion, the fibers would lose alignment. Therefore, to make a joint with a carbon fiber composite solder the solder cannot completely melt. The method used to overcome this problem involves using temperature and pressure [Ho and Chung 1991]. The composite-solder preform is placed in the assembly to be joined and heated to below the melting temperature of the solder while pressure is applied to the assembly. The pressure breaks the surface oxide on the metallizations of the components to be joined (and can be used as a fluxless process [Mizuishi et al. 1988]) and causes localized melting at the interface. A metallurgically sound joint is made in this localized region.

Results to date indicate that the use of composite solders does not improve the wetting behavior of solder alloys. Additional testing and further development is needed for composite solders to be used in electronic assemblies. However, compromises must occasionally be made in the ease of manufacture in order to improve solder joint reliability.

10.6.2 Mechanical Properties of In-Situ *Composite Solders*

The majority of the mechanical property data that exist for composite solders is for bulk material under isothermal conditions. Solder samples for testing and evaluation were prepared from 6 mm diameter extruded rods. The powders were first pressed to >95% density at room temperature, the die assembly was heated to 100°C and extrusion was performed at 100°C at an extrusion ratio of 16:1 to produce 6 mm diameter rods.

In practical applications all solder alloys experience reflow conditions, i.e., remelted and then air-cooled to room temperature. The extent of microstructural changes depends on the cooling rate and compositions of solder alloys. For the eutectic Sn-Pb solder alloy, a faster cooling rate results in a non-lamellar structure in which coarse lead-rich alpha phase particles are dispersed in tin-rich beta phase matrix. Slower cooling rates produce a lamellar structure and coarse grain size. For composite solders, intermetallic compounds such as Cu_6Sn_5 and Ni_3Sn_4 do not dissolve during reflow. The dispersoids tend to constrain grain growth in the Sn-Pb matrix.

Reflowed solder samples were prepared by heating the 6 mm diameter rods in an aluminum mold to 20–30°C above the melting point and cooling the die-sample assembly at rates comparable to that in service conditions. A representative temperature-time profile during heating and cooling of a solder sample is shown in Figure 10-9. The reflowed solder samples were subsequently hot isostatically pressed at 100°C and 70 MPa for 2–3 hours to close any voids formed during reflow.

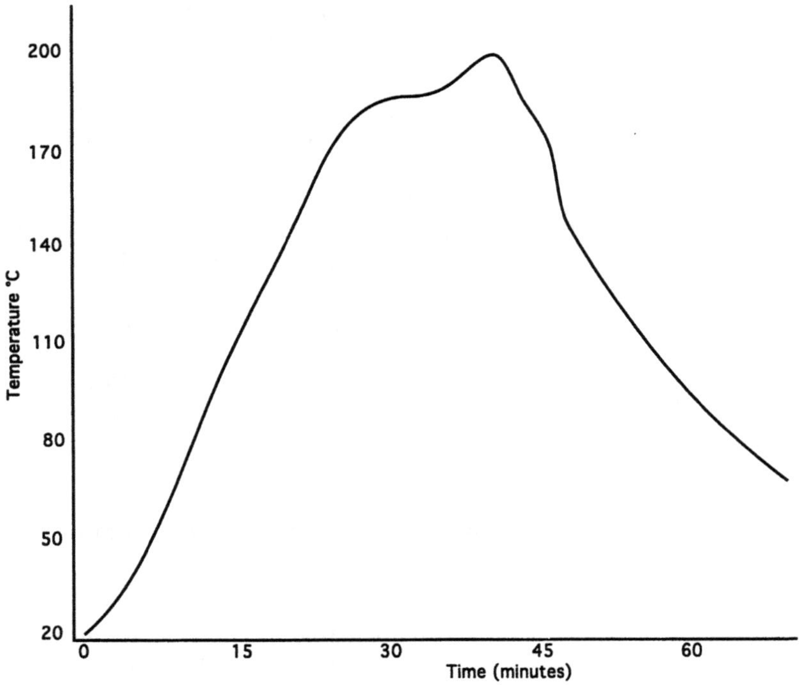

Figure 10-9 Typical temperature-time profile during reflow of solders

Table 10-2 shows the compositions, along with estimated dispersoid size and volume-fraction, of composite solders for which extensive mechanical property data are available. The volume-fractions of dispersoids shown in Table 10-2 are based on the assumption that copper and nickel completely combine with tin to form Cu_6Sn_5 and Ni_3Sn_4. The density of Cu_6Sn_5 and Ni_3Sn_4 were taken to be 8.28 g/cm^3 and 8.65 g/cm^3, respectively.

10.6.3 Mechanical Property Testing

In order to rank and compare the performance of different solder alloys, a comprehensive data base of the tensile, creep, and isothermal fatigue properties of bulk solder alloys was generated. Tensile tests were conducted at a strain rate of 2×10^{-3}/s on smooth tensile specimens of the geometry shown in Figure 10-10. Specimens were tested to determine Young's modulus, yield stress, ultimate tensile strength, percentage reduction in area at failure, and percentage plastic elongation to failure. Duplicate tests were performed for selected temperature and heat treatment conditions. Tensile creep tests were performed at 25, 75, and 125°C and the extent of primary creep, steady state creep rate, and creep-rupture time as functions of stress and temperature were determined. The stress exponent and activation energy values for creep were used to determine the operative creep deformation mechanism. Fatigue tests included isothermal low-cycle fatigue (LCF) and high-cycle fatigue (HCF). Stress-controlled HCF tests were performed in air at 25°C on smooth axial specimens in accordance with ASTM E 466 procedures. Tests were performed at four different stress levels between 0.4 and 0.8 of the yield stress to obtain S-N curves. The tests were conducted at a stress ratio of −1 and cyclic frequencies of 0.05, 0.5, and 5 Hz. Strain-controlled LCF tests were performed in air at 25°C in accordance with ASTM E 606 procedures. The specimens were tested at strain ranges of 0.5 and 1.0% under total strain control to obtain plots of peak stress as a function of the number of cycles. The tests were fully reversed and were conducted at strain rates of 0.001 – 0.0001 s^{-1}.

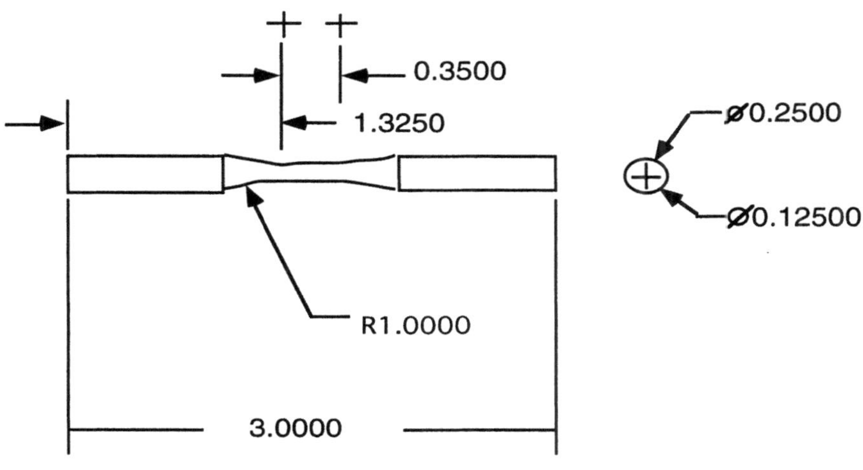

Figure 10-10 Geometry of specimens used for tensile, fatigue, and creep testing

Table 10-2 The Composition, Dispersoid Size and Volume Fraction of Dispersoids of Eutectic Sn-Pb-Based Solder Alloys

Solder	Nominal composition Sn% - Pb% - Cu/Ni%	Intermetallic particle type	Dispersoid size in diameter, μm	Volume fraction of dispersoids
6337	63 - 37	-	-	-
RS 6337	60 - 40	-	-	-
2.5% Cu	61.5 - 36.1 - 2.4	Cu_6Sn_5	0.33	6 %
5.0% Cu	60.0 - 35.2 - 4.8	Cu_6Sn_5	0.42	13 %
7.5% Cu	58.6 - 34.4 - 7.0	Cu_6Sn_5	0.50	19 %
2.5% Ni	61.5 - 36.1 - 2.4	Ni_3Sn_4	0.20	9 %
5.0% Ni	60.0 - 35.2 - 4.8	Ni_3Sn_4	0.25	18 %
7.5% Ni	58.6 - 34.4 - 7.0	Ni_3Sn_4	0.30	27 %

10.6.4 Tensile Properties

The tensile properties at 25°C and a strain rate of 2×10^{-3}/s of the commercial 63Sn-37Pb solder are summarized in Table 10-3, as well as rapidly solidified, extruded, and reflowed solders containing varying amounts of dispersoid forming ternary additions. The yield stress and modulus values of commercial 63Sn-37Pb solder are in agreement with those reported in the literature. However, the rapidly solidified and extruded 63Sn-37Pb alloy has significantly lower yield stress and modulus value because of the fine grain size resulting from rapid solidification. At high homologous temperatures, the flow stress increases with increasing grain size in contrast to the flow stress dependence on the inverse square root of the grain size generally observed at low temperatures. Upon reflow of the rapidly solidified 63Sn-37Pb alloy, yield stress and modulus increase but are lower than in the commercial 63Sn-37Pb solder.

Figure 10-11 shows the yield strength of composite solder plotted as a function of volume-fraction of dispersoids. It should be noted that the yield strength varies linearly with the volume-fraction of dispersoids irrespective of the type of dispersoid-forming element. Nickel is more effective than copper in forming large volume-fractions of dispersoids, hence, it is more effective in improving the tensile strength. Compared to conventional 63Sn-37Pb, dispersion-strengthened alloys have higher tensile strengths, but lower elongation and percent reduction in area.

Table 10-3 Tensile Properties of *In-Situ* Composite Solder Alloys at Strain Rate of $2x10^{-3}s^{-1}$

Solder	0.2% offset yield strength, MPa	Ultimate tensile strength, MPa	Elastic modulus, GPa	Elongation %	Reduction in area
6337	35	37	12	48	331 %
RS 6337	29	32	14	52	57 %
2.5% Cu	46	48	15	30	45 %
5.0% Cu	52	60	21	18	37 %
7.5% Cu	58	66	23	13	29 %
2.5% Ni	51	54	21	18	42 %
5.0% Ni	63	73	24	15	37 %
7.5% Ni	71	78	28	7	29 %

For dispersion-strengthened alloys, the strength increment from dislocation-dispersoid interaction is given by the modified Orowan equation:

$$\Delta\sigma = 2\Delta\tau = 1.06\,(\frac{Gb}{\pi L'})\ln(\frac{x}{\gamma_0}) \tag{10-8}$$

where G is the shear modulus (E/2(1+ ν), where E is the elastic modulus; the Poisson's ratio, ν, can be assumed to be equal to 0.35 for Sn-Pb alloys. The Burger's vector of the dislocation, b, is equal to 3.2×10^{-4} mm [Pink et al. 1981]. The dispersoid sizes and interparticle spacings (L') determined from scanning electronic micrographs are shown in Table 10-4. As Table 10-4 shows, the experimental values of yield strengths of nickel- and copper-containing alloys are in good agreement with the strength increments given by the Orowan equation.

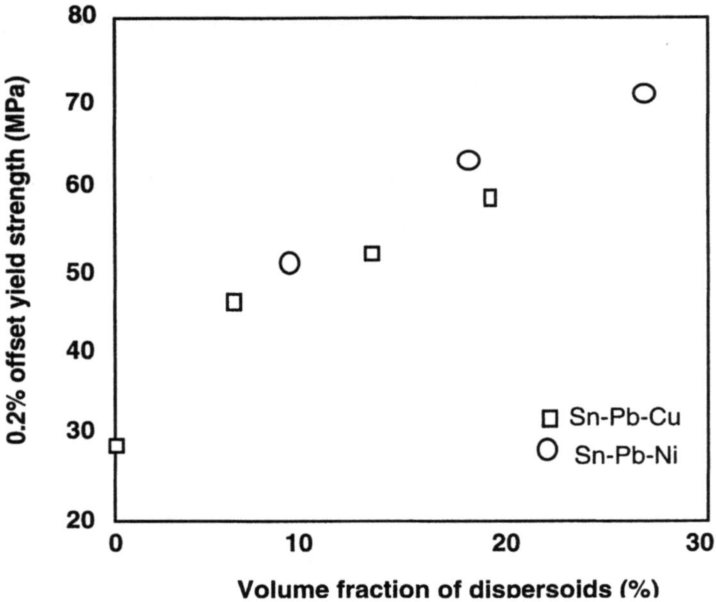

Figure 10-11 The yield strength of composite solders plotted as a function of volume-fraction of dispersoids

10.6.5 Creep Properties

Over the range of stress studied, the steady state creep rates at 25° and 75°C of composite solders and conventional solders are shown in Figures 10-12a-c. The steady state creep rates at all temperatures are significantly lower in nickel- and copper-containing composite solders than in the conventional solder. The increased creep resistance of the composite solders is due to the effectiveness of dispersoids in inhibiting grain boundary sliding and providing obstacles to dislocation climb. The steady state creep rate of 5% Ni is about three orders of magnitude lower than the conventional 63Sn-37Pb. However, the dispersoid-containing alloys have lower creep ductility. Rapidly solidified 63Sn-37Pb solder has lower creep resistance than the slow-cooled 63Sn-37Pb alloy.

The stress and temperature dependence of steady state creep rate is given by:

$$\dot{\varepsilon}_{II} = A\left(\frac{\sigma}{E}\right)^{n} \exp\left(\frac{-Q_c}{RT}\right) \tag{10-9}$$

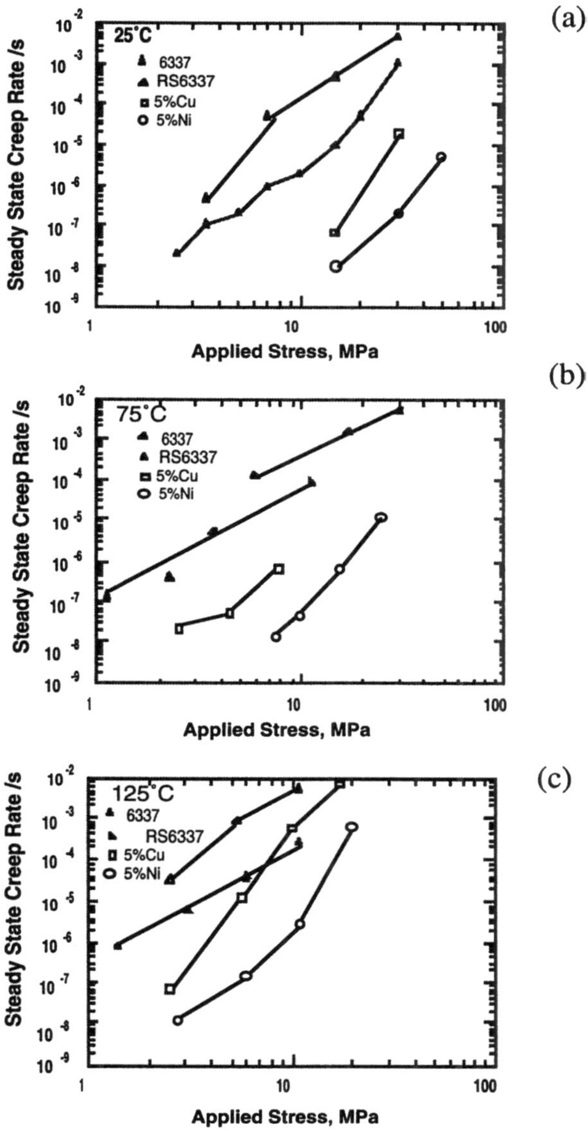

Figure 10-12 Steady state creep rate of solder alloys plotted vs. applied stress at: a) 25°C, b) 75°C, and c) 125°C.

Table 10-4 Interparticle Spacing and Dispersion Strengthening of Composite Solders By Modified Orowan Equation

Solder	Interparticle spacing, L; μm	Calculated strength increment, MPa	Experimental yield strength increment, MPa
2.5% Cu	0.34	9	11
5.0% Cu	0.26	12	17
7.5% Cu	0.21	16	23
2.5% Ni	0.17	17	16
5.0% Ni	0.11	26	28
7.5% Ni	0.08	36	36

where n is the stress exponent and Q_c is the creep activation energy. Values of stress exponents determined from the slopes of ln $\dot{\varepsilon}_{II}$ versus ln σ plots are shown in Table 10-5. The stress exponent varies with temperature. The stress exponents are higher in composite solders than in the conventional solder of eutectic composition. The values of activation energy, Q_c, determined from the plots of ln $\dot{\varepsilon}_{II}$ versus 1/T are shown in Table 10-6. The activation energy decreases as the applied stress increases. Activation energy for creep is higher in composite solders than in the eutectic solder alloy.

At temperature from ambient to 125°C, the activation energy values obtained in this study for conventional 63Sn-37Pb and rapidly solidified 63Sn-37Pb are very close to the activation energies for superplastic region II reported previously in the literature [Mohamed and Langdon 1975; Pink et al. 1976; Kashyap and Murty 1981] for Sn-Pb eutectic alloys. The activation energy for self-diffusion within grains and at grain boundaries for pure tin, lead, copper, and nickel are shown in Table 10-7 [Smithells 1976]. The activation energies for conventional 63Sn-37Pb and rapidly solidified 63Sn-37Pb are close to the activation energy for grain boundary diffusion of pure tin and lead. On the other hand, the activation energies for dispersion-strengthened solders are close to those of self-diffusion of tin and lead in the lattice.

Table 10-5 Stress Exponents of Solder Alloys for Creep at Diffrent Temperatures

Solder alloy	Stress exponent, n		
	25°C	75°C	125°C
6337	3.9	3.1	2.7
RS 6337	4.0	2.6	3.1
5.0% Cu	8.9	3.2	5.1
5.0% Ni	5.5	5.2	4.8

The dominant creep deformation mechanism for the solder alloys can be verified further by the stress exponent. It is well established that for materials exhibiting superplasticity, such as Sn-Pb alloys, the strain rate sensitivity ranges from 0.3 to 0.8 which corresponds to a stress exponent of 3.33 – 1.25. The most important mechanism in superplasticity is grain boundary sliding. For dislocation creep, either dislocation climb or dislocation glide, the stress exponent ranges from 3 to 7. Therefore, the dominant creep deformation mechanism for conventional 63Sn-37Pb and RS 63Sn-37Pb appears to be grain boundary sliding; for composite solders the dominant creep deformation mechanism appears to be dislocation creep.

The significant reduction of creep rate for composite solders is due to the presence of second-phase dispersoids which suppress grain boundary sliding and reduce the creep rate. The presence of second phase particles also increases the stress exponent in agreement with the high stress exponents for other dispersion-strengthened alloys Al-Fe-V-Si [Kim and Griffith 1988].

Table 10-6 Activation Energy (in kcal/mole) of Solder Alloys for Creep at Various Applied Stress Levels

Solder alloy	Activation energy			
	3.5 MPa	7 MPa	14 MPa	28 MPa
6337	15.4	13.4	11.5	9.5
RS 6337	14.5	12.9	11.3	9.8
5.0% Cu	40.0	32.9	26.2	19.4
5.0% Ni	26.8	25.6	24.4	23.2

Table 10-7 Activation Energy of Diffusion for Pure Elements [Smithells, 1976]

(a) Self – Diffusion in solid elements

Element	Q,kcal/mole	Temperature, C°
Sn	25.6 ± 0.8	178 ~ 222
	25.6 ± 1.0	160 ~ 228
Pb	25.65 ± 0.25	200 ~ 323
Cu	50.4 ± 0.2	698 ~ 1061
	47.12 ± 0.33	685 ~ 1062
	46.9	700 ~ 990
Ni	67.2	870 ~ 1404
	67.97 ± 0.18	980 ~ 1400

(b) Grain boundary self– diffusion

Element	Q,kcal/mole	Temperature, C°
Sn	9.55 ± 0.7	40 ~ 115
Pb	15.7	214 ~ 260
Ni	28.2 ± 2.0	850 ~ 1100

10.6.6 High Cycle Fatigue Properties (Stress Amplitude Controlled Fatigue)

High-cycle fatigue-life data were obtained from stress amplitude controlled fatigue tests. The test matrix included three frequencies (5, 0.5, 0.05 Hz) and three stress levels. The applied stress levels were selected close to 0.2% offset yield strengths of the alloys. The number of cycles to failure (N_f) was recorded for each test. The results are tabulated in Table 10-8 and are plotted in Figure 10-13. The 5% Ni solder alloy exhibits the longest fatigue life.

Table 10-8 Test Matrix and Results for High-Cycle Fatigue Life of Solder Alloys

	Cycles to failure, N_f						
	±34.5 MPa	±27.6 MPa	±34.5 MPa	±41.3 MPa	±27.6 MPa	±34.5 MPa	±41.3 MPa
Solder	5 Hz	0.5 Hz	0.5 Hz	0.5 Hz	0.05 Hz	0.05 Hz	0.05 Hz
6337	29,353	41,761	17,652	3,978	3,346	265	36
RS 6337		4,775	3,132	638	589	98	15
2.5% Cu			20,817	5,439			
5.0% Cu		116,384	32,055	9,736		12,264	4,126
7.5% Cu			25,705				
2.5% Ni			23,650				
5.0% Ni	234,210	285,120	167,153	53,794	55,836	20,570	9,245
7.5% Ni			106,650	36,869			

Because ambient temperature is 65% of the melting point of the eutectic Sn-Pb solder, time-dependent creep deformation can occur during fatigue and can contribute to degradation of fatigue life. The extent of deterioration caused by time-dependent deformation was assessed by studying the frequency effects on S-N behavior. The effect of frequency on fatigue life (shown in Figure 10-14) for different solder specimens indicates that frequency, dependence of fatigue life is significantly less pronounced in nickel- and copper-containing solders than in the conventional 63Sn-37Pb solder. At higher frequencies (5 Hz and 0.5 Hz), the fatigue life of the 5% Ni alloy is about one order of magnitude greater than that of 63Sn-37Pb. At the lower frequency (0.05 Hz), the fatigue life is about two orders of magnitude greater in the 5% Ni alloy than in 63Sn-37Pb. This renders the composite solders attractive for applications in which creep-fatigue interactions are pronounced.

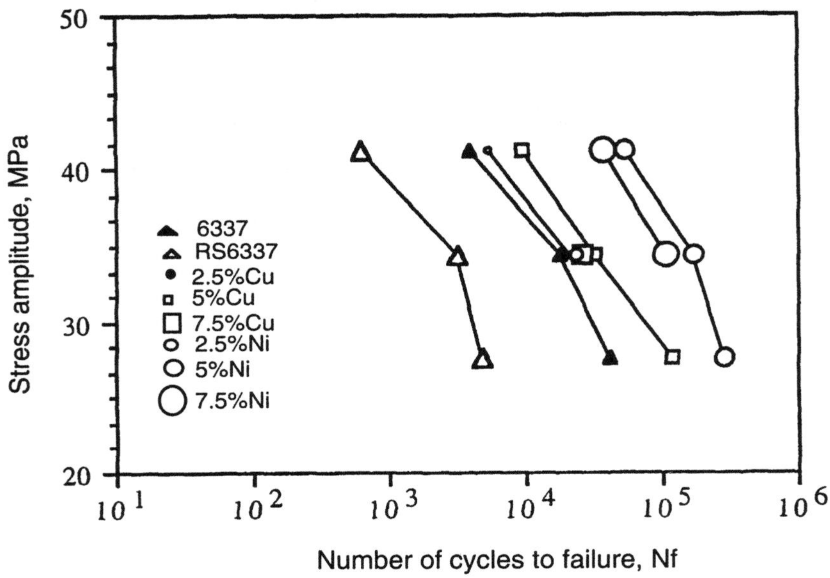

Figure 10-13 Stress vs. number of cycles to failure (S-N curves) at 0.5Hz for different compositions of solder specimens

The alloys can be ranked from longest to shortest fatigue life as follows: 5% Ni, 7.5% Ni, 5% Cu, 7.5% Cu, 2.5% Ni, 2.5% Cu, 63Sn-37Pb and rapidly solidified 63Sn-37Pb. This follows the same general trend in yield strength (see Table 10-3). The stronger materials tend to have longer high-cycle fatigue lives. The improvement of high-cycle fatigue resistance of composite solders can also be shown in Figure 10-15, where stress amplitude normalized by yield strength versus fatigue life at 0.05 Hz is plotted. At the same normalized stress amplitude, the composite solders have longer fatigue lives. Again, the fatigue life at a normalized stress level does not seem to depend on which element, nickel or copper, is added in the eutectic solder. The high-cycle fatigue life is affected by the yield strength, which is a function of volume fraction of dispersoids as has been illustrated earlier. The relationship between stress amplitude (SA) normalized by yield strength (YS) and fatigue life is given by the following equation:

$$\sigma_a / YS = 1.95 - 0.31 \log N_f . \qquad (10\text{-}10)$$

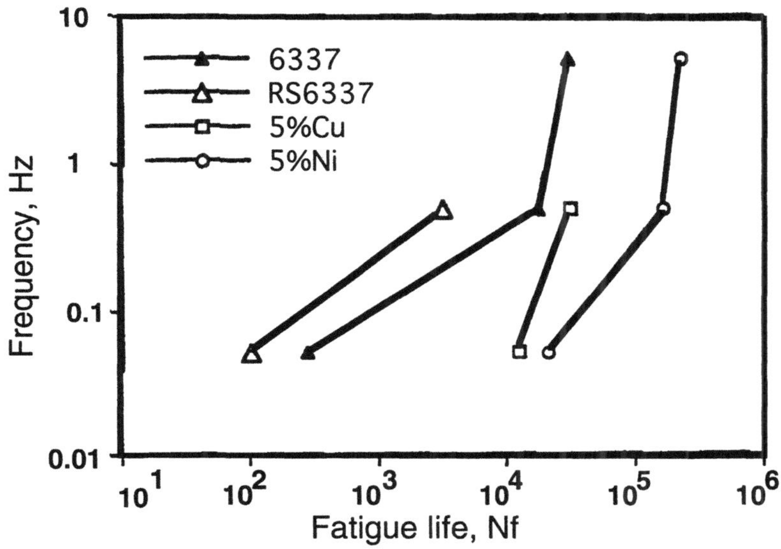

Figure 10-14 Effect of frequency on fatigue lives of solder specimens

10.6.7 Low-Cycle Fatigue Properties (Strain Amplitude Controlled Fatigue)

Fully reversed (alternate tension-compression) strain-controlled fatigue tests were performed at strain rates of 0.001 and 0.0001/s, and total strain ranges of 1 and 0.5%. Typical hysteresis plots obtained for different solder specimens are shown in Figures 10-16a-c. Results of constant strain amplitude, fully reversed fatigue tests performed at a strain rate of 0.001/s and the total strain range of 1% are listed in Table 10-9 and shown in Figure 10-17. There is significant softening in 63Sn-37Pb but the composite solders do not show any fatigue hardening or softening, indicating that they are more stable under reversed cyclic deformation than 63Sn-37Pb. This is especially desirable in the design of solder joints. Lower fatigue lives under constant strain amplitude conditions in copper- and nickel-containing alloys can possibly be explained by triple point void nucleation and growth in the presence of fine grain structure and dispersoids on grain boundaries, which result in stress concentration and early fatigue failure.

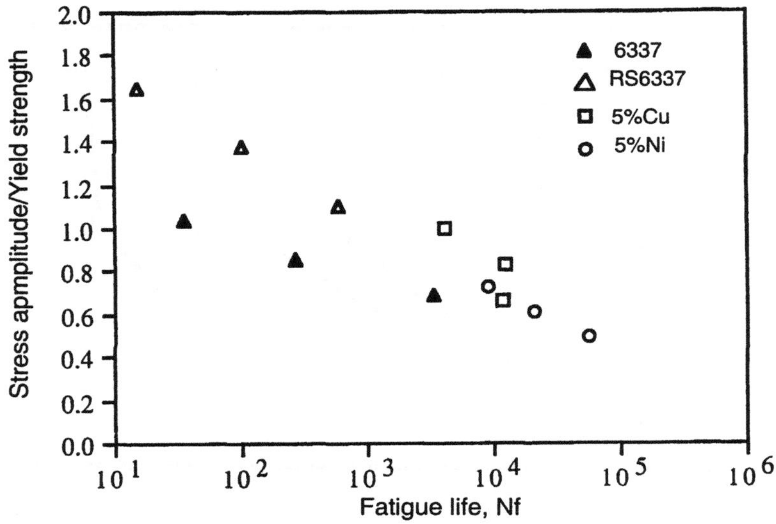

Figure 10-15 Normalized stress vs. number of cycles to failure at 0.05Hz for different compositions of solder specimens

The effects of isothermal strain rate and total strain on fatigue lives are shown in Figures 10-18 a-b. Apparently, fatigue life increases with decreasing total strain. Strain rate-dependent modulus and yield strength decrease with decreasing strain rate. The effect of strain rate on strain-controlled fatigue life does not seem obvious.

In general, high tensile ductilities result in improved low-cycle fatigue performance and high strengths result in superior high-cycle fatigue resistance. Optimum performance over the whole range of strains can only be obtained by the use of tough materials which have both high strength and ductility. Figure 10-19 schematically illustrates that the relationships of the resistance of a material to fatigue failure are similar to those used for improving resistance to tensile fracture.

Table 10-9 Test Results of Low-Cycle Fatigue of Solder Alloys

Solder	Fatigue life	Modulus. GPa	Yield Strength.MPa
(a) Strain rate = 0.001/sec (0.05 Hz), total strain = 1%			
6337 (9*)	981	28.1	40.0
RS 6337	222	23.6	25.0
RS 6337 - RF**	847	26.4	31.7
2.5% Cu	281	26.6	34.0
5.0% Cu	751	28.6	41.5
7.5% Cu	185	27.3	53.1
7.5% Cu - RF (2)	236	35.6	55.5
2.5% Ni	210	26.0	47.0
5.0% Ni (2)	162	35.7	56.5
7.5% Ni	185	40.4	66.0
2.5% Ni - RF	182	32.0	40.3
5.0% Ni - RF	43	42.3	59.0
7.5% Ni - RF (2)	117	46.6	69.5
(b) Strain rate = 0.001/sec (0.1 Hz), total strain = 0.5%			
6337	7759	24.6	30.8
5.0% Cu	4315	30.5	58.2
5.0% Ni	1395	31.4	67.5

Solder	Fatigue life	Modulus. GPa	Yield Strength.MPa
(c) Strain rate = 0.0001/sec (0.005 Hz), total strain = 1%			
6337	2095	19.9	20.0
5.0% Cu	460	25.5	33.5
5.0% Ni	73	28.9	52.3

The dependence of fatigue life on strain amplitude is given by the strain life relation:

$$\frac{\Delta\varepsilon}{2} = \frac{\sigma'_f}{2}(2N_f)^b + \varepsilon'_f(2N_g)^c \qquad (10\text{-}11)$$

where $\Delta\varepsilon/2$ is the strain amplitude, σ_f' is the fatigue strength coefficient, b is the fatigue strength exponent, ε_f' is the fatigue ductility coefficient, and c is the fatigue ductility exponent. A least-squares fit of the experimental data to the above equation results in the following strain life equations for different conventional and composite solders:

For conventional solder:

$$\frac{\Delta\varepsilon}{2} = 0.00216\,(N_f)^{-0.089} + 0.175\,(2N_g)^{-0.520} \quad . \qquad (10\text{-}12)$$

For copper-containing composite solders:

$$\frac{\Delta\varepsilon}{2} = (0.01060)\,(N_f)^{-0.167} + 1.352\,(2N_g)^{-0.189} \quad . \qquad (10\text{-}13)$$

For nickel-containing solder:

$$\frac{\Delta\varepsilon}{2} = 0.00653\,(N_f)^{-0.255} + 0.142\,(2N_g)^{-0.652} \qquad (10\text{-}14)$$

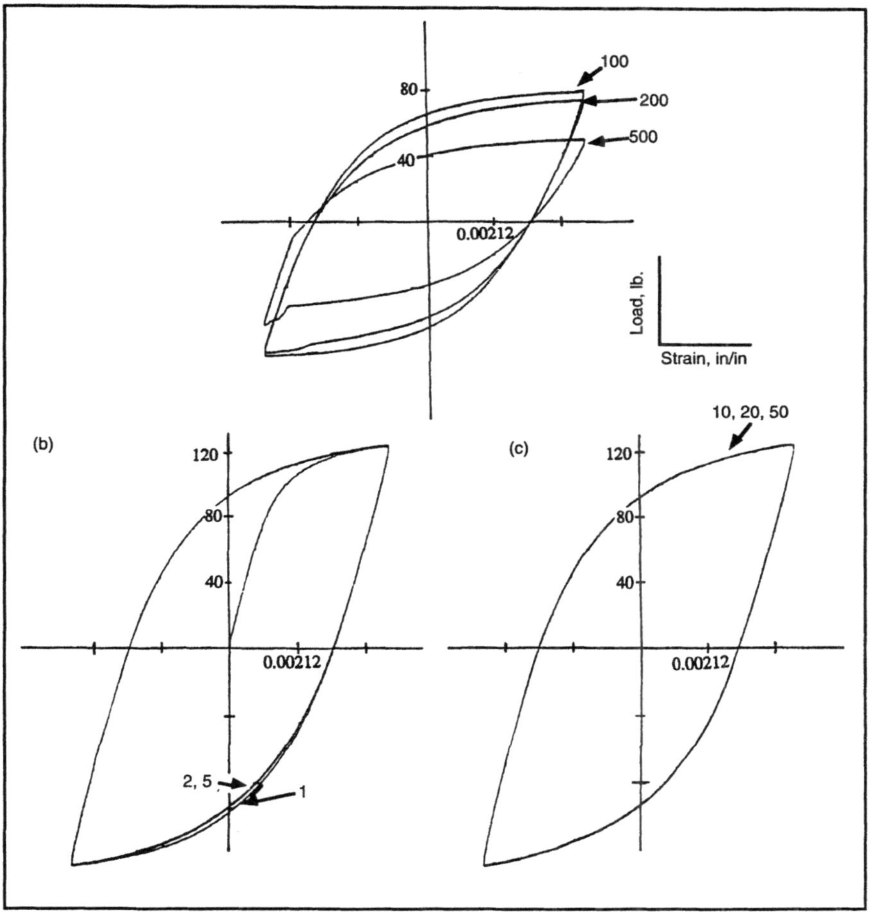

Figure 10-16 Hysteresis loops during low-cycle fatigue (strain amplitude controlled fatigue) of: a) 63WSn-37Pb, b) and c) 63Sn-29.5Pb-7.5Ni solder alloys.

The predicted reversals to failure and experimental data are shown in Table 10-10. Reasonable agreement is obtained, considering the scatter normally observed in fatigue experiments. Calculated reversals to failure are underestimated for 63Sn-37Pb and overestimated for 5% Ni. Note that all the coefficients and exponents used in these equations are not obtained from exactly the same strain rate or the same frequency. Further modification of these strain life constants must take into consideration the effects of strain rate and frequency. Also, the amount of data used in these calculations is sparse. Statistically significant conclusions require replicate tests and more experiments at several stress and strain levels.

10.6.8 Fractography

Figures 10-20a – 20b show the fracture surfaces of conventional 63Sn-37Pb and composite solder samples after strain-controlled fatigue tests. As shown in Figure 10-20a, conventional 63Sn-37Pb has experienced coarse intergranular fracture. Secondary cracks are also obvious on this micrograph. Fine intergranular fracture is found in rapidly solidified 63Sn-37Pb. For the composite solders shown in Figure 10-20b, fatigue failure is found to be fine transgranular, which is probably due to microvoid coalescence leading to a tension overload failure because of the fine grain structure and presence of dispersoids.

10.6.9 Creep-Fatigue Interactions

Since creep occurs in eutectic solders at ambient temperature, it is believed that the pertinent fatigue behavior of Sn-Pb solders can be best understood through creep-fatigue interaction studies [Weinbel et al. 1987]. Although the creep-fatigue interactions have not been studied for the composite solder alloys, it is worthwhile to present some of the related research and some ideas of how the creep-fatigue interactions could be studied.

Table 10-10 Strain Life Prediction of Solder Alloys

	1% Total strain		0.5% Total strain	
Solder alloy	$2N_f$ Calculated	$2N_f$ Experimental	$2N_f$ Calculated	$2N_f$ Experimental
6337	1524	1962	8990	15518
5% Cu	1670	1502	6676	8630
5% Ni	520	324	3058	2790

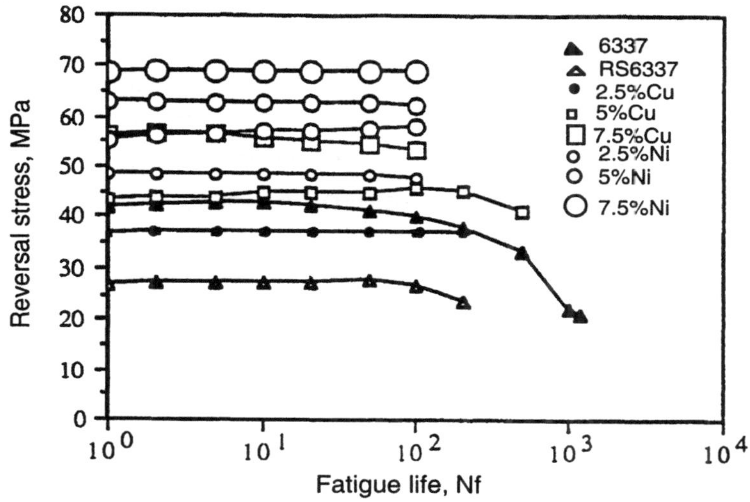

Figure 10-17 Peak reversal stress vs. number of cycles during constant strain amplitude controlled fatigue testing of solder specimens at a strain rate of 0.001 s^{-1} and total strain of 1%

Creep-fatigue interaction typically describes the cyclic application of a load at temperatures where time-dependent thermally activated processes can occur. For symmetrical fatigue tests with or without hold time, effects of frequency (or strain rate) and hold time on fatigue life are significant. It is usually found that, for eutectic Sn-Pb solder at ambient temperature, fatigue life increases as the frequency increases and as hold time decreases in isothermal fatigue. In thermomechanical fatigue, Frear et al. [1990, 1991] showed that as deformation rate decreases the fatigue life increases, and as hold time decreases fatigue life decreases. The evidence of creep-fatigue interactions is then obvious, because if there were no creep-fatigue interactions, the frequency, strain rate, and hold time would have no effect on the fatigue lives.

Creep-fatigue interactions have also been studied by cyclic creep tests. For cyclic creep tests, square-wave loading is employed, providing balanced time on-load and time off-load. The frequency effect and mean stress effect on the creep-fatigue behavior of eutectic tin-lead solder alloy at ambient temperature was also studied by Weinbel et al. [1987]. It was consistently found that the minimum creep (or cyclic creep) rate increases as hold time increases. Both the number of cycles to failure and the time to failure increase as frequency increases. The cyclic creep rate increases and the number of cycles to failure decreases drastically as the mean applied stress is increased.

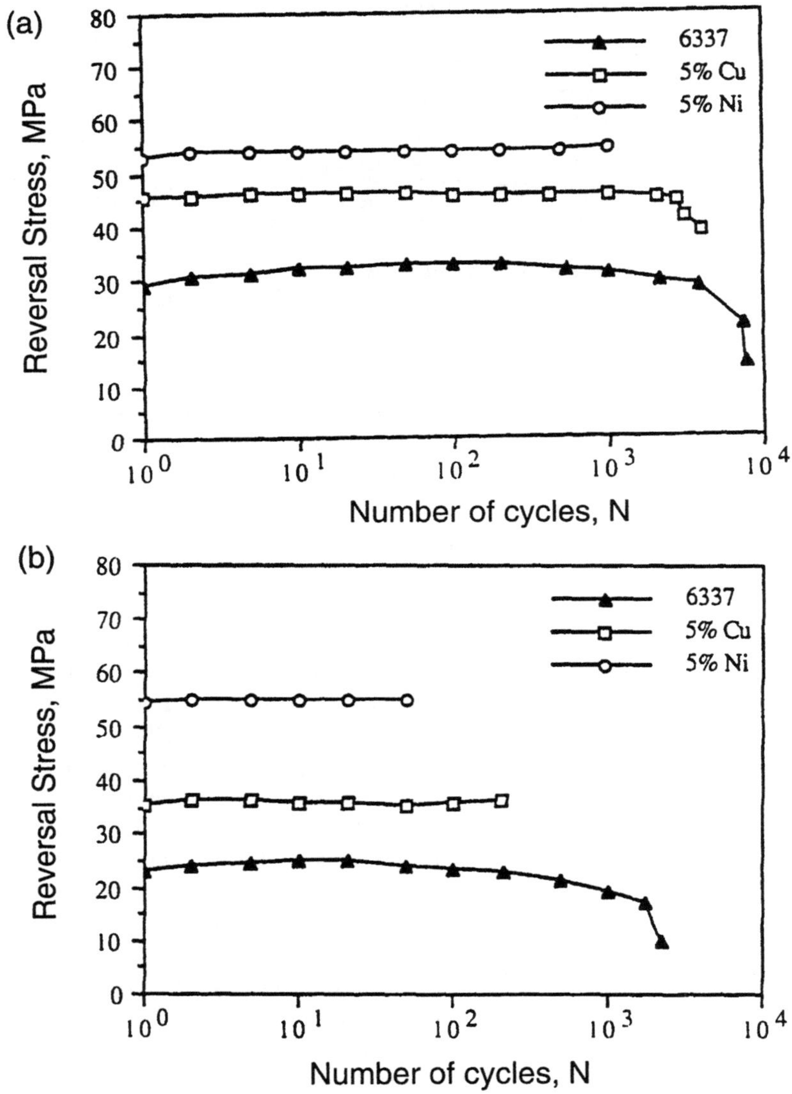

Figure 10-18 Peak reversal stress vs. number of cycles during constant strain amplitude controlled fatigue testing of solder specimens at: a) a strain rate of 0.001 s^{-1} and total strain of 0.5%, and b) strain rate of 0.0001 s^{-1} and total strain of 1%.

Knecht and Fox [1990] were able to isolate the contribution of creep damage from a symmetrical fatigue hysteresis loop. They consider the total shear strain rate as the sum of time-independent elastic, time-independent plastic, and time-dependent secondary creep strain rates:

$$\dot{\gamma} = \dot{\gamma}_{el} + \dot{\gamma}_{pl} + \dot{\gamma}_{cr} \tag{10-15}$$

where:

$$\dot{\gamma}_{cr} = C_0\left(\left(\frac{\tau}{\tau_0}\right)^2 + \left(\frac{\tau}{\tau_0}\right)^{7.1}\right) \tag{10-16}$$

The model for the time-dependent strain rate was derived from both the steady state creep rate of the grain boundary and the matrix. The stress exponents of grain boundary sliding controlled creep and matrix creep were determined to be 2 and 7.1, respectively. The activation energy is included in the coefficient C_o for a constant temperature. It is difficult to separate fast primary creep from time-independent plasticity; it was hoped that $\dot{\gamma}_{pl}$ might include such effects. For a stress-strain hysteresis cycle, the creep strain can be subtracted from the total strain at all stress values to obtain the "time-independent" strain (elastic plus plastic). In this way, the contribution of creep and fatigue deformation can be separated, as shown in Figure 10-21.

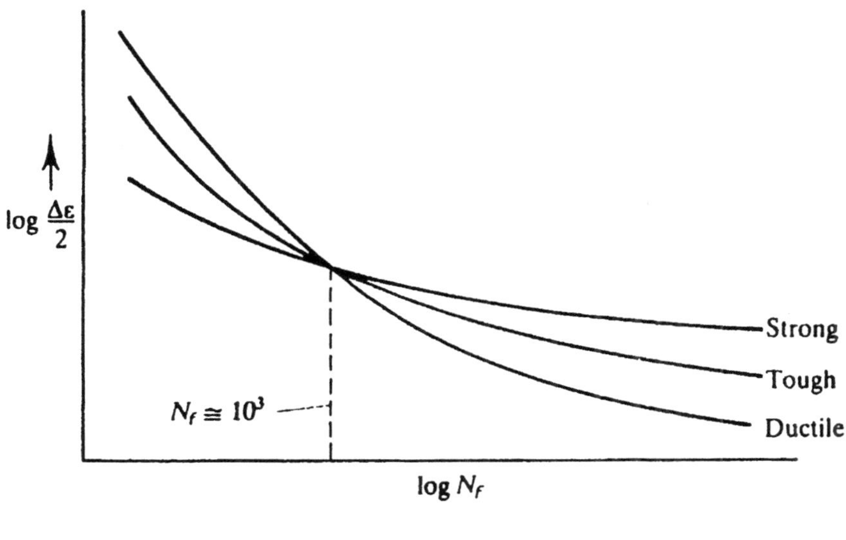

Figure 10-19 Schematic illustration of the relationship between the strain amplitude fatigue life and the nature of a material [Courtney 1990] **[Landgraf 1970]**

Creep-fatigue interactions have also been studied by crack growth under constant load and constant load with superimposed fatigue cycles at elevated temperatures. The linear elastic stress intensity factor, K, was used as a correlating parameter for these tests. A linear, cumulative damage model was used to successfully predict the total crack growth from the summation of crack growth contributions from time-dependent (creep) crack growth and fatigue crack growth.

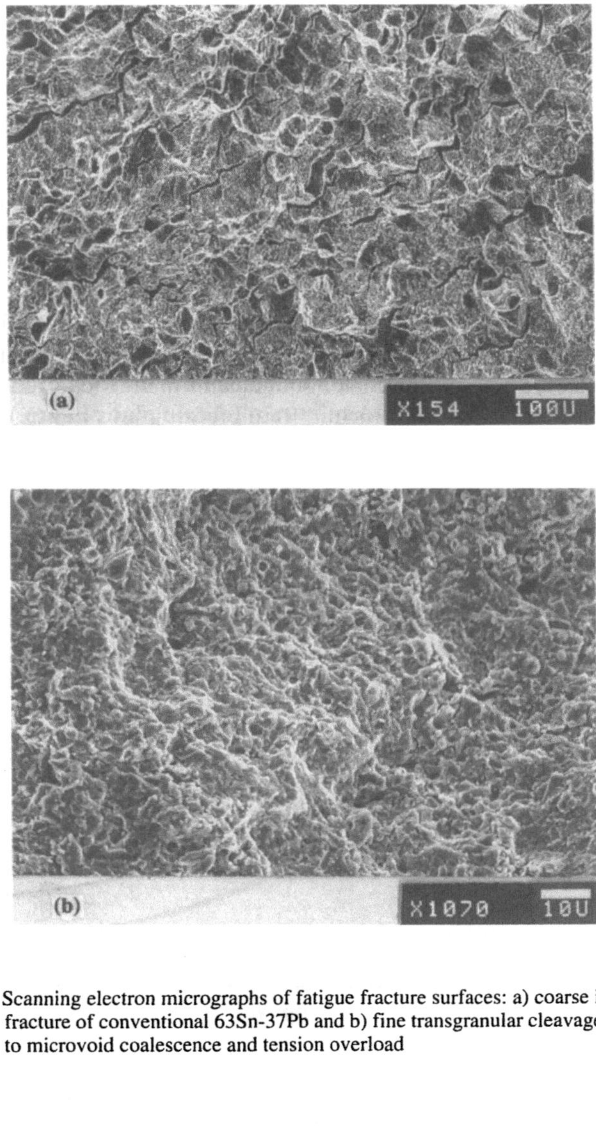

Figure 10-20 Scanning electron micrographs of fatigue fracture surfaces: a) coarse intergranular fracture of conventional 63Sn-37Pb and b) fine transgranular cleavage of 7.5%Cu due to microvoid coalescence and tension overload

$$\left(\frac{da}{dn}\right)_{total} = \left(\frac{da}{dn}\right)_{creep} + \left(\frac{da}{dn}\right)_{fatigue} \qquad (10\text{-}17)$$

In Eq. (10-17), the individual contributions of creep and fatigue are summed, and interactive effects (mixed-mode growth) are neglected. To account for the creep-fatigue interactions in hold-time tests, Eq. (10-17) has been modified by some mixed-mode correction factors, as shown in Eq. (10-18) to produce more accurate results [Mall et al. 1990]:

$$\left(\frac{da}{dn}\right)_{total} = \left(\frac{f_1}{6t}\right)\left(\frac{da}{dn}\right)_{creep} + \left(\frac{f_2 t_H}{t}\right)\left(\frac{da}{dn}\right)_{fatigue} \qquad (10\text{-}18)$$

where $f_1 = \exp(-\alpha t_H)$ and $f_2 = 1 + \beta \exp(-\alpha t_H)$.

Coefficients α and β are empirical constants to be determined from fits to the experimental data, t_H is the hold time in minutes, and t is the total cycle time in minutes.

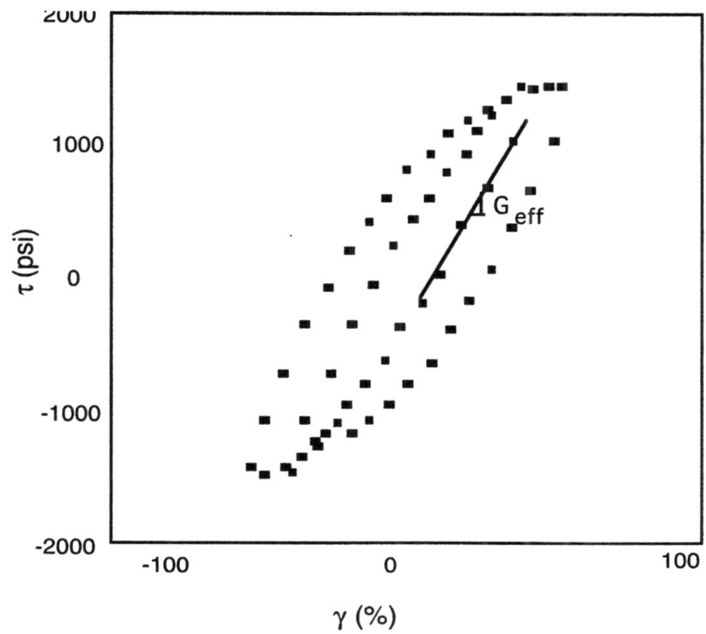

Figure 10-21 Plot of shear stress vs. shear strain hysteresis that shows how the creep contribution can be removed. The inner loop has had the creep contribution removed. (© 1990 IEEE)

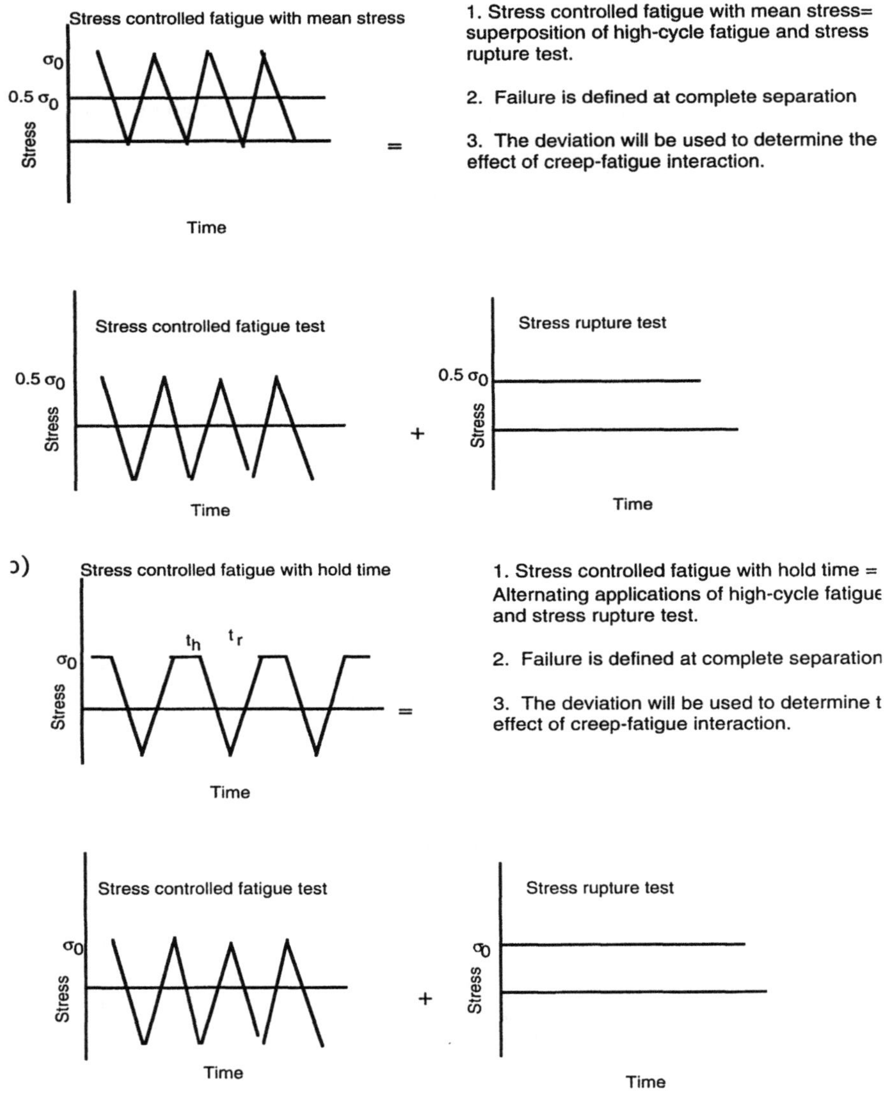

Figure 10-22 Creep-fatigue interactions studied by stress-controlled fatigue tests and stress rupture tests: a) Stress-controlled fatigue with mean stress as the superposition of high-cycle fatigue and stress rupture testing, and b) Stress-controlled fatigue with hold times as the alternating application of high-cycle fatigue and stress rupture testing.

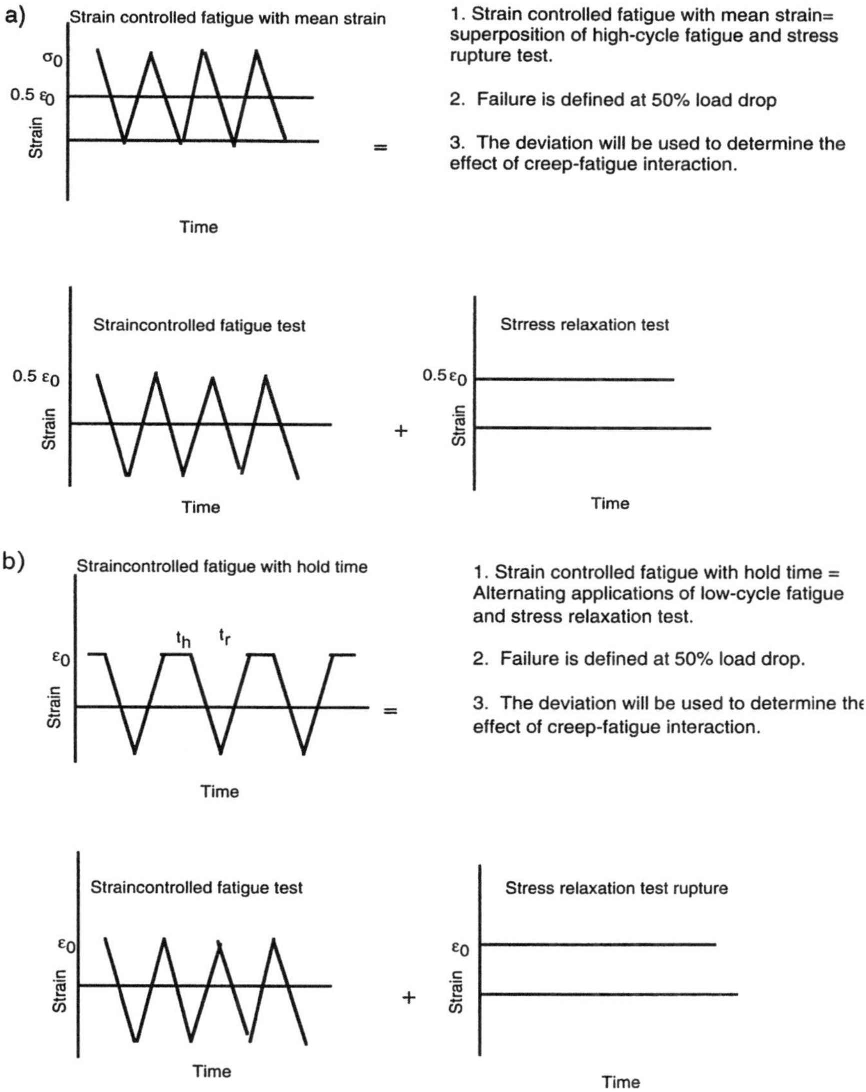

Figure 10-23 Creep-fatigue interactions also studied by strain-controlled fatigue tests and stress relaxation tests: a) Strain-controlled fatigue with a mean strain considered as the superposition of low-cycle fatigue and stress relaxation, and b) Strain-controlled fatigue with hold times considered as the alternating application of low-cycle fatigue and stress relaxation.

To investigate the additivity of creep damage and fatigue deformation and evaluate the effect of creep-fatigue interactions, the following test methodologies are suggested. Stress-controlled fatigue with mean stress can be considered as the superposition of a high-cycle fatigue and stress rupture test, as shown in Figure 10-22a. Stress-controlled fatigue with hold time can be considered as the alternating applications of a high-cycle fatigue and stress rupture test, as shown in Figure 10-22b. The deviation can be used to evaluate the effect of creep-fatigue interactions. Strain-controlled fatigue with mean strain can be considered as the superposition of a low-cycle fatigue and a stress relaxation test, as shown in Figure 10-23a. Strain-controlled fatigue with hold times can be considered as the alternating application of a low-cycle fatigue and stress relaxation test, as shown in Figure 10-23b. The deviation can then be used to determine the effect of creep-fatigue interactions.

10.7 SUMMARY

Composite solder development is in its infancy. Preliminary results indicate that fine stable dispersoids can be produced by suitable alloying additions and rapid solidification processing. The dispersoids improve tensile strength, creep resistance, high-cycle fatigue life, and stress-rupture life of solders. However, the dispersoids may not be beneficial for uniaxial ductility, plane-strain ductility, stress-rupture ductility, low-cycle fatigue, and toughness. An important goal is to develop a methodology to improve the ductility of dispersion-strengthened alloys, or to strengthen the grain boundary of solder alloys. Because fatigue failure of eutectic solder is driven by the formation of heterogeneous coarsening bands in the plane of shear, the formation of recrystallization bands can be prevented by homogenization so that the shear deformation does not concentrate in some particular slip band. Alternatively, the growth of bands can be prevented by interposing microstructural obstacles which limit their extent. The metallurgical modifications that have potential to improve the fatigue resistance of eutectic solders are grain refinement and superplasticity, alloy additions to suppress the heterogeneous coarsening bands, and second-phase dispersoids to suppress growth of recrystallized bands. The wetting behavior of composite solder alloys is a critical feature that has not been investigated. For a composite solder to be useful, manufacturability (wetting) and reliability (fatigue resistance and mechanical properties) are equally important. Experimental work must be performed to evaluate the wetting behavior of current composite solder alloys. Furthermore, the development of future composite solder alloys must focus on improved fatigue life and wetting behavior that is equivalent or superior to that currently used with near-eutectic Sn-Pb alloys.

REFERENCES

Ackroyd, M. L., and C. A. Thwaites. 1975. *Metals Technol.* pp. 73-85.

Argon, A. S., Editor. 1969. *Physics of Strength and Plasticity.* Cambridge, MA.: The MIT Press. p. 127.

Bangs, E. R., and R. E. Beal. 1975. *Welding Res. Supp.* Oct. 1975. p. 377s.

Betrabet, H. S., and S. M. McGee. 1992. *Proc. NEPCON '92.* Anaheim, CA. pp. 1276-77.

Betrabet, H. S., S. M. McGee, and J. K. McKinlay. 1991a. *Scripta Metall.* 25:2323-28.

—. 1991b. *J. Mater. Sci.* 27:4009-15.

Clough, R. B., et al. 1992. *Proc. NEPCON '92.* Anaheim, CA. pp. 1256-65.

Courtney, T. H. 1990. *Mechanical Behavior of Materials.* New York: McGraw-Hill. p. 577.

Embury, J. D. 1985. *Metall. Trans.* 16A:2197.

Frear, D. R. 1989a. *Proc. 39th IEEE Electron. Comp. Conf.* 39:293.

—. 1989b. *IEEE Comp. Hybrids and Manuf. Technol.* IEEE CHMT-12. p. 492.

Frear, D. R., D. Grivas, and J. W. Morris Jr. 1988. *J. Electron. Mater.* 17:171-80.

Frear, D. R., J. B. Posthill, and J. W. Morris Jr. 1989. *Met. Trans.* A. 17A:1325.

Frear, D. R., and W. B. Jones. 1990. *Proc. NEPCON West '90.* Anaheim, CA. pp. 1324-52.

Frear, D. R., W. B. Jones, and K. R. Kinsman. 1991. *Solder Mechanics: A State of the Art Assessment.* Warrendale, PA.: TMS Publications. pp. 191-237.

Frear, D. R., W. B. Jones, and N. R. Sorenson. 1991. *CREEP: Characterization, Damage and Life Assessment.* D. A. Woodford, C. H. A. Townley, and M. Ohnami, Editors. ASM International. pp. 341-50.

Glazer, J., T. Kramer, and J. W. Morris Jr. 1991. *J. Surf. Mount Technol.* 4:15.

Grivas, D. 1978. *Deformation of Superplastic Alloys at Relatively Low Strain Rates.* Ph.D. Thesis. Berkeley: University of California.

Ho, C. T., and D. D. L. Chung. 1991. *J. Mater. Research.*

IBM. 1986. *IBM Technical Disclosure Bull.* 29:1573.

Johnson, C. C., and J. Kevra. 1989. *Solder Paste Technology: Principles and Applications.* Blue Ridge Summit, PA.: TAB Professional and Reference Books, Inc.

Kashyap, B. P., and G. B. Murty. 1981. *Mater. Sci. and Eng.* 50:205-13.

Kelly, P. M. 1972. *Scripta Metall.* 6:647-56.

Kim, Y. -W., and W. M. Griffith. 1988. *Dispersion Strengthened Aluminum Alloys*. p. 181.

Knecht, S., and L. R. Fox. 1990. *IEEE Trans. CHMT.* 13(2):424-33.

Lee, S. N., and D. S. Stone. 1990. *Proc. 40th IEEE ECTC.* Las Vegas, NV. 40:491.

MacKay, C. A. 1981. IPC-TP-380. *IPC Fall Mtg.* Dallas, TX.

Mall, S. E., A. Mall, and T. Nicholas. 1990. *J. Eng. Mater. and Technol.* 112:435.

Marshall, J. L., J. Sees, and J. Calderon. 1992. *Proc. NEPCON '92*. Anaheim, CA. pp. 1278-83.

Mei, Z., R. Hansen, M. C. Shine, and J. W. Morris Jr. 1990. *Proc. ASME Annual Winter Mtg.*

Mizuishi, K., M. Tokuda, and Y. Fujita. 1988. *Proc. IEEE Comp. Hybrids Manuf. Technol.* IEEE CHMT-38. pp. 330-34.

Mohamed, F. A., and T. G. Langdon. 1975. *Phil. Mag.* 32:697-709.

Nakamura, et al. 1986. Japanese Patent No. 61-16523.

Pinizzotto, R. F., et al. 1992. *Proc. NEPCON '92.* Anaheim, CA. pp.1284-98.

Pink, E., R. Kutschej, and H. P. Stuewe. 1976. *Scripta Metall.* 10:759-62.

—. 1981. *Scripta Metall.* 15:185-89.

Sastry, S. M. L., et al. 1992. *Proc. NEPCON '92*. Anaheim, CA. pp. 1266-75.

Schmidt-Thomas, K. G., and S. Wege. 1986. *Brazing and Soldering*. 11:27.

Shine, M. C., and L. R. Fox. 1988. ASTM STP 242. *Amer. Soc. Testing and Mater.* pp. 588.

Smithells, C. J. 1976. *Metals Reference Book*. 5th Ed.

Steen, H. A. H., and G. Becker. 1986. *Brazing and Soldering*. No. 11. pp. 4-11.

Tribula, D. 1990. *A Microstructural Study of Creep and Thermal Fatigue Deformation in 60Sn-40Pb Solder Joints*. Ph.D. Thesis. Berkeley: University of California.

Tribula, D., D. Grivas, D. R. Frear, and J. W. Morris Jr. 1989. *ASME J. Electron. Pack.* 111:83.

Tribula, D., and J. W. Morris Jr. 1990. *ASME J. Electron. Pack.* 112:94.

Wasynczuk, J. A., and G. K. Lucey. 1992. *Proc. NEPCON '92*. Anaheim, CA. pp. 1245-55.

Weinbel, R. C., J. K. Tien, R. A. Pollack, and S. K. Kang. 1987. *J. Mater. Sci.* 22:3901-06.

Wild, R. N. 1972. *Welding Res. Supp.* 51:521s.

Wolverton, W. M. 1987. *Brazing and Soldering*. 12:33.

Index

MIX
Papier aus verantwortungsvollen Quellen
Paper from responsible sources
FSC® C105338

If you have any concerns about our products,
you can contact us on
ProductSafety@springernature.com

In case Publisher is established outside the EU,
the EU authorized representative is:
Springer Nature Customer Service Center GmbH
Europaplatz 3, 69115 Heidelberg, Germany

Printed by Libri Plureos GmbH
in Hamburg, Germany